When Roger Met Patty

1 Minute of Film ... 47 Years of Controversy

by William Munns

The Patterson-Gimlin Film, the Hominid seen in it, the remarkably intense debate it has provoked, and the solution to the mystery.

First Printing

Cover design by William Munns

ISBN-13: 978-1500534028
ISBN-10: 1500534021

Table of Contents

When Roger Met Patty

Introduction

The 1967 Patterson-Gimlin Film is often described as the second most analyzed home movie in history (second only to Abraham Zapruder's filming of John Kennedy's assassination in 1963). Two men, Roger Patterson and Bob Gimlin, were exploring the northern California woodlands in October of 1967, and Roger carried a 16mm movie camera loaded and ready to film something on a moment's notice. He was looking for some evidence of the creature called "Bigfoot", because there had been a trackway of footprints found nearby two months before. On October 20, a little after noon, he and Bob were on horseback with a pack horse in tow, going north on the west shore of Bluff Creek, and they rounded a curve of the creek with a log jam on their left obscuring what was ahead. Past that logjam, they had their famous encounter, which Roger filmed. In his film, we see a humanistic figure, apparently covered in a dark brownish/black fur, walking hastily away from the men and at one point looking back at the camera, and continuing to briskly walk away, until Roger's camera runs out of film. In the 47 years that followed, that film footage has been the eye of a controversial hurricane of debate, and a single film frame of that figure looking back has become the international photographic icon for the mystery called "Bigfoot" or "Sasquatch".

This collective document is an analysis of the famous Patterson Gimlin "Bigfoot" film, but from a whole new perspective. There is a powerful urban legend, believed by many people, that this film has been proven to be a hoax with a man in a fur "creature costume". It is not true, but still, it is a formidable legend to challenge, and can only be done so by a professional who actually designs and builds such costumes. So a major part of this book will be devoted to both the makeup artist craft itself, and the applied knowledge of this craft to the analysis of the Patterson Gimlin Film (PGF).

A proper investigation requires that one must be open to all paths of investigation that the data leads to, and once I started this analysis effort, I found that issues of cameras, lenses, editing and film copy process also were relevant issues needing analysis. So my effort expanded into those areas, and I had some professional experience in these filmmaking skills as well as my makeup expertise (described in Chapter Three).

But this analysis effort has been personalized because there is a powerful skeptical community who believe Bigfoot does not and cannot exist, and so as soon as anyone offers an analysis to the contrary, they set forth on a mission to attack the person, personalize the discussion and then try to discredit the person who does the analysis, instead of keeping the discussion on track with the ideas and the scientific analysis itself. We see this increasingly in every walk of life, this politicizing of ideas and personal attacks on people who offer an idea that the challenger doesn't agree with.

I have seen my life examined microscopically, not by people with any intent to admire what I have accomplished, but rather by people who simply are hoping to find some arguing point to discredit me or belittle my accomplishments. Everything good or admirable they read about me simply does not stick in their minds. It is discarded as useless to their agenda of criticism. They want "dirt".

Because this personalization has been so intense, this book must be as much a personal story of my efforts to stay centered on a path to real knowledge and understanding, as it is a story of how the mysterious Patterson Gimlin film is rightfully analyzed. While the film's analysis is a unique endeavor, the story of one trying to achieve a factual and reasoned solution amid an environment of partisan gamesmanship and "win at all cost" opposition is a story I suspect many readers can relate to in their own lives and activities.

In my effort to understand this film and the controversy around it, I have become part of the controversy. Much of what I present herein is derived from my research work. While authors do commonly take the position of observing or reporting on a matter from an outside perspective, and remove themselves from the discussion or dispassionately reference themselves in the third person, I can't remove my research from my presentation of the controversy. So I made the decision to simply and factually acknowledge that I am part of this controversy and my analysis can be best explained in the first person form. Much of my insight is derived from my life experience and my eclectic resume of career and technical skills acquired from about 45 years of professional endeavor and self-taught education. So a personal and first person approach is what I feel is the most appropriate form for this book. I have chosen to simply write a book that I feel is a truthful expression of what I have learned and considered in this curious seven-year adventure, a book I will always be proud to have written, a book I would actually like to read myself if another had written it.

Reception to this book will be intense and diverse. Only time will tell if it endures as a reasoned analysis, a scholarly text. But I will have expressed my thoughts, ideas, and reasoning, in a form that I feel is a positive accomplishment, and so I am at peace with my effort and my decisions.

Bill Munns

Dedication

This book is dedicated, first and foremost, to the actual participants of the Patterson Gimlin Film Mystery (one deceased and one still living) because they have seen their lives transformed (or cursed, in some cases) by this film. It has made them famous and forced them to adjust their lives to the controversy of the film. In a sense, they have lost a significant portion of their lives to this film, so powerful is its impact, and so turbulent its controversies.

It defined Roger Patterson's life, literally becoming his legacy, but it also put his life under a public microscope, and as his life was not that of a "regular guy" with a steady job, every arguable flaw or mistake, every irresponsible or thoughtless action he may have done, is documented, debated and publicly examined. People intent on showing the film to be a hoax have vilified Roger with an intensity that still astonishes me. So first and foremost, this book is dedicated to Roger, because he did not hoax this film. Whatever else in his life you may wish to criticize him for, as being flawed or fake, this film is honest and what he claimed to film was true.

Second in dedication is Bob Gimlin. When he agreed to accompany Roger on this expedition to search for Bigfoot evidence, he could not possibly have known that the encounter in the Bluff Creek forest would devour his life like the shark in JAWS. This film would make him world famous, it would make total strangers hate him, it would provoke people to impune his personal integrity again and again, and it would be something he could never erase, never ignore, never resolve. People examining his life obsess with second-guessing his decisions, his actions, his words and his demeanor. But those who do are fools, because they judge Bob without full knowledge of his thoughts, his personal decisions and private considerations. He has attempted to maintain some semblance of privacy in his life, despite being something of a "public figure" by virtue of this film, and his public actions and statements are driven by choices, considerations and factors in his private life, which he has every right to try and keep private. So to judge Bob is to indulge in petty character assassination, because to defend himself, he would need to surrender his privacy, and every human needs some privacy just to be human.

In October, 1967, something happened to Bob that was so rare, so bizarre, and so powerful in changing the course of his life, that we can never judge how well he has managed to endure this experience. How well would we, if fate put us there instead of him? But while Roger passed away less than five years after the filming, Bob has endured 47 years of this film's grip on his life. And yet the film has defied resolution during those years, holding his life, his reputation, in suspense. So with this dedication, I hope that it gives Bob some comfort and some resolution or closure to the controversy. He saw something real that day, 47 years ago. It was not a guy in a fur costume. He's always known that. Hopefully, now more people can share what he has always known.

Third in dedication is Patricia Patterson, Roger's widow. When Roger packed up the truck to go with Bob Gimlin in early October, 1967, headed for Bluff Creek, CA. Patricia was a housewife with three young kids and no financial stability in their life. Her problems were the problems of millions of women, to keep a household going and raise some kids despite financial insecurity. She also was struggling with the difficulty of seeing her husband endure the crude radiation treatments of the era for Roger's cancer (Hodgkin's Lymphoma). Then he came home from his trip and announced to the world what he had filmed. And Patricia's life went from being challenging and understandable, to bizarre and unprecedented.

Mothers the world over deal with comforting their children who may be bullied or mocked at school, for being too fat, too skinny, too dumb, too smart, or too unattractive. But Patricia had to comfort her kids who were bullied or mocked for having a father who claimed to have filmed "Bigfoot". She watched the media circus Roger engaged in with the assistance of his brother-in-law, Al DeAtley as they four-walled the movie around the country. She watched his cancer return and lead to his death in 1972. She found herself alone, with three kids, no estate or assets, except a bizarre piece of film that TV producers would pay good money for rights to broadcast. How does a woman deal with such a situation? What friend do you ask, what course do you take, what professional do you hire, to help you make the kind of decisions she was now faced with, trying to manage this one utterly unique and controversial asset?

Since Roger's passing, 43 years ago, Patricia has been approached by people who try to prove the film is real, others who are determined to prove it a hoax, and media producers who want to license and exploit the film for program content. She has been poorly served by some of these people, and outright deceived by a few. Now, her own health is declining and she deserves some closure on this film which looms over her life. So this book is dedicated as well to her life, her challenges, and hopefully, some closure to this film controversy that has caused her so much grief and aggravation.

Finally, I would like to dedicate this to my father, Howard Munns, not because he is connected to the PGF, but because of his inspirational guidance while I wrote this PGF analysis. My dad was a talented artist who never got caught up in the "pretentious artist" mindset. He was a man who simply liked the truth, in his art, in life, in nature, in science, and in people. He instilled in me a love of truth above agenda. I apply that love to this analysis effort. I have only one goal here, to follow the data and see if I can find out the truth of this most astonishing and unique film. I think, if he were alive today, that my father would be pleased to read this book.

Bill Munns

Additional Acknowledgments

Many people have helped me along the way. But they, collectively, are to be commended for the contributions they made to my understanding of the PGF, and my research effort.

Thank you, Chris Murphy, John Green, Jeff Meldrum, Peter Byrne, Bill Miller, Daniel Perez, Doug Hajicek, Robert Hackett, Paul Graves, Tom Yamarone, Mike Rugg, Alton Higgins, Byran Brown, Kathy Strain, Bruce Harrington, Al Guinn, Steve Streufert, Robert Leiterman, Bart Cutino, All the 2012 Bluff Creek site survey team.

Research Team:
Models: Traci Truehill, Virginia Ma, Carol Hannan, Gail Cook, Ellis Hooper, Louise McCartney, Ahsley Guarrasi, Cynthia Dallas, Joe Foley, Partick Caberty, Bryan Betts

Crew: Pat Loudermilk, RN, Chelsey Jensen, Iris Jauregui, Mathew Rubene, Shahrooz Raffi, Nadja Hoyer-Booth, Bouka Riley, Matt Thomas

Makeup team: Lee Joyner, Kelsey Boutte, Robin Rebbe, Thirati K, Emily Parks, Katherine Howard, Cody McEwan, Benjamin Peter, TJ Loza, Tyler Meehlies.

A special acknowledgment of appreciation must go to Michele Cestone, who has shown me a remarkable vote of confidence. The research work which took my effort to a new level of factual certainty could not have been done without the generous support of the Cestone Foundation.

And a special appreciation to my wife, Jakie, for keeping up with me all these years.

Chapter One - Forget What You think

One of the great mysteries of the 20th century is found on less than 24 feet of 16mm color film, taken in a remote wilderness area of Northern California in October, 1967. Roger Patterson, a man with an independent and improvised life, a man who has a multitude of reasons to question his integrity, emerged from that wilderness with what he claimed was actual footage of a real "Bigfoot" type creature. He died less than five years later (in January 1972) of cancer, and went to his grave unwavering in his testament that the filmed footage is something real.

His companion of record on that expedition, Bob Gimlin, is still alive today and still insists the footage is real, and the filmed "creature" is not any kind of fake or hoax. He has endured unrelenting attacks on his personal integrity over the years by skeptical or doubtful people who passionately believe that nothing like a "Bigfoot" can or does exist, and that he is merely perpetuating an epic fraud upon the public. Total strangers with one fanatical intent, to prove him a liar, have examined every documented detail of his life, in search of the proof of deception they so passionately believe exists.

Fig. #1-02 shows Bob Gimlin during a 4-hour oral history interview, made on April 2014. Insert is Bob back in 1967, holding two plaster footprint casts.

Another man, Bob Heironimous, was a longtime friend of Roger's, a man who loaned Roger one of the horses they rode on that fateful expedition, and a man who acted as a cowboy in some of Roger's prior effort to make a film about the "Bigfoot" story. Heironimous now claims that he is indeed the "man in an ape suit" and that he is what we see in this footage. His claim was revealed over 25 years after the filming, after Heironimous first denied any such involvement, and after talking with an attorney. His life likewise has been subjected to intense scrutiny, and people doubting him have put his life and words

1

under the same microscope that skeptical people have subjected Bob Gimlin's life and words to, in search of cause to discredit the man's claim.

Fig. 1-02

Bob Gimlin April, 2014 from Oral History Interview
Insert - Bob in 1967, with foot casts.

Each living man has his supporters and his detractors, like fans of two rival football teams going into the Super Bowl.

A few scientists and academic scholars have tried to authenticate the film as showing something biologically real, and their efforts thus far have been far from conclusive. Many critics and skeptics have tried to expose the film as a hoax, and failed as well. Many veteran Hollywood Makeup Effects artists have pronounced the film a fake, yet these Hollywood critics have not made any effort to take their opinion beyond a sound bite catch phrase and into a formal analysis and proof of the fakery they claim.

Roger Patterson found the public reception to his film footage and claim rather odd. Many scientific, academic, and media entities shunned the footage and wanted no part of what they perceived as a hoax. But with the assistance of his Brother-in-law, Al DeAtley, Roger took the film and his story directly to the general public, with theater showings around the country, and the public reception was both profitable and more sincerely curious about the prospect of the film being real.

The camera original footage was passed from Patterson to a film company, American National Enterprises (ANE) and they made a documentary ("Bigfoot: Man or Beast) in 1971 before Patterson passed away. It included the footage Patterson took. With Roger's death in 1972, that company kept possession of the camera original, until that company went bankrupt a few years later and their physical company assets were auctioned off. It is reported that one buyer of assorted office furnishings found a small reel of camera original film in that collection of office furnishings, and thus acquired the famous film by accident. This new owner had rightful physical possession but not did not own licensing rights. This new owner, generally not identified publicly, put the film in a film storage vault in Southern California. There were some reported attempts to purchase this film from the new owner, and rumors of million dollar prices are usually connected, but all that can be factually said is that no sales were made.

With Roger's passing away, Bob Gimlin felt that he had some rightful claim of ownership to the film, being the second man there that day and participant in the filming event. His claim was taken to court, and the judge made a curious ruling, giving Bob Gimlin 51% ownership in the film, while awarding 49% ownership to Roger's widow, Patricia Patterson. But this judicial ruling further specified that the licensing rights (and future revenue) would be divided so Mrs. Patterson retained film licensing rights (for any media showing the film itself), but Gimlin would have rights to the image licensing of still images from film frames, for publication. Bob Gimlin would then sell his 51% claim of ownership to researcher Rene Dahinden for a token sum (alternately reported to be $1 or $10).

An investigator of all things strange, the late Jon-Erik Beckjord, was obsessed with the film and thought not only that it was real, but that there was evidence in the film of a connection between the "Bigfoot" creatures and aliens. Beckjord took one famous frame of that film, printed that frame image, and then added a multitude of red circles to indicate suspicious things in the landscape. What he lacked in reason, he made up for in imagination.

Researcher Rene Dahinden pursued a relentless and partially successful attempt to take over control of the film, and once he acquired Bob Gimlin's 51% legal ownership of the film rights, Mr. Dahinden convinced the film storage facility that was holding the precious camera original footage, to check it out to him and one Bruce Bonney, for further analysis. The camera original was never returned to that film vault and its current whereabouts is unknown. Dahinden is reported to have parted ways with Bruce Bonney in such a bitter divide that there were rumors he threatened to kill Bonney. Mr. Bonney now lives as a recluse in Arizona and will not talk to anyone about his experience, even though Dahinden has been dead for over a decade. His formal analysis, called "The Bonney Report", has been seen by some people but currently is unobtainable to any current researchers, so we cannot know what was in that several hundred page document.

A researcher with background in astrophotography, M. K. Davis, began to study the film and strongly advocated the film's authenticity and the reality of the filmed subject, with new computer image analysis tools and processes. But he then went down a path of tabloid sensationalism, by extending his analysis of this film to supposedly prove that there was a secret massacre of a group of these Bigfoot individuals (which he likened to a form of Native Americans) by evil lumber industry power barons who wanted to keep the woodlands free of this rare species and open for logging.

I think we can say with some measure of assurance that this film does not have a dull or common history.

If you decide to look further into this question of the film's authenticity or the claims of a hoax, you will likely find more questions than answers. In many respects, and with many specific events of the story, people did not (indeed could not) anticipate how intensely this film would be studied and debated, so documentation for many events was poorly kept. The smattering of documentation we now have is often used to mislead more so than enlighten.

This lack of documentation is most infuriating with the issue of how, when and where the film itself was processed (generally referred to as "The Processing Timeline" in debates), because the film was a Kodachrome type film and processing options are far more limited than with other color films or any black and white films. Questions about the processing timeline lead to suspicions the filming was not actually done on October 20, 1967 as reported, but was done days or weeks before, allowing time for hoaxers to review the footage and edit it to remove anything that would have exposed the alleged hoax activity. The only problem with this theory is there is absolutely no evidence of editing the original (although some individual copies have been edited for TV program presentations over the years). Appendix Three details this splicing investigation.

If you have any kind of respectful professional credentials and credibility and you go on record with an endorsement that the film may be real, there are a small army of internet analysts who will search your life and every word you have said on record for some cause to discredit you and thus, discredit your appraisal of this film. They will find things about your life that you, yourself, may have forgotten and could never find documentation of. They will take every word said or written by you or about you and subject it to the most rigorous scrutiny, and will go to unbelievable lengths and tortured semantic explanations to try and show the world you are a fringe "believer" of Bigfoot, and not a person of reason and logic. By their reasoning, you cannot be logical, rational or credible, because you do not agree the film is a hoax, and they will search your life obsessively until they find what they can claim is proof of such.

If you bring up the topic of this film in a forum where factual and technical topics are generally discussed (such as my experience in a Cinematography Forum), you will meet total strangers who will find the whole topic of this film so laughable that they will ridicule even a serious and matter-a-fact attempt to discuss it.

Fascinated with paradoxical situations and puzzles? This is a beaut. With 47 years of astonishing brainpower on both sides of the debate, why can't a conclusion be shown and accepted? Why can't the critics get their act together and develop one formal and agreeable proof of a hoax? Why can't Bigfoot researchers find a body to show the world such creatures exist? Why is there absolutely nothing in all the photographic material showing a "Bigfoot" which even comes close to this one strange film. Most hoaxed films and videos fall apart within days and even hours of their public debut. But the PGF has endured for 47 years. Most footage that doesn't fall apart is sustained by the vagueness of the image data, so often inconclusive that a new word was invented just for these vague images. They are called a "blobsquatch" (for an unrecognizable "blob" instead of a verifiable sasquatch). The PGF, by comparison, has a splendid wealth of highly detailed frames revealing the anatomy of this subject.

This film is often described as "the gold standard" for photographic evidence that Bigfoot is a real entity, but where are the lesser nuggets of gold, where is the silver, or even the bronze or copper? Compared to this one curious "golden" film, everything else is just lead or scrap iron. How can one golden example exist, yet nothing else is even marginally second best?

There is nothing simple about this film. There is nothing common about its colorful history. And there is nothing in truth even remotely matching the urban legends that have arisen about it. It inspires irrational thinking on both sides of the argument, which sadly overshadows the more rational attempts of analysis by responsible people on both sides of the controversy. And just about everything you may think you know about this mysterious film is probably wrong.

And then there is a quote attributed to the dying Roger Patterson, who offered the rather cryptic remark that "I am probably the worst person this film could have happened to". I suspect that Roger did see clearly how his own life actions and choices were being used by people to judge the film.

We do know many people continue to judge the film by judging Roger himself, even long after his passing. It is the standard skeptical direction of argument. So we may ask, would the film have been received differently if Roger had a steady job, if there wasn't an arrest warrant issued for theft because he had kept his rented 16mm camera far longer than the rental agreement specified, if the film's processing had been done in a more routine and documented fashion, if the subject ("creature") in the film had done something more monstrous than simply walk away with a brief lookback to the camera for a mere second?

Everything about this film is debated and disputed by somebody. Even the most basic and longstanding standard of what constitutes quality of evidence is sometimes turned upside down, so that empirical evidence (the actual film image data) is downgraded, and hearsay remarks are upgraded to sterling integrity. An amazing amount of research into the lives of Roger Patterson and Bob Gimlin have been undertaken, most often by people intent on proving the claim of hoax. They find "suspicion" in every discrepancy, however minor or benign, and elevate that suspicion to a "red flag" and then use the implied "where's there's smoke, there must be fire" reasoning to say that there must be a hoax because of all the red flags, the mountain of suspicions.

Yet for all these red flags, no one can actually explain the claimed hoax in a single coherent manner. Indeed, many of the suspicions directly contradict other suspicions. But like a criminal defense strategy in a trial, you can try to win by simply creating enough "reasonable doubt", and if you take that approach, it does not matter if some of these "reasonable doubts" contradict other reasonable doubts.

Roger went into the woods specifically looking for footprint evidence of the existence of Bigfoot, based on a trackway sighting two months before in the area. He was prepared to spend weeks in the wilderness, and had his camera ready at all times, to film anything of interest. But the first 3/4 of the 100 foot roll of film in his camera was just mundane segments of a man on horseback leading a smaller pack horse through the woods. Evidence of Bigfoot eluded the men for weeks, and so Roger just shot the horse and rider footage as possible filler for his intended documentary. Would a man intent on staging a hoax put weeks into the mundane wandering around and filming uneventful filler footage, leaving only a minute or less of footage in the camera to capture the eventual hoax he is believed to have staged?

First Reel "Horse and Rider" footage

Fig. 1-03

Fig. #1-03 shows the Men on Horseback footage, sample frames from the eight individual filmed segments. In each case, the rider just sits on his horse, or walks the horse with the white pack horse in tow. This footage accounts for about 75% of the 100-foot reel.

Personally, I would think a better plan for a hoax would be to film some mundane impressions in the ground that sort of resemble footprints, and some broken or disturbed tree saplings and branches, to suggest some form of wildlife has been active in the area, and then film the staged encounter. Why? Because he could say "we found some curious things, and that encouraged us to keep looking." But behind the scenes, planning a hoax, if the first 76 feet of film is intentional, and you screw up the finale with your guy in the costume, you'd have to load another roll of film and go back and redo all that horse and rider footage again to get the first 76 feet used up so the creature could be the finale where the film runs out. Complicated. Setting up some mundane ground impressions and tree sapling or branch breaks would be easier to redo in preparation for "take two" of the guy in costume.

Does this horse and rider footage seem like the pattern or logic of a hoaxer? If it does not, be prepared for a skeptic to tell you the action (that I personally find persuasive as a real event) is just part of the meticulous plan of hoax. If someone seems sincere, he is pretending to be, to fool you. If an event seems spontaneous, it was cleverly planned to look spontaneous. If an action seems illogical to plan, that is sure proof of the brilliant mind which planned these deliberate illogical actions to throw off any suspicion of hoax.

But if we piled together all these "deliberate acts" which were deceptively engineered to hide the hoax, we'd have a plan more complicated than anything the old Mission Impossible TV series ever devised, and they strained credibility far beyond its limits.

In an interview with both Roger and Bob shortly after the filming was announced, the two men described that their agreed policy, if they should encounter a real Bigfoot, was to shoot it only if they felt their lives were threatened. So according to Bob Gimlin, he did have his rifle at hand and poised to fire if need be, but as the subject just walked away from them, there was no threat that would justify shooting the thing.

A few days before Roger died, while speaking with researcher Peter Byrne, Roger reflected upon his life and his choices in regard to this famous film that had defined his life and legacy (for better or worse) and he told Peter that in retrospect, seeing the aftermath of his film's public reception, if he could go back to that day in October, 1967, he would have shot the thing and brought out a body instead of a reel of film.

If one is to believe the film is a hoax, and Roger's friend Bob Heironimous was in the creature costume pretending to be the "thing", it certainly is an odd regret for a dying man.

The debate, as to whether what we see in that film is real or not, has raged with unrelenting determination for 47 years, although the debaters tend to change over time. The old guard get tired of the lack of conclusion, and a new guard steps up to recycle and refresh the same talking points.

For many years, the prevailing theory, (that the film is a hoax), described the ape suit as fabricated by Hollywood Makeup Artist John Chambers (famous as the designer of the makeups for the "Planet of the Apes" movies), but that theory has fallen into the chasm of disinterest. New claims by a costume maker, Phillip Morris, are the current rage, even though Mr. Morris and his company make a much lower level of costume realism than John Chambers and the Hollywood Makeup group. The difference between the average "Morris" Suit (that was sold then for $495 to anyone who wanted one), and what we see in the famous footage is now fashionably explained by saying

11

Roger Patterson was a really talented guy and an experienced saddle maker, so of course he could take a stock average gorilla costume and modify it into something so unique that for many years most Hollywood Makeup Artists would swear that only the great John Chambers could have made it. It's an interesting shift of theory, that the idea goes from "only the best guy in Hollywood could have done it" to "the talented rodeo cowboy and saddlemaker fixed it up on his first try from a mediocre Halloween/carnival grade commercial costume"

This shift in theory should make the film all that easier to debunk once and for all, yet it continues to defy any conclusive and singular hoax explanation. Veteran Makeup Effects Professionals should be able to easily explain the work of an amateur (if that's what it is), yet they cannot.

Another curious wrinkle in the "Roger as costume-maker" and the "many Hollywood Makeup Artists say it's a fake" combo so loved by current skeptics is that none of these Hollywood makeup artists (so often quoted as saying the film's a fake) will endorse the "Roger as costume-maker" and the "Morris gorilla costume modified" scenarios. So one theory they love is never endorsed by the experts they love to quote for support of their epic hoax argument. That is one of the more glaring contradictions they quietly shy away from and hope nobody notices.

But, in fairness, flawed claims occur on both sides of the debate. And it started with the first scientists and academic people trying to explain this film. They tried to explain the film as best they could, but focused their efforts on issues that actually prove nothing. The most glaring example of this is "the walk".

It has been described over the years that the filmed "creature" exhibits a walking action technically categorized as a "compliant gait". This type of walking requires that the knees be bent somewhat throughout the walk step cycle, and that such causes a "gliding" sort of walk. Some analysts went so far as to say humans couldn't replicate the walk. Others will argue with conviction that humans most certainly can, and that the late film comedian, Groucho Marx, did exactly that kind of curious walk in some of the early Marx brothers movies. I personally never understood the concern for this issue and it seems to have fallen by the wayside as a form of "proof" the filmed subject is not human. But for many years, an enormous amount of effort was invested in this claim and attempts to show why it proves something.

Similarly, a great deal of debate has raged on about apparent muscle motion in the body, and is has been argued that creature costumes of the time could not replicate such muscle motion effects. Now, on the one hand, I (as a long time professional Creature and Makeup Effects Designer) can agree that creature costumes of the time had no capacity for creating any form of believable muscle motion (they could be sculpted to look like muscles but still would not move like real muscles). But on the other hand, I never saw anything in the film which I found really convincing as evidence of real anatomy muscle motion, as some analysts claim. The reason I hesitated to endorse the claim of seeing real muscle motion is that I see indications of fatty tissue and a body that is "bulking up" for the coming winter, and so between the subcutaneous fat deposits and the fur covering (both of which are on top of real muscles), these structures have more impact on what we see of the body surface shifting highlights and shadows. As I did not see any previous analysts factor this into their appraisals of the body, I could not endorse most claims of seeing real muscle motion.

On the other side of the fence, proponents that the film is a hoax have invested immense effort in poor choices of argumentative reasoning as well. The most egregious flawed argument is the one which claims "Bigfoot does not and cannot exist, so the film must be a hoax, even if we can't prove the specifics of how the hoax was done." This laughable claim falls apart on two counts. First, no fact or determination is accepted by default, in scientific matters. Any determination must be proven. Second, there is absolutely nothing "impossible" about a Bigfoot creature existing, especially as seen in this film. Its descriptive form does not defy any of the laws of physics or biology, not its size, its anatomical form, its mammalian features, or its physical motions and activities. People trying to shore up this argument usually try and compare Bigfoot to truly fanciful creatures like fairies, aliens or winged unicorns. People already biased to deny any possibility Bigfoot exists welcome the argument because it confirms their bias in a colorful and elegant way. But it will fail as an argument if rationally analyzed.

Proponents that the film is a hoax also love to use the "backstory", the pile of suspicions based on Roger Patterson's unconventional life, as "proof" the film is a hoax. They say he is a proven hoaxer and conman, so the film is simply his ultimate "con", his most successful hoax. But they cannot prove Bob Gimlin is a conman of any kind, although they do try with inventive interpretations of "discrepancies" in his life and remarks. In doing so, they lose all perspective on human frailty and the simple fact that we humans are not perfect, we may not describe an event with exactly the same words on different occasions, we may has concern for our privacy and may not admit with bold candor every detail of our past to inquiring strangers, we may have errors in our memory of events. But to a proponent that the film is a hoax, any human frailty is evidence of deliberate deception, and gets blown way out of proportion to try and sell the hoax idea. Chapter Ten addresses the backstory with detailed consideration for why it is flawed.

So could it be that both proponents that the film is real, and proponents that the film is a hoax, have wasted decades of time and effort chasing false goals of proof? Oddly, yes, that is the case. Nobody on either side actually took a proper look at the full inventory of evidence, and put the evidence into a proper grading as to quality and reliability, so they could pursue the real truth of this film, instead of arguing passionately and unsuccessfully for false or simplistic illusions of truth.

It would be easy to ask, where did they go wrong? And that would be the wrong question. The rightful way to look at this is to start with this question: If you have the most remarkable and unique film evidence of something which may be a real but unknown species of North American primate, but may be just a guy in an ape suit, where do you go to determine exactly what it truthfully is?

And the answer is, ... nowhere.

Overlooked by one and all up to now is the simple fact that there was not, and still isn't, any formal established procedure for making such a determination. The second consideration is more pragmatic. A great solution would be to assemble a custom panel of experts across multiple scientific and technical disciplines to do a formal analysis. But, who's paying for it?

There is another factor that more-or-less doomed the early analysis efforts from ever achieving a proper and conclusive result, and that was technology. When we look at the computer graphics and image analysis tools we employ today, tools which did not exist then, we can see that the technology of the time simply could not result in a proper analysis for the truth of this mysterious figure in the film.

So it is often overlooked that the early analysis efforts were doomed from the start, by a lack of full understanding of the problem, a lack of the full range of necessary skills, and a lack of technology.

Now let us examine what did occur, and why those efforts to resolve the mystery would not succeed.

From the first showing of the developed film on Sunday, October 22, 1967, in the home of Roger's brother-in-law, Al

DeAtley (the man who arranged for the film's processing but now cannot remember where he had it done) two veteran Bigfoot researchers, John Green and Rene Dahinden, were present, and they remained connected to this film's legacy from then on. Although Rene passed away, John is still with us, although in his 80s now and less active because of health concerns. But these two men helped start the film's analysis with good intentions and a reasonable choice of where to turn to. They went to the academic and scientific community looking for experts on real biological species, hoping that these experts would validate that the creature seen in the film was biologically real.

But in retrospect, this was somewhat naive, because biologically knowledgeable scientists and academic scholars may verify that the filmed subject appears to be something real, but such people were utterly unqualified to falsify the alternative, which is demanded of a rigorous scientific proof. These men were not qualified to show or prove that what was seen was not a clever costume worn by a human performer. They were not special effects makeup artists.

As there are no full transcripts of these early days and preliminary analysis discussions, we will never know how this issue was considered, how much they assumed, how much they knew of the matter. It is unlikely that they even recognized how vital it would be for a person (or several) skilled and experienced in the design and fabrication of "creature costumes", in order to do a truly comprehensive analysis. This is not to say that they neglected the idea, so much as to say that they were not prepared to carry out the idea successfully. To carry in out successfully, they would need to understand Hollywood, and few people outside the industry did, or even do today. It is a world unto itself, quite removed from the scientific and academic worlds.

Now John Green would actually take the film to Hollywood, and ask for experts to analyze the film, but John was a

Journalist by profession, and Bigfoot investigator by avocation, and so while his intentions were excellent, his choice of experts was wrong. He went to Disney, and there was a well established perception at that time that Disney had the most brilliant effects technicians in the world, a perception cultivated by the Disney empire to promote their world famous flagship theme park, Disneyland, and its new satellite park, Disney World in Orlando, Florida. But Disney's genius effects technicians were the robotic designers at W.E.D. the "Imagineering" people who pioneered the parks amazing audioanimatronic show figures.

In John Green's book, he describes the discussion as follows:

(quote from 1978's "Sasquatch: The Apes Among Us," page 129:)

"Ken Peterson, a senior executive, told that their people [Disney] has already studied the film [PGF], and that if they wanted something like it in one of their own movies they would not attempt to film it, they would draw it. None of their mechanical creations for Disneyland was sophisticated enough to walk free, they were all attached to a base at some point. The only way to imitate the Patterson creature would be with a man in a suit."

John, having no film industry experience, could not have known that they were the wrong people to ask. Disney's expertise, as noted, was with robotics and cel drawn animation, not elaborate creature suits worn by actors. But the man was correct in advising that the suit technique was the only prospect for explaining the film as a staged event.

What was needed was a brilliant makeup artist who did creature costumes, and Disney was, frankly, a poor choice to find such. Disney Studios was never at the time a strong studio for films with state of the art creature effects. For them, an elaborate "creature effect" was turning Tommy Kirk into "The Shaggy Dog". Nothing Disney did came close to Universal Studio's rich history of creatures from the 50s and 60s. MGM also was a powerhouse in that category, with the recent works of "The Time Machine" and "The Seven Faces of Dr. Lao" (For which Makeup Artist Bill Tuttle received a special Academy Award). And 20th Century Fox had just stepped up into the major league of makeup effects with its production of "Planet of the Apes" (which would earn John Chambers a special Makeup Academy Award), released a few months after the Patterson Gimlin film was taken. But Disney was not the studio to find a great "creature" makeup artist.

So the error of thinking that Disney was a place for genius effects people was understandable for a Hollywood outsider like John or Rene to make, but it was an error none the less, and contributed nothing to the early analysis of this curious film.

In the book, "Sasquatch" by Don Hunter and Rene Dahinden, p.122, they describe Roger Patterson also taking the film to Hollywood, specifically Universal Studios Special Effects Department, where they asked if the studio staff could duplicate the film. They don't mention a specific person Roger talked to, but quote the effects technician as appraising the film as follows: "We could try. But we would have to create a completely new system of artificial muscles and find an actor who could be trained to walk like that. It might be done, but we would have to say that it would be almost impossible."

It is curious that the people at both studios were top professionals of their time and would know all the processes and options of that time. Apparently neither dismissed the film as a "cheap fake" and the people at both studios should have been knowledgeable enough to know a cheap fake if they saw one. But in recent years, the "it's a cheap fake" accusation has become the criticism of choice for makeup artists. Now granted, these modern makeup artists may know about new technologies that the 1967 studio people didn't know (because said technologies didn't exist then), but can any of these contemporary artists claim they know more about what could be done in 1967 than the studio professionals in 1967? That contradiction is generally ignored by people advocating a hoax, when they use contemporary artists opinions to support the hoax claim. Why are the professional appraisals of then and now so contradictory?

Returning to the ideal person to go to, if you had to choose one man of that era, that would have been Stuart Freeborn, in England, who had just completed his groundbreaking ape suits for 2001 A Space Odyssey. But how was John Green to know that? All the publicity was telling the world about Hollywood and the Planet of the Apes. The movie was becoming a franchise series of films, each trying to dazzle the world with more incredible makeup effects. And the makeup artists in Hollywood were loathe to even acknowledge how amazing Stuart Freeborn's work was (I was there, so I speak from experience here).

So the best man for the job was likely someone John, Rene, or Roger would not have known anything about. And asking around Hollywood would not have likely produced a reference to Freeborn's qualification. So while these attempts to get the expert opinion of Hollywood Special Effects Wizards was right in theory and well-intentioned, it simply was not to be and the film's analysis effort was clouded by that fact.

But could even Stuart Freeborn have done a proper and definitive analysis? Not likely, because the analysis technology of the time, his era, was inadequate. His opinion, however, would have been quite fascinating to hear.

This brings to a point one of the most misunderstood factors in the whole PGF Legacy. How do we properly analyze this film? What tools and techniques do we use, what expertise do we need to draw from, and what can the film data actually prove , as compared to what the film data may falsely suggest if mis-applied?

Perhaps the most profound misunderstanding in the 47-year analysis of this film was the under-estimation of the need for proper foundation knowledge of the best evidence, the film footage. It is film, it has been copied, and there are suspicions it was edited. Any analysis demands a person who has worked with film, cinematography (to know cameras, lenses, other equipment, and the results of film image exposures based on the use of said equipment), editing (to appraise the claims of the film being edited, and if such supports the claims of hoax), and film lab processing and copying.

Were experts with this background brought in, and given the necessary film copies, scans and data to do a thorough analysis? Sadly, no, never before.

Even the makeup artists often consulted in the 70s 80s, 90s and the present, what were they given, what did they see? Did they have full access to the best image frame scans and film data plus the computer imaging software to do a proper analysis? Sadly, no.

No one is to blame, because this mystery is unique and simply nobody was prepared for the process of doing it right. So once I jumped off the deep end into this pool of evidence, and began to try and identify the underlying issues, the foundation issues, I realized what was lacking. It's been seven years working to assemble the largest film image database ever held by a researcher in the film's 47 year history. And it has taken a lot just to sort out the real issues from the distractions.

And that is why this analysis is unlike anything done before in the history of the PGF.

I understand in depth how creature suits and effects are designed and created.

I understand film, cinematography, editing and film lab processes.

I have listened to all the skeptical claims of hoax, and graded them as to worthiness to investigate.

I have listened to all the proponent claims the PGF is real, and graded them as to worthiness to investigate.

I have studied other documentary footage Roger Patterson also filmed before the famous Bluff Creek PGF was taken, to better understand his use of filming equipment and his use of film types and lab processes.

Finally, after 47 years of good intentions hampered by failed efforts and inadequate technology, the right tools for the job are finally in place, and a true solution is possible.

If you love a truly great mystery, this is it.

And now is the time to solve the mystery.

Fig. 2-01

Chapter Two - Just The Facts

An introduction to the evidence and prospects for analysis

To begin, there's good evidence and not-so-good evidence. The good evidence is referred to as empirical evidence and it has some factual or tangible nature which makes the analysis or evaluation of it a potentially repeatable procedure, and good science relies heavily on the characteristic of repeatability in analysis. The not-so-good evidence is testimonial and anecdotal evidence, what people said, or recollected, or heard somebody else say. Amazingly enough, I've seen people so desperate to try and prove the PGF is a hoax that they have turned this fundamental concept of evidence upside down, throwing out the good evidence and embracing the not-so-good evidence. They disregard the film image data evidence, which is truly empirical and fine, and cherish the recollections of people interviewed 25, 30 or even more years after the fact, and regard those recollections as solid word-for-word perfect and reliable.

So I begin this chapter with my fundamental position on this issue. The film image data is the best evidence, and the solution is found in that best evidence. The "backstory", the pile of circumstantial, hearsay, anecdotal, and long-after-the-fact remarks and recollections of people are crappy evidence, and proves nothing because it's junk more so that fact. It most likely would never be admissible in any formal evidentiary proceeding (one that actually had rules of evidence that were enforced) or it would be torn apart on cross-examination. People desperate to prove the PGF is a hoax rely on it because it's their only hope of winning the argument and prevailing with their hoax belief (and the hoax idea is every bit as powerful a "belief" system as all the accusations of Bigfoot advocates being slaves to an irrational belief system). I will address this in detail in Chapter Ten, so I will for now stay focused on the good evidence, the empirical data which actually can prove something.

But allow me to illustrate two circumstances where testimony and empirical evidence conflict. Both oddly enough have to do with a PGF research endeavor where men went to the filming site the following June, 1968, and one man walked the approximate path of the PGF Hominid while the other man filmed him. The man filming, John Green, has a very clear recollection that he filmed Jim McClarin's walk with a Keystone K-50 movie camera which John recalls purchasing specifically for that event, a camera which John still owns today. But a photograph of John Green at the site with Jim McClarin beside him as John lines up the shot shows John is using a Revere "Arte Deco" model movie camera, not the Keystone K-50 camera he owns. John's recollection, as clear and unequivocal as it is, disagrees with the photograph. This photo of John Green and the Revere Camera was printed in Daniel Perez' "Bigfoot Times" dated October 2012.

Fig. #2-02 shows a Keystone K-50 model camera, the type John Green thought he had used, and below, is a sample of the Revere "Arte Deco" model camera Green is photographed actually using on site with Jim McClarin. There is simply no confusing one with the other.

Jim McClarin recalls that day that there were only four people on that site excursion, himself, John, John's son, and a friend of the son. He also recalls only one camera there that day. But the photographic evidence, from the archives of researcher George Haas, show a second camera was there, a Bolex with a 12-120 zoom, and film of a second walk by Jim McClarin is known to exist and has been scanned for study, and the second walk was taken by a different camera than the one John Green used to film the first walk. John's film itself verifies he used a 50' magazine camera (the Revere) and the other walk footage verifies it was taken by a 100' daylight load camera (such as a Bolex). The Haas archives also have numerous photos of that

24

event and show 11 people at the site that day (actually 10 people in the picture and person #11 taking the photo).

Fig. 2-02

Above, Keystone K-50 camera John Green recalled buying and using to film Jim McClarin

Below, Revere "Atre Deco" model camera that photographic evidence shows John Green actually using to film Jim McClarin, with Jim in photo as John lines up the shot. There is no mistaking one for the other camera.

So which do we consider the best evidence, the actual photographs and film image frames, or the recollections of the men? With no disrespect to either John or Jim, we must use the photographic image data as the more empirical, reliable evidence and believe it. The men are sincere and likely do actually feel their recollections are accurate, but the empirical data tells a different story. The recollections are in error. Rarely do we have such an opportunity to compare recollection and empirical data so succinctly as this instance, but recollection is known, studied and described as flawed in multiple studies or appraisals. More on this in Chapter Ten.

Fig. 2-03

Jim McClarin Walk #1 (start)

Walk #2 (start)

Fig. 2-03 Shows a frame sample from the first and second walks, and the camera aperture shape is distinctly different on

the two, irrefutable evidence two different cameras were used, and that walk #2 was not filmed with a 50-foot magazine type camera, as Walk #1 was.

The process of forming a "proof", a conclusion or determination, should ideally rely upon facts, empirical evidence and repeatable methodological analysis. That is the strongest proof. But we live in a world that is not perfect and so some proofs are simply not as pure and irrefutable as others. For example, a mathematical proof is about as pure and irrefutable as it gets. A physical or chemical proof is pretty solid, but can potentially be contaminated and ambiguous. Proofs in the "social sciences" (anthropology, sociology, psychology, etc) are less solid and definitive than physical science or mathematical proofs, but still, they are regarded with scientific and academic merit.

In law, like science, there are levels of exactitude in a "proof" or judgment. In criminal cases, the law requires the highest form of proof, "beyond a reasonable doubt", while in civil disputes the law only requires a preponderance of the evidence (the stronger case, or over 50% certainty) for the judgment. The higher standard requires you to rise above any reasonable doubt to prevail while the lower standard simply requires you to present the more reasonable argument, even if doubts linger.

Realistically, in any analysis of the PGF, its hominid figure and the solution to the mystery, there will similarly be levels of proof. Some topics or issues will have solidly empirical evidence and analytical methods to achieve solidly factual proofs (the identity of John Green's camera, noted above, as one example). Some issues will rely upon a level of argumentative reasoning, where the person making the judgment draws upon personal knowledge or experience to help arrive at a conclusion. There is nothing wrong with such judgments. It simply would be appropriate for the person making the judgment to acknowledge that it is more argumentative than a pure empirical proof. And I will aspire to follow this method in my discussion of issues, topics and controversies. I'll lay out the facts, and if they alone can form a conclusion, I'll state so. But if after the facts are laid out, an interpretation of those facts is needed, then I will explain the reason for my conclusion based on my judgment.

Finally there is the matter of proofs of inclusion or exclusion, and these types of proof are often confusing to many people. This type of proof can potentially prove one side of an argument conclusively beyond any doubt, but cannot prove the other side of the argument with equal certainty. An example is the above described comparisons of Jim McClarin's two walks filmed in 1968 at Bluff Creek. If the two films had an apparent similar camera aperture, we may conclude that the camera filming each walk was the same or a similar type camera, but we could not say with absolute certainty it was "The Same Specific Camera" (not just the same make and model but the same individual and specific camera). But if the camera apertures are different (as they actually are), we can say with absolute certainty that the two walks were Not Filmed With The Same Camera. So one argument, that there were two cameras, can be proven conclusively beyond any doubt (and that is the case here) but we cannot prove with equal certainty that if the apertures look the same, that only one camera was used. We could only conclude "there's no evidence of two cameras".

These "Inclusion/Exclusion" types of proof are valuable, but we must realistically recognize when they may prove only one argument with certainty, and will only fail to disprove the other argument. One is proof with certainty, and the other is actually inconclusive. But for the issues where a specific proof would exclude a real spontaneous encounter to a certainty, thus proving a hoax to a certainty, (or the opposite) are the strongest forms of proof we should explore. Whenever they come up in the discussions to follow, I will try to make it clear which claim (hoax or real) can be proven to a certainty, and which cannot, by that specific point of analysis.

The Evidence:

The film image data, in the form of both scanned individual frames from the various 16mm films associated with the PGF event, and the various still photos taken by researchers in various investigations of the event, are the best evidence. That said, let's now see what that filmed and photographic image evidence actually is.

1. The PGF itself, 23.85' of 16 mm color film footage, a verified 954 individual film frames, showing the subject figure (whom we call "Patty") walking away from Roger's camera as Roger chases after her, until he runs out of film. So we have 954 individual images here. But we also have multiple copies of this footage, some full frame, some optically printed to zoom in, repeat print (for slow motion effects on projection), some freeze framed, and printed at least three different times by three different printing systems. These varied copies also have evidentiary value in some of the issues, so while they are in one sense the same frames, the varied printing methods are valuable for comparative study. I was the first researcher to appreciate the value of studying various copies, and I accumulated the largest image scan database in the PGF history, one of the factors which allowed me to produce a unique and definitive analysis no previous researcher could accomplish.

Fig. #2-04 shows one example of studying multiple copies, and it grades each copy as to image clarity and detail, and only by having multiple copies to compare can such be done. Never before had a researcher done this kind of analysis, but it is vital to insuring that we work with the best photographic evidence.

2. The rest of the "First Reel" has some research value in the overall investigation because it helps rule out hoax or fraud. This other first reel footage is either Roger or Bob Gimlin on horseback, with a white pack horse in tow, the one man being

filmed by the other man. This horse and rider footage accounts for the remaining 76 feet of the 100-foot roll.

Fig. 2-04   Quality Studies.

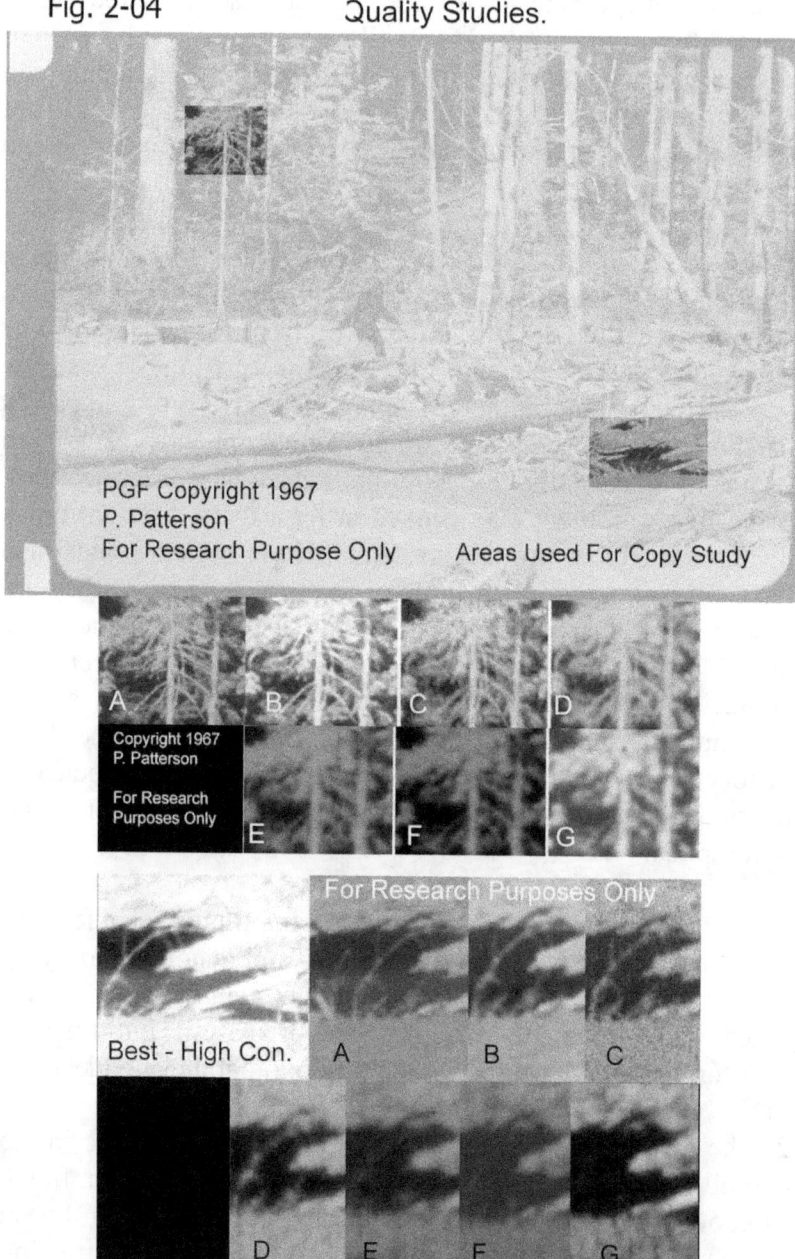

PGF Copyright 1967
P. Patterson
For Research Purpose Only   Areas Used For Copy Study

Copyright 1967
P. Patterson

For Research
Purposes Only

A B C D
E F G

For Research Purposes Only

Best - High Con.   A   B   C

D   E   F   G

3. The second reel footage is less assured and actually causes more controversy than it clears up. The trackway footage is more likely second reel, the material of Roger holding two cast footprints standing by a tree may be second reel, the shot of Roger pouring a plaster cast of a footprint is disputed as being second reel, and the majority of whatever else is on that second reel is unknown today. So second reel footage has little evidentiary value in relation to the central question of the authenticity of the film, but the missing parts are like darkened culture dishes where suspicions are spawned. More conspiracy claims arise from the second reel material than do legitimate evidence for the authenticity of the hominid. Only the trackway footage actually does support the authenticity. But debates on what the second reel material proves are ferocious and imaginative. (See Chapter Twelve - Loose Ends)

4. Roger's other documentary footage, before the October 1967 Bluff Creek event, can tell us some interesting things about Roger's skill as a cameraman, and what equipment he used. They also document some of his activity before the famous event.

5. Various still images made from the PGF 16mm footage are useful for some specialized forms of analysis. The 4x5" transparencies Roger himself had made by Kodak are the image quality benchmark standards for appraising the camera original image. The 12 known Cibachromes made by Rene Dahinden and Bruce Bonney are excellent for study of "Patty", but there are concerns about some image artifacts in them, especially the "hand" in the Cibachrome 352, which is not factual. The Noll frame set, taken at 3K resolution with a microscope by Researcher Rick Noll from a 2x zoom-in copy in John Green's archives are an excellent study set of frame images. And then there are the complete frame scans which I have made over the last 6 years, numbering in the thousands.

6. John Green's filming of Jim McClarin walking a path close to the PGF Hominid path in June 1968 is an excellent research

reference, and the second walk by Jim (taken by a cameraman not yet identified) is also useful, especially when the two walks are compared.

7. Rene Dahinden had someone film him on the Bluff Creek site in 1972 while Rene held up an 8' measure bar and various takes had Rene in various positions along what Rene assumed was the PGF Hominid path. This footage has some research value.

8. Various still photographs by Rene Dahinden and Peter Byrne at various times visiting Bluff Creek are also useful for study of the environment, and give us a perspective on the setting itself.

As you can see from this list, there's a wealth of photographic image material to work with for study and analysis. The challenge is in understanding how to use and analyze the photographic image data. In "Abominable Science" author Daniel Loxton makes the curious claim that the PG film cannot speak for itself. Actually, the film is remarkably eloquent and informative about what it can say about the truth of the event. Author Loxton simply passed off his own ignorance as the status of photographic analysis process. It did not speak to him, because he does not know the language. So we will disregard his appraisal and analysis as naively uninformed (that's the kindest way I can put it).

How much image data is there? Here's a fairly thorough summary:

Frame Inventory Reference Image Files   131 photoshop files
PGF Copy 1 scans   1003 4272x2848 JPEGS
                                        1003 RAW files
PGF Copy 2 scans   35 JPEGs
                                        35 RAW files
PGF Copy 3   960 JPEGs
                                        960 RAW files

PGF Copy 4		214 JPEGs
		214 RAW files
PGF Copy 5		46 assorted format image files
PGF Copy 6	frame captures	224 image files
PGF Copy 7		
PGF Copy 8		5849 assorted format image files
PGF Copy 9		18 scans and image files
PGF Copy 10		542 JPEGs
PGF Copy 11		1 image of filmstrip
PGF Copy 12		959 JPEGs
		959 RAW files
PGF Copy 13		951 JPEGS
		951 RAW files
PGF Copy 14		605 JPEGs
PGF Copy 15		18 JPEGs
PGF Copy 16		278 JPEGs
PGF Copy 17		87 assorted image file formats
PGF Copy 18		1060 JPEGs
PGF Copy 19		1512 JPEGs
PGF Copy 20		1368 JPEGs
PGF Trackway Green copy		206 JPEGs
	ANE Trackway	123 JPEGs
PGF Roger casting tracks C8		350 JPEGs & RAWs
Roger Patterson Documentary footage		804 JPEGs &RAWs
First Reel Horse and Rider footage		5149 JPEGs & RAWs
Green/McClarin Walk 1 and other site work		1272 JPEGs
McClarin Walk #2		468 JPEGs
72 Dahinden Survey Scans		216 JPEGs

Total Image Files   **28,571**
(Munns Research Archives)

Aside from these listed 28,571 image files, there are additional files for comparative study of 16mm film, splicing studies, camera edgecode studies, camera ID studies, film date code studies, lens studies, film image quality studies, human anatomy studies, and mammal fur image studies, derived both from actual historical home movies and older 16mm documentary professional productions, and new PGF research work I performed with research grants. Additional scans and image files do continue to accumulate, so the exact number at the time of publication is likely a bit higher still.

There are also the holdings of other researchers. Daniel Perez has a remarkable archive of material as well, many images and documents exclusive to him, from the estate of George Haas. Some of these are stills taken when Green filmed McClarin walking at Bluff Creek, June 1968.

So all together, we are looking at an image database of about 400 gigabytes of empirical data with analysis potential, about 35,000 images. The Haas Archives could expand this number once they are fully inventoried.

Now let's get acquainted with the film and some descriptive terms I'll use often.

The PGF - The common abbreviation for the Patterson-Gimlin Film, the 23.85' piece of footage showing the hominid Roger chased and filmed.

"Patty" the PGF's hominid subject figure (so named because she has apparent female breasts and so Roger's last name, Patterson was shortened and feminized to "Patty" for her)

Bluff Creek - the specific film site where the event occurred and Roger filmed.

PAC Copy family - the copies made from the true full frame contact prints Roger himself had a lab make in late 1967. PAC is my abbreviation for Patterson Archive Copy.

Green/Dahinden Copies - This copy family derives from the Ektachrome master Canawest labs made for John Green and Rene Dahinden in early 1968 from the camera original Roger loaned them. These were made on an optical printer (not a contact print) and so even the full frame version is slightly cropped. Also contain a 2x zoomed-in version, showing Patty better, slow-motion copies (each original frame printed two, three, or four times) and a freeze frame on VFC 354 (erroneously called frame 352), the best lookback frame.

ANE Copies - These copies were made from the original for the "Bigfoot Man or Beast" documentary American National Enterprises (ANE) released in 1971. They also used an optical printer, made full frame, zoom in, slow motion and freeze frame effects in their copies and used them in the documentary.

The PGF has six filming segments, where Roger's camera was switched on and off. These will be described in far more detail in Chapter Six, How the Event Occurred, but here's a brief reference description of each.

Segment One is frame 01 to 094. Patty is seen full body, and from a back view as she walks away.
Segment Two is frames 095 to 190. Patty is seen sometimes blocked by foreground objects, and her lower legs obscured by terrain, and in a back view.
Segment Three is frames 191 and 192. This is likely a camera trigger slip, and Patty is not verifiable but may be seen, but the creek Roger must cross is seen in part.
Segment Four is frames 193 to 233. This segment has very few sharp frames because Roger is running around a branch structure and charging forward toward Patty as he films.
A lot of this is just the camera pointed down at the ground.
Segment Five is frames 234 to 686. This is the most important segment, because it includes the lookback by Patty and also because Roger has planted himself to hold his camera steady as he pans to follow Patty walking across the terrain. Most of the good body images come from this segment.
Segment Six is 687 to 954. This is where Roger has moved again to get around some trees blocking his view, and films Patty's distant walk away toward the woods, and then Roger

runs out of film. But Patty is very far away and so images of her have little anatomical detail for analysis.

The Lookback - This is about one second of film, where Patty is walking at a near profile path for a moment and she looks back at Roger's camera to see what he's doing. As Roger is holding his camera steady at the time, these film frame images are the sharpest ones in the set. But Patty's legs are partially blocked by foreground clutter and debris so we don't see her feet well. This is the only time we see Patty's breasts, or her face.

Now, how does a researcher analyze this incredible volume of image data? To begin, what are you trying to study or determine? A vague question like "Is she real?" needs refinement and focus on specific and articulate study goals.

For Example:

If you want to study the breasts and any motion of them, you need the lookback sequence, and ideally the 4x zoomed in copy set from PGF Copy 8.

If you wanted to study the head shape, again the lookback segment is ideal.

If you want to study overall anatomical proportions of her, select frames from segments One and segment Five would be best, which show her body in a variety of walking postures and varied arm positions.

If you want to study the contours of the back, segments One and some early frames of Segment Five show the back well.

If you want to study the landscape where the event occurred, or issues of splicing the film, you need frames from a PAC Family copy, which is true full frame contact printed. The optically printed copies in the Green/Dahinden group or the

ANE group are not true full frame. Photogrammetry studies of the Bluff Creek area need to be done with PAC full frame copies, plus the other site film frames from Green and Dahinden, and stills from Dahinden and Byrne. The Perez Archives also have some fine site photos from different perspectives that can be useful.

If you want to study the interaction of Patty's walk away and Roger's pursuit of her, together in the context of the terrain, the PAC full frame copies are best.

The best set of closeups for any image stabilizing of Patty's walk from beginning up to the end of segment Five are the Noll frame scans, from a 2x zoom copy in John Green's collection. If you just want the lookback segment, the Copy 8 4x zoom in set are the best.

If you want to study elapsed time for tree shadows at Bluff Creek, you need a PGF copy set of frame scans plus the two McClarin walks (to show elapsed time between his walk #1 and his walk #2) and the Rene Dahinden 1972 Site survey footage, to see how tree shadows shift over the elapsed time as he changes his position for each of his filming segments.

So what film frames or photographs one uses must be matched with the specific goal of analysis. And this is where all researchers before me could not succeed. They did not have the multiple copies and an understanding of the value of each so they could use the best image data for the intended analysis goal. My database was the first to inventory this body of image data and see where each copy or frame scan set was most useful. One of the surest ways to get a false conclusion was to use the wrong image data for an analysis.

Example: For awhile, there was a crazy idea being advocated by some that there were multiple creatures at Bluff Creek that day, and men with rifles or guns, and the men were shooting these creatures. This is commonly called the "massacre theory". One of the cornerstones of this was a white flare-like spec on the film, at frame 613, on one of John Green's copies. But it was apparently scanned for TV viewing and the TV conversion to 30 frames per second blended some frames and one such frame blend made a half intensity flare. Analysts looking at these TV frames thought they were seeing a gun muzzle flash increase and then disappear, but all they were really looking at was a single light flare on one copy, and the flare wasn't on any other copy, so it wasn't on the original film. And the TV blend made it appear as if it was on two frames, one true frame and one conversion frame blend. This shows the value of being able to compare various copies to see what is true original content and what is an image artifact of one specific copy. The "massacre theory" fell apart with this analysis. It has no merit.

Fig. 2-05 shows the one frame from one of John Green's copies that has a light flare, and the same frame from another copy, which does not. No PAC copies have the flare either, so it is isolated to one copy. But researchers with just that copy as their reference (and likely just a tv scan at that because of claimed second frame with a partial flare, which is created by a TV conversion that blends frames to get the frame count up to 30 frame/per/second) and they used this flare to concoct a rather absurd claim that men were shooting ast the creature in the film.

If this body of image data evidence seems to be intimidating in its volume, you can start any research or study of this film mystery by looking at various books and studies done on the topic.

The few scientific papers on the subject can be found on an online scientific journal named the "Relic Hominid Inquiry" (or RHI). Two of those were co-authored by myself and Jeffery Meldrum, Professor of Anthropology at Idaho State University. http://www.isu.edu/rhi/

Fig. 2-05

Light flare on one PGF Copy from John Green Archives mistaken for gunshot "muzzle flash" which briefly resulted in the "massacre theory" (the idea hunters were shooting at Patty and other claimed subjects at the scene.

However, there is only one subject, Patty, and the flare is just an anomaly of one specific copy, and is not on the original or other copies.)

Left below is from Green Copy 13. Right is the flare, on Green Copy 12.

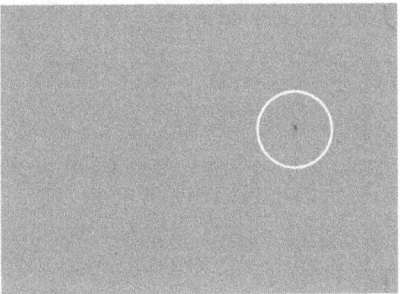

Using a Tonal Inversion overlay (described in Appendix Six) the flare is isolated (bottom image)

One paper is on the film image evidence itself, and is a foundation paper which simply evaluates the factual evidence potential of the film image data, and addresses issues such as questions about splicing, image graininess, motion blur, film resolution, copy types and processes, etc. That paper does not address the authenticity of the hominid subject. Rather, it simply studies the quality of the evidence other studies use in making such determinations.

"Analysis Integrity of the Patterson Gimlin Film image"

http://www.isu.edu/rhi/pdf/ANALYSIS%20INTEGRITY%20 OF%20THE%20PATTERSON-GIMLIN%20FILM%20IMAGE_final.pdf

The other paper co-authored by Meldrum and myself is specifically on studies of anatomical features such as apparent adipose tissue deposits on the body of the PGF Hominid in relation to real human anatomical features and costume structures.

"Surface Anatomy and the subdermal adpose tissue features in the analysis of the Patterson Gimlin Film Hominid"

http://www.isu.edu/rhi/pdf/Munns-%20Meldrum%20Final%20draft.pdf

Another essay authored by Brian Keith deals with studies of movie ape costumes in relation to the PGF Hominid.

Essay: "The Patterson Gimlin Film: What makes a hoax absolutely genuine?"

http://www.isu.edu/rhi/pdf/Keith_rev.pdf

Critics dismiss all these papers as unscientific and useless, but when you ask them to show any scientific papers with differing or opposing conclusions, those critics shift into an array of evasive claims, like they don't have to prove anything, or all burden of proof is with proponents, or that the proof of hoax is contained in internet forums where the PGF is discussed, and these posted comments by anonymous people of no verifiable credential or accomplishment constitutes a splendid and ironclad "proof" of the hoax. But ask one of these anonymous people to write up a formal paper, put their real name and verifiable credentials on it, and submit it for some type of academic publication, and you will only get evasions, excuses, and hostility.

As a skeptical text, the best one can find is David Daegling's book, "Bigfoot Exposed" (2004), and the author does have academic standing at a university and he does approach the topic with generally good academic discipline. He err's in one discussion of movie ape suit technology (noted in more detail in Chapter Four), but the author does try to appraise the topic with reasonable consideration of the evidence. And even he concedes, on page 106, about the PGF, *"since at this writing thirty-six years have passed, and definitive proof of a hoax has not surfaced."*

"Definitive proof of a hoax has not surfaced!" That's the best you'll get from any kind of reasonable scholarly writing.

"Abominable Science" (2013) fares even worse, in terms of scholarly discipline and conclusive results. Co-Author Daniel Loxton, who wrote the PGF section, ignores the film image data with his naive opinion that the film cannot speak for itself, to open the door to the horribly unreliable recollections and anecdotes which a dedicated skeptic must rely upon to try and make a case for hoax. And the best he could do is compare the PGF to an earlier eyewitness account of a Bigfoot encounter by a man, William Roe, and then try to tie the two together and say *"If Roe's report is a hoax, we would be compelled to conclude that the Paterson-Gimlin film is also a hoax"*.

Given author Loxton cannot prove the first "IF" (that Roe's story is a hoax), he is admitting he cannot prove the PGF is a hoax either. He may as well have said, "Since William Roe's report is not proven to be a hoax, we cannot conclude that the Patterson-Gimlin Film is a hoax, either." That's what he's actually saying, but he hasn't the courage to admit such.

Loxton's book is discussed in greater detail in Chapter Ten, "The Swamp".

So, if you want to do scholarly research on this question, " is the PGF a hoax?", the pickings are slim for resources. You have a few academic papers supporting the film's authenticity (noted above), and none concluding a hoax. You have the above two skeptical texts, plus Greg Long's "Making of Bigfoot (and often abbreviated MoB), the skeptics' "New Testament", and many books by Chris Murphy, Grover Krantz, and others on the affirmative side advocating Bigfoot does exist.

What encourages me the most is that people open to the prospect that the PGF is real, and that something called "Bigfoot" may exist, are trying to approach the matter with scholarly discipline (regardless of how you appraise their success or failure toward that end) while the people adamant that the PGF is a hoax and there is no Bigfoot take the low road with the crappy evidence and present their arguments in academically undisciplined fashion. So I would anticipate that eventually, the people who are in fact trying to do science and follow academic discipline are the more productive in their analysis and research, and thus more likely to get to the truth.

Now, let's meet Patty.

As previously noted, the subject figure seen in the Patterson-Gimlin Film footage is commonly referred to as "Patty", a nickname derived from Roger's own last name, Patterson, and feminized because she has apparent well-endowed female breasts. There is occasional confusion in the coincidence that Roger's wife is named Patricia, and also called Pat, but there is no intended or implied association between Pat Patterson and the figure known as Patty. Patty can also be referred to more clinically as the PGF subject figure, or the PGF hominid.

Fig. #2-06 is the finest image we have of Patty.

In terms of basic foundation analysis of Patty, I personally like the process where I list the options I consider reasonable, and see which can be then excluded for some stated and logical reason, for cause. I proposed it in various discussion forums and the idea was met with surprising objection or derision by the skeptical brethren (who claim to be "critical thinkers", but seem to have only understood the adjective and not the noun), and I can't imagine why. It seems a nice, logical and methodological way to approach a mystery in search of a solution.

So, thinking out loud here, this is how I personally work the question of what Patty may or may not be:

Because she appears fur covered, I'd first consider known animals, either native species, like a bear, or exotic species, like chimpanzees, orangutans, or gorillas. But we have excellent anatomical references on all known animals and Patty does not match any of them. If anything, I find her more humanistic than animalistic. So I exclude known animals for that reason. I don't know of anyone arguing the affirmative for other known species, so I see no reason to go further with this option.

Fig. 2-06

"Patty" in VFC-2 Frame 354 (mistakenly but popularly called frame 352).

I'd second consider the film subject figure might have been faked by some special effects process, like stop motion animation or cel drawn animation, composited into footage of Bluff Creek. But these techniques require a locked down camera and a constant level of clarity, and the PGF footage is hand held, the camera moving constantly, and has some intense motion blur in sections. Composited animation effects had never been successful at the time (1967) with hand held, moving and occasionally motion blurred footage (they can do it today, with CGI processes, but couldn't in 1967). So these options I can exclude for cause.

As John Green noted in Chapter One, Disney people did consider a robotic solution, as they were one of the industry leaders in humanistic robotics at the time. But no one of that era had ever made a bipedal robot self-contained that could walk freely through a natural terrain. So they excluded that option and I concur. It could not be a robotic figure.

Nobody is advocating these options so I see no reason to explore them further.

Some Bigfoot enthusiasts advocate a paranormal entity as an explanation, but I don't find any paranormal entity credible and so I personally will not explore a paranormal option. Others are free to disagree. Curiously, many skeptics think the description of Bigfoot IS (my emphasis) a paranormal entity, and they often use a strawman tactic to select the most outrageous fringe description of Bigfoot to set up their strawman they can easily defeat. I simply define Bigfoot as another hominid possibly co-existing with us humans, and I find nothing paranormal about that definition or that possibility. I'm sure there will be lively debate on this opinion of mine.

So, by my thinking, we are left with three real world options. Either Patty is a human in a fur costume (a fake), or she's a real human naked female with a medical condition called hypertrichosis (excessive body hair) and a deformed skull, or she's a hominid as yet unclassified or recognized by the biological sciences co-existing with us. At this level of consideration, the human in a fur costume is by far the most common and probable, and so it is reasonable for people to gravitate toward that option. And that is precisely why so much of this book discusses the concept and processes of designing and fabricating such fur costumes, plus the process of making (or faking) a film. The option must be fully explored and considered.

Now, in terms of the essential evidence of Patty we see in the film footage, we can say with confidence that the subject figure does appear to be fundamentally humanistic in general form, upright bipedal posture and locomotion, limb proportions somewhat different from either known humans or known apes, a head with a conspicuously flat cranium after the brow ridge, a head that seems to have a somewhat pointed back top (when seen from the lookback face position) and a body covered in a dark brownish/black fur. But as to the color, keep in mind that color film does sometimes shift the color of something photographed and the copy process can also alter colors somewhat, so don't be surprised if you see some images of her looking near black, and some images of her looking almost an auburn brown. That's the color balance of the film copies creating the discrepancy. Her soles of the feet seem pale. We don't see much detail of her hands, except they occasionally bend and straighten, or seem to ball up in a fist. We don't see much facial detail we can reliably analyze.

So in summing up this chapter, the good evidence is the film image data, it is plentiful, it is of excellent photographic quality, and we have sufficient data to eliminate suspicions of falsified image data. What you see is what Roger filmed and Bob Gimlin witnessed in real time (with allowance for brief camera starts and stops between segments) that day at Bluff Creek, CA.

And in terms of analysis, we must give the human in a fur costume every consideration for being validated or affirmed.

We can only dismiss that option if we find that costumes simply cannot accomplish what we see in the film.

Finally, I do want to give warning to any aspiring researchers. The PGF is a profound mystery and not for the dilatant. Any real understanding of it requires a substantial commitment of time, effort, thought and analysis. But I can assure you that the mystery of the PGF is worthy of such a commitment of time and effort. It is a splendid body of evidence of a truly unique encounter and a subject which sooner or later, the anthropological world will need to come to terms with.

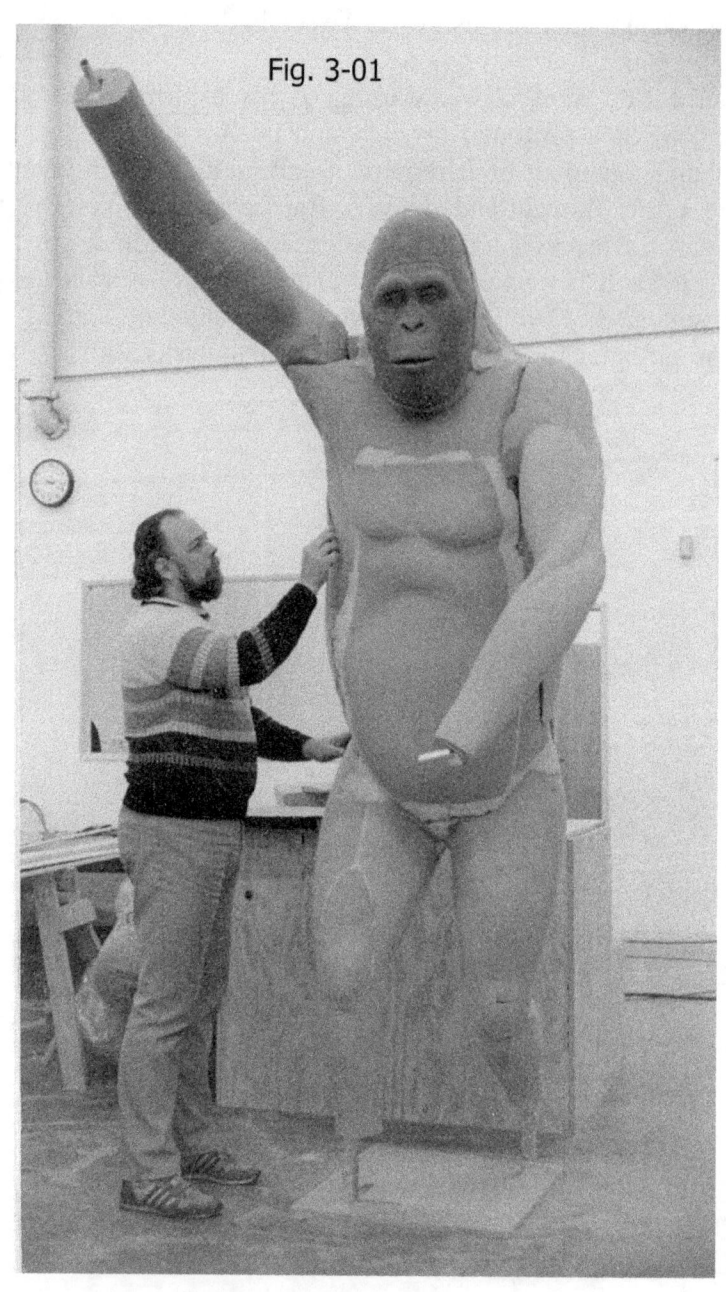

Fig. 3-01

Chapter Three - Standing at the Intersection of Hollywood and Sasquatch

An introduction to My Background in Relation to This Film Study

In the fall of 1967, I was a second year theater and film major at Los Angeles Valley College, and among my classes for that fall semester was Basic Stage Makeup. In the same fall of 1967, Roger Patterson was exploring the Pacific Northwest woods and forests, with hopes of documenting the existence of the creature commonly known as "Bigfoot". In late October, I was learning how to cut medical paper tape with a curved half-almond opening on the bottom side, to attach to the eyelid of a Caucasian actor, simulating the epicanthic fold of a Asian person's eye, so I could work on the Theater Arts department stage production of "The Chinese Wall". Roger was with Bob Gimlin, at a woodland area called Bluff Creek, in Northern California, with a 16mm camera, investigating some large footprint trackways that had been recently reported in the area.

On Friday, October 20, 1967, Roger filmed something which has been a controversial cornerstone of cryptozoology ever since. That something he filmed may have been a human actor wearing a fur suit. Or, it may have been a primate-like creature of unknown scientific identification, and generally referred to as Sasquatch, Bigfoot, or "Patty" (in the case of this film's female creature). Roger went on to try and convince the world he'd filmed a real Bigfoot, until his untimely passing less than five years later. I went on to work a variety of film crew positions on student films, independent films and documentaries, before specializing as a professional makeup artist for films and TV, and focused on the special makeup effects aspect of this job. And 41 years after that day in October, 1967, my career as a "creature guy" (a specialized makeup artist who does prosthetic and special effects makeup, including creature body suits) would intersect with Roger Patterson's filmed legacy, the 16mm filmed documentation of Patterson's Creature.

Before I got to that intersection, it seems that fate or destiny made sure I was equipped and prepared.

In the fall of 1966, Los Angeles Valley College was starting a film studies program, and I started college that same semester. I loved film and wanted to pursue a career making films, so I enrolled in the one film production class they offered for the first time that semester. Other films schools (at UCLA and USC quite notably) had been running their film studies for quite some time and so they had settled into a system whereby student projects were done on Super 8mm film to start, graduating up to 16mm for more advanced student projects. The Valley College program, being a new and experimental class, went off on a totally different track. The whole class was going to make a short film, and in 35mm color, with sync sound, like the major studios did. Now 35mm was unheard of for student film projects, but the class was an experimental one so the staff figured "why not?" The instructor had some industry connections to offset the costs of such a project, a bonus for the idea.

For filming at nearby CBS Studio Center, they used a standard 35mm BNC Mitchell with a 1000' roll of film, and the people filming all the CBS TV programs didn't want to be filming a good scene and have the magazine run out in the middle of it.

So common practice was whenever they ended a shot and checked the magazine footage, if it had under 200' left, they'd just pull the magazine and load a new one with a fresh 1000' load. The camera department would take the magazine, cut the film loop that normally threaded into the camera's film gate, and thus separate the exposed film footage on the take-up core from the unexposed raw film on the feed core of the magazine. Both were unloaded in the darkroom, the exposed film was sent to the lab for processing, and the unexposed film was wrapped back up in black camera paper and put in a film can, and marked as a "short end" and its approximate footage (120', 90' etc.).

The CBS people usually didn't like to use the short ends, and the head of the CBS camera department was friends with the instructor of the class at Valley College. So he agreed to give us all these short ends as our filming stock. He also agreed to loan us some 35mm cameras, usually Arriflex II, occasionally with "blimps" (sound isolation shells) and assorted lenses, tripods, etc. for weekend shooting, as long as we checked the stuff out on a late Friday and had it back first thing Monday morning.

So we had our 35mm camera equipment and our 35mm film, and the teacher used other industry connections to get processing lab discounts. And so we were set to make the only 35mm student film in the country, when UCLA first year film students played with Super 8mm.

The teacher decided that the whole class would do one 20 minute short project and a writer/director was selected. He had some history with the theater arts department and had written and directed some plays there already. The rest of us in the class would be his crew. I had no special film crew skills then, and so I thought my best shot for a crew position was as First Assistant Director (the "AD"), because the job was mainly planning, scheduling and just sort of coordinating the various crew activities on set. After the duties were taught to us all, and a sort of skill test was devised, I took the test and didn't get the job of AD. I got the one above it, Unit Production Manager (UPM). But while the AD is the lead man on the crew as filming is done, the UPM is the behind the scenes, before and after planner and organizer, as this project divided the tasks. It wasn't the job I had hoped for, but in retrospect, it was a blessing I now appreciate.

One of my tasks was to get the camera equipment and short ends each week from CBS Studio Center and return the camera equipment early Monday morning and deliver the weekend's filming rolls to Kodak for processing. So to make sure I get the right equipment, I had to learn what everything was, what attached to what and how. Thus began my education in filming camera equipment, and my education in film itself, processing labs and the like. I got to the point where I could easily switch lenses on the camera, change the motor from a wild motor (variable filming speed) to a 24fps sync motor, and I even learned to load film in a magazine blindfolded (which is essentially what you are doing with a camera black bag, because your hands are inside doing the work and your eyes are outside seeing nothing of the work), and I knew the film processing timeline for getting the film developed.

When I took up analyzing the PGF and realized there were issues about the camera, lenses, and film processing, I was on familiar ground, well prepared to study the issues.

A second curious thing happened on that first film project with prepared me well for my later studies of the PGF. CBS didn't have sound equipment for us to borrow, so we had to rent it. The premium sound recorder of the day was a Nagra 3, the pro recorder used in the industry at that time. But our budget was limited so our sound crew ended up with something considerably cheaper, what I vaguely recall as being called something like a "StellaVox". It could record decent dialogue sound, and it did have a sync cable to hook up to the camera, but we never got the sync pulse signal to properly regulate the two devices. As a result, when we had our film developed and a work print made, and had our dialogue sound transferred to a 35mm mag stripe stock, and both were loaded into a movieolla for editing, the lip sync was wildly off, when the scene marker (also called a "clapper stick") was synced up at the start of the shot. Somebody had to do lip syncing of the audio track. That somebody became me.

Once we knew our picture and sound tracks were out of sync, the teacher looked for volunteers to learn how to run a movieolla and do the lip sync work, a very tedious task. I was fascinated by every part of the filming process, so I essentially volunteered for any job that was open, and I volunteered for that one too. They showed me how it was done, I picked it up quickly and so on weekends I was helping organize the filming of the project, and on mid-week days I was in the editing room syncing up the last weekend's filming. The task involved cutting the mag stripe sound track (a magnetic stripe on clear 35mm film base stock) to either shorten it up or slug it out longer with added pieces, to shift the timing of words so each line of dialogue timed correctly for the picture of the talking person. To do it well one needed to learn how to watch a mouth moving and figure out what sounds you hear matched the lip motion. My brain was wired to make the connection and so I did well at it.

30 years later, I would invent a digital character lip sync software and get my first patent awarded for that invention (called "Totalsynch"), and here was where it started, in that editing room on the college campus, lip-syncing dailies on this student project in the fall of 1966. Once I was in the editing room syncing dailies, helping the director cut the picture was a natural continuation of the work, and that lead to setting the film up for printing, negative cutting, and sound effects editing. And 40 years later, when I started my analysis of the PGF and realized that there were issues with the film, various copies and questions about splicing and editing, I came to the PGF task well prepared. I was a "hands on" film editor.

The next semester, the teacher decided to go with one major project, again a 35mm short film the whole class would do, but also do a few 16mm projects others in the class could do as well. I got one of the 16mm projects and was on my way as a writer/director. I had a small crew of fellow students, including a cameraman who was the lead cameraman on the prior semester's 35mm project. But on our first weekend of filming,

53

he canceled on me that Saturday morning as we were getting loaded up to drive to the beach, my chosen filming location. Without a cameraman, what could I do? Some people would desperately call in another cameraman. Others would fold it up for the day. I just figured "how hard can it be to run a camera?" and so I became my own cameraman. And off we went to film at the beach.

And so, as Roger Patterson was renting his K-100 camera and filming his documentary footage in May, 1967, I too was filming with a 16mm camera and loading and unloading 100' rolls of film on daylight reels. So as soon as I got immersed in the PGF debate and heard people debating how Roger had reportedly unloaded his first reel under a poncho and loaded up the second reel for filming the trackway, I had a funny sense of Deja Vu, "been there, done that" with 100' daylight reels many a time. These people debating that aspect of the PGF story were unsure if the described camera unloading was true or false, possible or not. I'd done it many a time and knew from hands-on experience that the described activity was perfectly common, easy and truthful. What a difference some real-world hands-on experience makes, when analyzing this PGF mystery.

I would go on to do a second 16mm film of my own (writing, directing, photographing, and editing both), and work again on another 35mm major class project as UPM, during my first two years at Valley College. On my own projects, I became a skilled cameraman, using Arriflex, Eclair NPR, Bolex, Beaulieu, and Auricon cameras on various sets, as well as becoming a skilled still photographer with a Nikon F camera. I had my second edition (1966) ASC Manual (the "Cameraman's Bible", a cinematographer's reference manual put together by the American Society of Cinematographers), which I would oddly turn back to, 40 years later, to research issues of film stock, cameras, lenses and such for the PGF analysis. I had spent ample time in an editing room doing essentially everything, cutting picture and sound, syncing dailies, setting up A/B Rolls of Picture for optical printing, setting up sound reels for mixing, cutting negative, the works.

Having mastered the various crew jobs, I started getting requests to crew on other people's student and independent film projects. The illustration of me beside an Eclair NPR camera is when I was one of three cameraman filming singer Rod McKuen's concert tour, "Sounds on a Summer Night" in 1971. We filmed at various locations around the USA.

Fig. #3-02 was taken on our tour around the country. We filmed a concert at Kent State, Ohio, at the Red Rock Amphitheater outside Denver, and at the Santa Monica Civic Auditorium, as I recall. The photo was likely taken on the road to one of those concerts, where a sign or billboard referenced either the location or the event.

As I went into my third year at Valley College (which was a bit ironic because it was traditionally thought of as a "two year college", but they kept adding new film courses so I kept attending), I started doing more outside film assignments on independent and documentary projects. One stands out in my mind because of a curious PGF connection. A professor at UC Irvine was doing a short showcase film of his own and needed a sound man. I had mastered on-set sound recording as well (having taken the class in Production Sound Recording), using the Narga 3 (we had given up on the StellaVox after that first project, and moved up to the Nagra), and had done some film lighting, and so I often got bundled as a sound/lighting man on set, and was recommended to this guy as such. All went well for his filming and then he needed some help in the cutting room, so I volunteered for that. I did a first cut of his picture and sound for him.

Fig. 3-02

One of our filming locations was the home of a wealthy man in Newport Beach, and the home had an enormous "Trophy Room" of stuffed animals the man had hunted and mounted as trophies. One was a standing polar bear. The scene we had filmed in that home was a guy who had met a girl in a club, they partied, they went to her home together and had a wild night. He wakes up in this massive trophy room bewildered at where he is. In the editing, the director began to wish he had some footage of that standing polar bear so when his actor wakes up and looks around, we cut to his point of view, the polar bear, and he expresses his bewildered state of mind. But that filming day, we didn't deliberately film the polar bear or any room trophies by themselves.

A film editor looks at every single piece of footage in preparation to do his job, and I had remembered that on one reel, when they loaded the camera and ran it to check how smooth the film loop was in the film gate, the camera had been pointing at that polar bear, and by sheer luck, had captured a few seconds of good clean footage of it. Our camera was an Eclair NPR, and took a 400' core load, instead of the 100' daylight loads of the Arriflex I had used myself. The Eclair NPR is uniquely designed so its film gate is actually the merged parts of the Magazine face and the Camera Pull-Down mechanism plate. So this camera is unusual in that almost 100% of the footage used to test the film loop and smoothness of the threading becomes potentially usable footage, depending on the lens settings.

So we actually had polar bear footage, taken when the magazine film loop was tested for smooth running, but there was one problem, a minor one. The guy is to the right, looking screen left, and the polar bear in this footage was standing facing screen left as well. So I could use the polar bear shot, but the edit would suggest the guy was looking at the polar bear's back, not the front and face. The solution? Flip the film, left right, which was easily done when the film stock was double perforation film (in those days most was, unlike the single sided perf film more common in more recent years). So I simply found the clip, flipped it over so the image reversed left to right, and delivered exactly what the director wanted. The guy looks to his left, and then we see this huge polar bear seeming to be looking back at him.

Why is this so memorable? Once I began participating in internet forum discussions of the Patterson Gimlin Film, I found the James Randi Educational Forum (JREF) was intensely discussing my activities on the Bigfoot Forum where I started talking with others about the PGF. So I joined the JREF to at least respond to their sometimes wild and erroneous assumptions and gossip about me (see Chapter Thirteen for more on this). And over on the JREF, I was discovering the fuller inventory of skeptical suspicions about the film. One of those was the fact that one documentary showing the PGF had flipped the first few frames of the footage, before running the rest of it in normal left-right orientation. And these JREF skeptics found this flipped frame incident incredibly suspicious. They ranted and gossiped about it at great length, and were sure this was one more solid piece of proof of a hoax. And I was astonished at their unbelievable ignorance of what a flipped film frame really means. It means nothing!

I recalled the editing on that film, the polar bear segment, and how I flipped it for a better left-right visual look between the bear and the guy. I recalled many long days and nights in the editing room, trying to finish a film cut, and being so sleepy that I occasionally flipped a piece of film by accident and then had to go back and open the splices and flip it right, correcting my sleepy error. Yet here were people imagining that they were intelligent, knowledgeable, and critical thinkers, behaving like complete idiots and talking about flipped film frames as being suspicious when not a one of them had even a meager amount of real-world hands-on knowledge and experience with film editing. They were ranting away about some flipped frames being suspicious and that must be evidence of a hoax, when a few flipped frames on one copy of the PGF edited for a TV show means absolutely nothing in terms of the PGF being a hoax when filmed years before. Did they honestly think the TV show's film editors were in on the hoax?

Delusional skeptics continue to point out that one incident where the first few frames of one copy of the PGF, shown on one documentary program, are flipped left-right before restoring the rest of the footage to correct left-right format. They continue to think it's suspicious. They continue to behave like paranoid fools, because they are so desperate for any oddity or irregularity to hang their hoax claim upon. Critical thinkers? Certainly not! Just fools who know nothing about the realities of film handling, editing and splicing, and form conclusions based on their own appalling ignorance of the procedure. And the JREF is sadly a magnet to such fools, who swarm there like flies around a trash can.

So in the fall 1968 semester at Valley College, I was taking whatever new film studies course was offered for the first time, while doing these small film project jobs on the side, continuing to develop my crew skills as a cameraman, lighting man, sound man, editor, and such. The second semester, starting in January of 1969, they added a new course that quite literally changed my life. It was the Advanced Stage Makeup course to be taught by Michael Westmore, who was Assistant Supervisor of the Makeup Lab at Universal Studios, and nephew of the legendary Bud Westmore, the Studio's Makeup Department Head. Suffice to say, the moment the course was announced, I signed up for it.

Mike's course taught us the basics of making prosthetic facial appliances, like the Planet of the Apes used. We learned how to make an impression of a person's face, how to cast a plaster positive head, how to sculpt plastilina clay into the changed facial feature (a different nose, maybe, old age jowls, or even Planet of the Apes mouths, noses and eyebrows), make the negative molds and finally cast the actual facial prosthetic appliances and apply them as part of a makeup. I was fascinated by this process, and once the specific steps and materials were revealed to me, I started doing more facial casts and experimental prosthetic makeups on myself, my sister, and

59

my friends. I wanted to excel at this skill, and when you consider that within a year, I was doing a professional monster makeup of the "Blackenstein Monster" (yes, a parody of the Frankenstein Monster, on a black man) for the little cult classic "Blackenstein", I guess you can say I succeeded in that goal.

Fig. #3-03 shows my work making the monster for the movie, "Blackenstein".

But the big break was not that film. It was when Mike announced to his class in May that there was an opening for a makeup artist to help with the Studio Tours Makeup Show, for the coming summer season, because the guys currently doing it were studio makeup apprentices (Steve Abrams, Rick Sharp, and Charlie Sorkin), and they would be finishing their apprenticeships soon. The studio had decided not to take in any new apprentices in the formal union apprentice program, so outside help was sought for the makeup show (think "cheap labor", non-union newcomers). Along with 12 others from the class, I tried out for the job, and I was the first one hired. Three others from the class would also be hired as the summer season ramped up.

Up to that point, it was my ambition to be a writer/director, and I just did crew work for the experience and the chance to meet industry people, waiting for my "big break". So I wasn't planning a career as a makeup artist. But we go through the doors that open up for us, and sometimes that door leads us down a path we did not plan or anticipate. Such was the case here, and so I began my career as a professional makeup artist, eventually specializing as a Makeup Special Effects and Creature Guy. In the Summer of 1969, I was 20 years old, and working in the Makeup Department of a Major Hollywood Studio.

Fig. 3-03

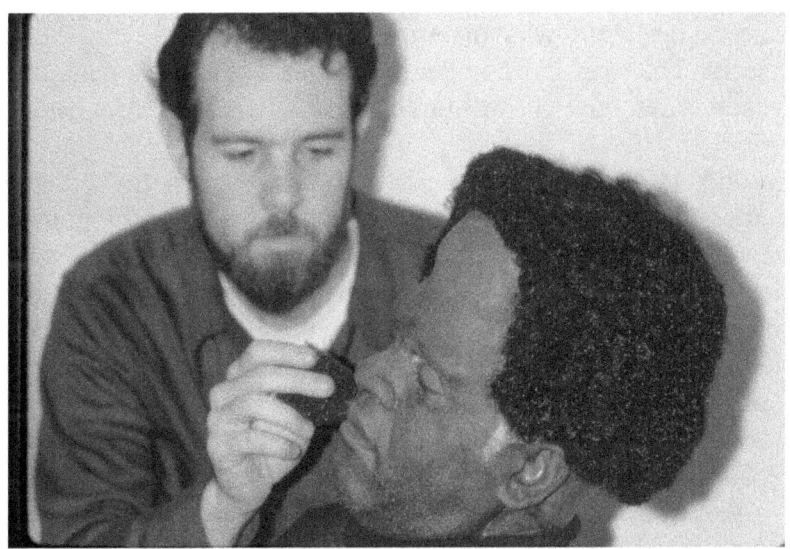

Making the monster for the movie "Blackenstein"

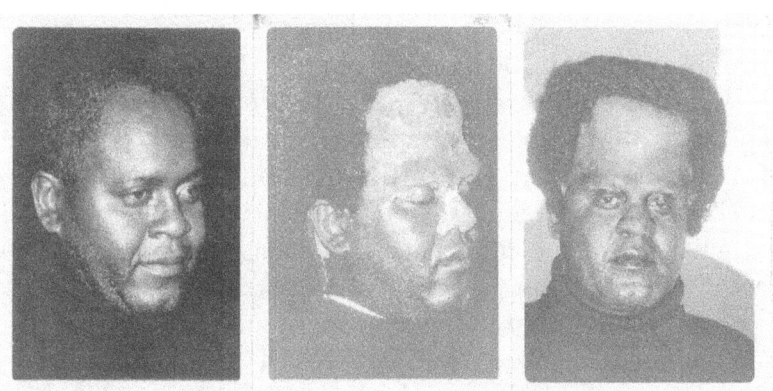

I was learning what the finest studio makeup lab was doing, at the studio with the greatest legacy of famous creature creations and full creature costume suits. I was learning what the materials of the 1960s were and were not capable of. I was learning exactly what was needed to do an analysis of the famous Patterson Gimlin Film's creature, learning how suits were designed and what materials were available to build them. It seems Fate was indeed opening a door leading to a path, and decades later, the PGF was on that path.

Around 1970-71, a fellow makeup artist from my college studies told me about a makeup school he was teaching at. It was called "Elegance International, Inc" (today it's generally identified as E. I.) and was primarily a salon and beauty makeup school. But the owner wanted to expand into motion picture makeup. I met with the owner and agreed to develop a course outline for motion picture makeup and teach the first class. Once that began and was successful, we talked about expanding the program to include a course in makeup prosthetics and lab work. So I developed a course outline and taught the first one of those courses as well. In the fall of 1973, I was promoted to director of the school, and continued as director and supervising teacher for 6 years.

There is a wonderful adage that goes *"We learn some from our teachers, more from our peers, and most from our students."* I don't know the source, but can emphatically attest to the truth of it, because I found I learned a tremendous amount of new material on the makeup arts in my years as a teacher. As much as I gave to my students, I gained as much or more knowledge myself. An important part of my own development during that time was when I chose to embark on a program of recreating many classical makeup masterpieces. I felt these classical makeups were excellent teaching examples, but I couldn't find enough photographic documentation on them to create good step by step teaching visuals. My solution was to recreate the makeup completely, and film or photograph every step of my work, so a teaching document of the process, in detail, could become a permanent part of our instructional inventory.

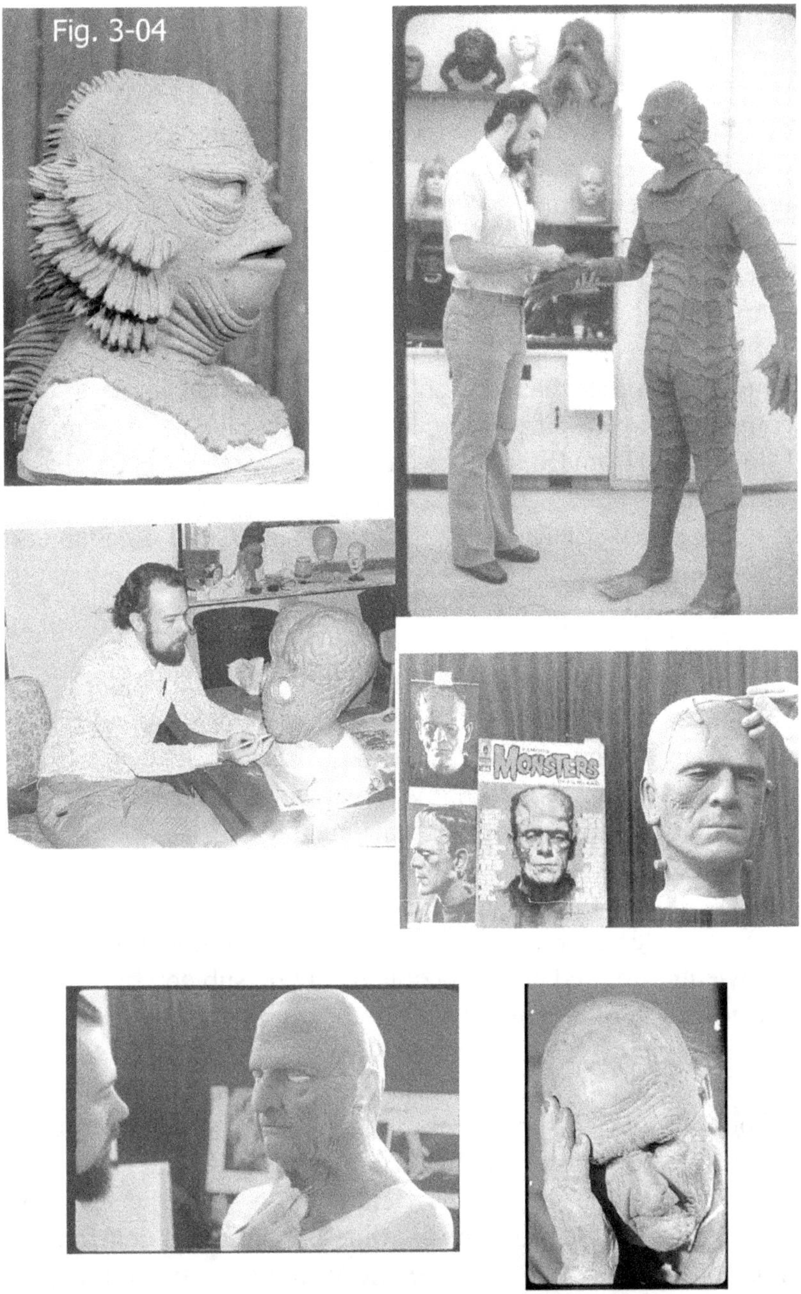

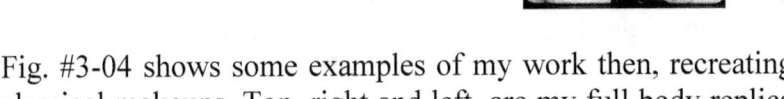

Fig. #3-04 shows some examples of my work then, recreating classical makeups. Top, right and left, are my full body replica of the "Creature from the Black lagoon". Middle left is my sculpted head of the Metaluna Mutant from "This Island Earth" (I did a complete body as well), and middle right is my replica of the classic Boris Karloff Frankenstein monster originally done by Jack Pierce. Bottom left and right are my attempt to replicate Dick Smith's classic aging of Dustin Hoffman for "Little Big Man". Bottom left shows the sculpture, and right shows the complete makeup, with a student, Shawn MacInroe, as my model.

Among the other makeups and creature effects I made were: a complete Morlock from the 1960 Time Machine, a complete Chewbacca from Star Wars, a replica of the Cecily Tyson aging by Stan Winston and Rick Baker for "The Autobiography of Miss Jane Pittman", a "Close Encounters" lead alien head, a replica of Dick Smith's Mark Twain makeup on Hal Holbrook, assorted makeups by John Chambers, etc.

In doing so, I could work through the techniques and challenges of the original work, and compare my skill level with that of the master makeup artist whose work I was recreating. And I filmed the work, in Super 8mm format, and edited the film into a teaching documentary.

This work provided my foundation for my later professional work as a "creature guy" and also was the foundation for my knowledge applied to analyzing the PGF subject figure with consideration of whether it's a fur costume or not.

In 1979, another school lured me away from the makeup school, but in this new school, I was the student, not the teacher. A veteran wild animal trainer named Ralph Helfer had an animal compound called "Gentle Jungle" and he was very successful in Hollywood at the time. And he had developed a three phase course training people in the care and training of exotic animals. I loved animals and since the course was on each Sunday for 8 weeks (my day off from the makeup school), I enrolled in Phase One. The Phase One course was essentially an introduction to care and handling of exotic animals, a chance to learn more about them and touch, pet or otherwise just get close to them. Elephants, chimpanzees, lions, tigers, jaguars, servals, cougars, bears, emus, pythons, tarantulas, eagles, and llamas were all part of this course. It awakened in me a powerful sense of discovery, the fact that I could actually get so close to these animals.

Fig. #3-05 shows me with Dandy, a great lion, top left, me walking Sultan, a great tiger, top right and middle left, me working on a chimpanzee named "Doc" for the movie "Brainstorm", and bottom, me taking an impression of the tusk pegs of an elephant named "Misty" for a commercial where she needed tusks put back on.

As soon as it ended, I signed up for Phase two, which was actual hands on animal care. Those who knew me then were somewhat mystified that I'd pay money to learn how to clean an elephant barn (which was one of our lessons), but I loved it. We were also taught how to handle the animals, and that to me was magical.

So I completed that and signed up for Phase Three, actually learning to be an animal trainer. By this time, Ralph had gotten to know me and my makeup work, and he occasionally had requests for his animals to be altered in appearance. I was glad to offer my advice, such as, how do you paint a chimpanzee green to look like a chimp from outer space? How do you make an elephant look like a woolly mammoth? How can you make a Giant Irish Elk (Megaloceros)? Can you make a chimp look like a gorilla? Can you make an orangutan look like a pre-human "missing link"? Can you put a large scar on a grown male lion's face?

Fig. 3-05

This experience with Ralph, and many of his best trainers (who later went on to have their own animal compounds and services for movies), Boone Narr, David MacMillian, and Sled

Reynolds, brought me into a phase of my life that was quite unique even among makeup artists. I developed techniques for doing special effects makeup on live animals, and Ralph hired me to do such special makeup effects for a movie he was producing, called "Savage Harvest."

Fig. #3-06 shows me putting a makeup wax scar on the forehead of a lion named "Bear" while trainer John Gillispe watches the lion. On that show, I did the scar also on another lion, Dandy, and made lion prosthetics to interact with actors safely, like maul actor Tom Skerrit.

So in late 1979, I closed the chapter of my life as a teacher and school director, and returned to private practice as a movie makeup artist. Where this experience helped me in the PGF research 25 years later was that this experience put me in very close contact with chimpanzees and orangutans, and by close contact, I mean actually putting makeup on them. I studied their anatomy in meticulous detail, and eventually made full body suits for chimps to look like gorillas, a first in Hollywood. This also gave me a unique perspective of ape suit designs and construction, which helped me in my later PGF analysis.

So from 1979 to 1997, I was working freelance as a movie makeup artist doing prosthetics and other special effects work mainly. But when you work freelance, you don't actually have much say in the types of jobs you get. You generally take what's offered, but that isn't necessarily what you want to do personally. I wanted to do superbly realistic animals, but I was offered comic book characters, zombies, and aliens. So I decided that I needed to showcase my realistic animal skills and find customers who had more interest in such work. So I began gravitating toward the museum and theme park fields, and exploring a crossover of makeup artist techniques applied to wildlife art, making perfectly realistic replicas of animals.

Fig. 3-06

Fig #3-07 shows one of my wildlife art pieces, a full-scale baby orangutan. I sold two of these to private clients. Each one has the head and body hair punched in, one hair at a time, a meticulous process, but necessary to replicate real baby orangutan hair patterns.

Fig. 3-07

During this phase of my career, I would make quite a few dinosaur figures, some full scale, some scaled down, make dodos, a panda, a lot of bald eagles, a giant tortoise, and a lot of varied primates, especially gorillas. I would go to the World Taxidermy Competition and enter my work in the newly established contest category of Wildlife Re-Creation, and come home twice with "Best-In-World" honors (1988 and 1992), and subsequently judge and lecture at future contests (1995 and 1997). I also started a series of instructional magazine articles in "Breakthrough Magazine", one of the premiere Taxidermy and Wildlife Art magazines in America. Over the years, I wrote close to a dozen instructional articles on Re-Creation techniques.

Where I feel this experience is relevant to the PGF analysis is that it challenged me to study the appearance of real wildlife with a high degree of detail, especially the outer appearance of same, it challenged me to explore material techniques to replicate same, and it challenged me to expand my understanding of ways to fabricate fur costumes.

In 1997, I transitioned into computer graphics, because the CGI field was rapidly taking more and more jobs that had formerly been done by makeup FX people. Simply put, CGI was the future. I could not foresee, 17 years ago, that the CGI skills I would acquire were an excellent preparation for the digital image analysis work needed to understand the PGF, but now, in retrospect, I can see that these skills gave me a unique vantage point for appraising the PGF evidence and potential for analysis. I won't dwell on this phase of my career, here, but a resume and descriptive career essay is in the Appendix of this text, and one may refer to if you wish to know more.

Figure #3-08 has two samples of my recognition as a computer graphics artist. The magazine Design Master, published in Australia, ran a six-page feature article about my work with a software called "Bryce". The lower panel shows my recognition for an animation of lighting and my Hanging Gardens of Babylon model set to music.

Summing up, while my makeup effects work seems the obvious and immediate thing that qualified me to examine the question of whether such a creature costume is seen on the PGF, my background in cinematography, still photography, film editing, film processing pipelines, copying and printing, computer graphics and digital image processing also factor into my analysis and are skills that I have a solid background in. Most creature effects makeup artists don't have such filmmaking or CGI skills. As far as I know, no one who previously did research and analysis on the PGF had this unique combination of skills and background to bring to the analysis effort.

I would like to note as well that however much I knew when I started this analysis work in 2008, I was constantly challenged to learn more to try and solve the various elements of the overall mystery. So these past seven years have been a remarkable educating experience for me as well.

For this reason, I feel my background is both unique and ideal to make a final determination about what it is we see in this most mysterious film footage.

Now is probably as good a point as any to correct one common misconception, to address one common false accusation. My critics love to try and ignore my professional accomplishments, ignore my hands-on experience, and dismiss my analysis of the PGF by saying I'm a "Believer" of Bigfoot, so my findings are not factual or reasoned, but are manipulated to satisfy my supposed bias, my "belief". In other words, they say I believe Bigfoot is real, so I distort facts and analysis to support that belief. So now I'd like to set the record straight on this "believer" issue, because skeptical people use it to ignore or belittle my professional capabilities.

Since I was a kid in the 50s I was curious about all things nature, and mysteries like the "Abominable Snowman" fascinated me. I grew up on "creature movies" and so fake movie "creatures" were also fascinating. I was in college when the PGF was announced and the first photos were released, the footage shown on TV for the first time. I paid attention, I was aware of it back then, but had no particular belief it was real. I paid far more attention to the apes in the two movies "Planet of the Apes" and "2001 A Space Odyssey" and about learning how each one was accomplished. I loved special effects and 2001 had pushed the envelope in many ways with bold and innovative photographic effects (courtesy of FX genius, Doug Trumbull).

As my makeup skills continued to develop in the early 70s, and I began teaching makeup at the first private school training makeup artists to do movie makeup work, I embarked on a series of teaching projects, ambitious showcase efforts of classical past makeup achievements and some hypothetical projects as well, to push my own skills to new heights, and then document the effort as teaching aids and examples. I did a Bigfoot concept design as a 1/6th scale model, and later build a full size standing creature model for the school.

Fig. #3-09 shows me in the process of sculpting this model Bigfoot creature, just as a showcase piece in case I might be offered a job making such for a movie.

Did I believe in Bigfoot? Not really. I knew it was a topic people discuss, I knew that with the PGF publicity, sooner or later people would be making movies with "Bigfoot" creatures in them, so my activity was a logical part of my chosen profession. I did real creatures, fantasy creatures, prehistoric creatures, and comic or cartoonish creatures. I didn't know if anything like a Bigfoot was real, and had no ambition to investigate it further. I just wanted to know how to create a great one for a movie, if such a movie offer came along.

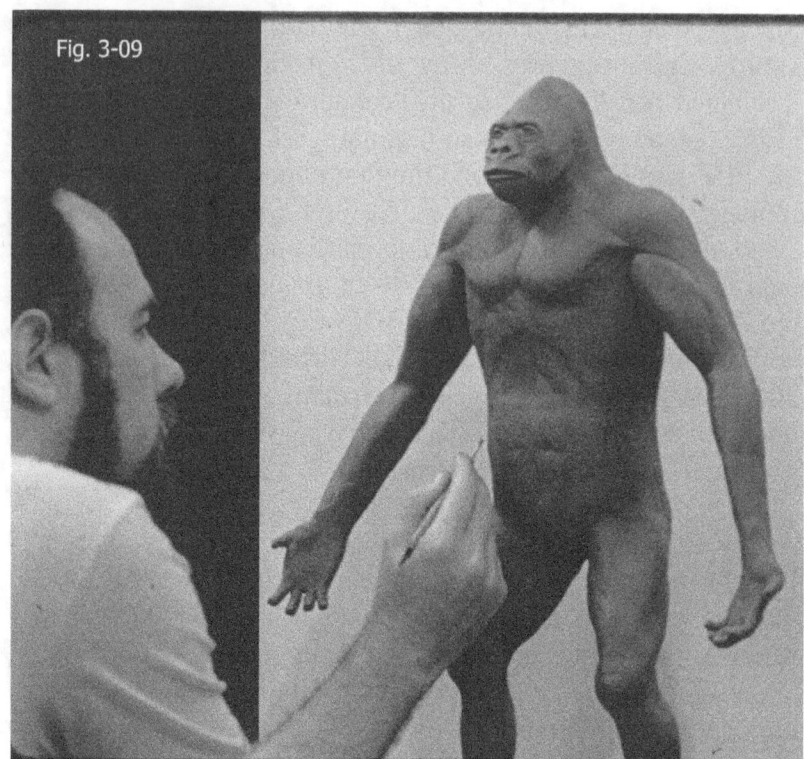

Fig. 3-09

In the late 80s, when I was working for a theme part robotics company, Creative Presentations, one of my duties was to try and develop ideas for robotic creature exhibits we would build ourselves and rent out to museums, the way Dinamation and Kokoro were renting out their robotic dinosaurs. So aside from designing a dinosaur show, I looked at other potential exhibit topics and thought a cryptozoology exhibit would be a crowd-pleaser, a profitable venture.

So I revisited the PGF as one of the exhibit sections, along with a Yeti section, a Loch Ness monster section, a Krakan section, a Roc section, and so on. I didn't believe any of them existed. I simply developed the exhibit with a specific concept for each creature. The Loch Ness exhibit focused on the radar and sonar technologies that were being used to search for the creature. The Yeti segment focused on the possible connection between the current rumored yeti and the pre-historic ape

Gigantopithecus. The Krakan segment simply wanted to convey the immense size of this fabled creature. The Roc segment focused on the intriguing idea that if you took a real egg of the Elephant Bird (Aepyornis maximus) and imagined it was a raptor instead of a ratite, and scaled it to a modern eagle's egg, how big would such a raptor grow? And sure enough, it would grow to have a wingspan of 35'. So the exhibit speculated that maybe the legend of the Roc came from early explorers finding elephant bird eggs on Madagascar and mistakenly thinking they were raptor eggs instead of ratite eggs, and imagining how big an eagle would hatch and grow up from that egg.

It was during that time that I made my famous Gigantopithecus figure and was photographed standing beside it. That photo would become an internet sensation years later and especially on Bigfoot forums and websites. Many people would wonder if anything about my design was influenced by a belief in Bigfoot, but absolutely nothing about it was. My associate, Dr. Russell Ciochon, even said going into it that he was adamant that everything we do must be scientifically responsible to known animals and prehistoric fossil evidence. It was not intended in any way to encourage a belief in Bigfoot. And I was fine with that because I didn't believe Bigfoot existed. I wasn't as hard line as Russell, saying that Bigfoot does not and cannot exist. I just was open to the prospect, curious about looking into the idea with an open mind. But if I did finally commit to a conclusion about Bigfoot, any such conclusion must be derived from a factual basis and solid evidence.

Fig. #3-10 Me and my full figure of Gigantopithecus. There is some confusion about the fossil ape's expected posture. All presumption of the time was that the ape was quadrupedal, but I felt seeing him standing would show the immensity of his size, so I suggested we pose him as if he was reaching for some type of fruit in a tree or high vine, and that is why the one arm is reaching upward. The pose was not meant to suggest we thought the ape was bipedal in his walk behavior.

Fig. 3-10

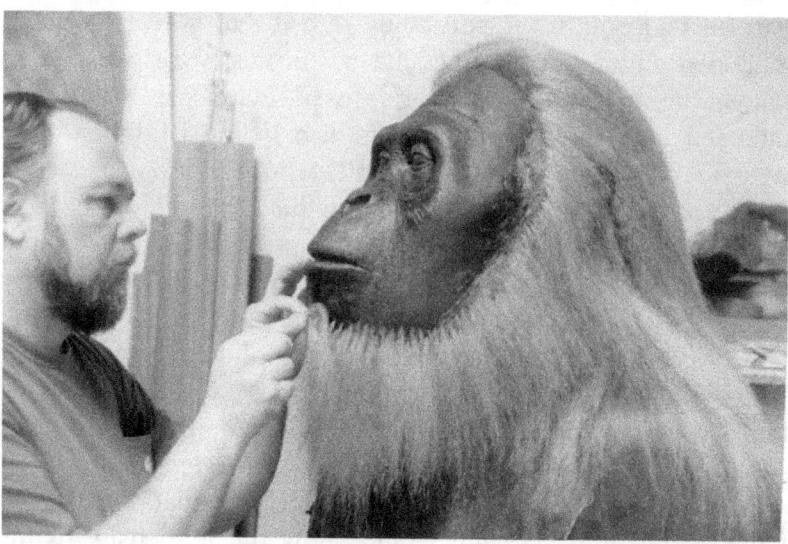

In the year 2000, I was interviewed in a newspaper, and offered my opinion that I felt the PGF hominid was not a hoax because of the way the surface of the back of her neck did not compare with any fur costume neck I was aware of from that era (the 60s). So my conclusion, that it was not a fake, was derived from a factual consideration of costume technology. My conclusion was the result of where the data lead me. When I entered the internet forum Bigfoot world in 2008, and found the wealth of image data available, I decided to put all options on the table and renew my analysis of where this new data would lead to. It lead me to a firm conclusion that the PGF Hominid is not a costume. But again, it was the data which lead me to the conclusion, not any "belief". That has always been my position, and remains so today. Belief has nothing to do with my conclusions. I have far too great a love of science and reasoning, and find irrational belief systems unacceptable.

But the "believer" accusation is a skeptical ploy to discredit anyone who shows plausible reasoning for why the general concept of Bigfoot may be real, and to discredit anyone who concludes that the PGF may be a real biological entity and not a hoax. People who don't like this book, and my analysis and conclusions, will no doubt rally to the cause and keep yelling "Believer", in the hopes somebody will buy into their accusation. No doubt some people will. Others will actually appraise the facts and information I lay out herein, and form their own conclusions as to whether my analysis and conclusions are the result of "belief" or the result of following factual data to a reasoned conclusion.

This personal and professional odyssey to find an answer to the mystery of Patterson's Creature has been an adventure, an education, and a gratifying experience, because of what I have learned in the process. And one of the many things I have learned is that anything I say or do will assuredly offend somebody, but I will not hesitate to present my thoughts and analysis for fear of that offended person. Because I am sincere in presenting what I truly feel is factual and responsible analysis, I will continue to present my ideas with confidence and enthusiasm.

Fig. 4-01

Chapter Four - The "Creature" Business

An Introduction to the Special Makeup and "Creature" Craft

Before we begin with the actual PGF analysis, the reader needs to be acquainted with the craft and profession of special makeup effects, and the subdivision of it that covers creature effects, costumes and fur body suits. A basic commentary of some material, as it relates to the PGF, will be included here, but developed in greater detail in a later chapter when the PGF is studied more fully. There will be many terms used which are common professional terms, but have different meanings in general usage, so the reader should be acquainted with the terms and concepts of the craft.

Makeup, masks, puppetry, and costumes have a legacy in classical theater dating back throughout recorded history. The practices were fairly simple, but the results were imaginative and entertaining. So with the advent of moving pictures, the well-established techniques of the theater were used for character makeups and costume effects needed for this new media called "the movies".

But as the motion picture industry developed, the more creative and artistic practitioners soon recognized two vital differences between the film medium and live theatrical stage performances. One was the intimacy of the camera, the "close-up", which could inspect every detail of an actor's face as if the audience were an arm's length away, whereas a stage audience rarely was closer than 20-30 feet to the performer. This intimacy of the camera required makeup to be toned down and applied with more precision and attention to detail.

The second difference was that the photographic process, accompanied by the film editing process, had a remarkable potential to achieve effects that simply were not possible or practical on stage during a live presentation. Thus was born "special effects", illusions using the technical processes of photography and editing to accomplish results unheard of in the theater.

One of the simple but effective effects was the "transformation", wherein a person somehow transforms their appearance while we watch. It could be done in one of several ways, most commonly a simple cross-dissolving two footage segments in the film lab, the actor's "before" appearance fading out as the actor's "after" appearance fades in.

A person could get older or younger, change from a human to an animal or a monster, an injured person could miraculously heal, any transition the story needed could be accomplished. And it could be achieved with far greater realism than any comparable stage effect or illusion, in close-up.

Image compositing could also put real people into fantasy settings, make people giants or smaller than dolls, make ghosts and apparitional spirits. So this advent of the film special effects process both excited the audience and encouraged film makers to develop stories with creatures, monsters, and fantasy beings.

During the silent film era (before sound was effectively attached to the projected image in 1926), one man lead the field in character makeup effects. Lon Chaney, famously known as "The Man of a Thousand Faces", was an actor with a fine background in stage character makeup, and he saw the motion picture medium as a great opportunity to push the makeup craft to new levels of innovative technique. He became famous for his inventive transformations of himself into his many film characters. The most famous is likely his "Phantom of the Opera" character, where he actually had hooks pull his nose and lips upward and out of shape before applying the nose putty to resculpt the nose, and used well-crafted false teeth to achieve a horrifying, unforgettable face.

Fig. #4-02 shows Lon Chaney as the Phantom. For its time and the techniques available, it still stands today as one of the great makeup achievements.

Fig. 4-02

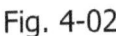

During his era, the actor did his or her own makeup, and few actors were anywhere as skilled as Mr. Chaney. But with the advent of the "talkies" (movies with a sound track, so we could actually hear the actors speaking), the movies grew tremendously in popularity, more actors were hired for this increased production, and their makeup skills, particularly for the movie circumstance, were modest at best, and stage oriented. The makeup artist as a crew member was born out of the need for makeup skill consistently satisfactory for the camera. The makeup artist began replacing the actor doing his own makeup.

The Westmore family of wigmakers and makeup artists were one of the more highly publicized practitioners of this new field, makeup artistry. The father, George, his seven sons, and even the sons' children and grandchildren, built a unique legacy within the makeup profession that endures today. Frank Westmore's book "The Westmores of Hollywood", does a fine job of explaining the family legacy, and is highly recommended for anyone wishing to explore this further.

But among makeup artists, a lesser known man is the one we look to for inspiration. His name was Jack Pierce. It was he who transformed the young actor Boris Karloff into one of the most famous monster faces of all time, the original Frankenstein's Monster in 1931, using rubber latex and cotton to build up the headpiece to reshape Karloff's head, and then detail it with scars, hair, and the famous electrical terminal bolts in the neck.

It was Jack Pierce who transformed both Henry Hull and a few years later, Lon Chaney Jr. (yes, son of the "Man of 1000 Faces", above) into a werewolf, with extensive body hair meticulously applied to the actor's head, face and torso. Those two classic makeup effects would inspire the generations to come, makeup artists wanting to get into the "special makeup effects" specialty.

Ape costumes would see some fascinating sophistication in the early Tarzan movies, with costumes to look like chimpanzees. It certainly helped the costume designers that real chimpanzees were available and used in the film, so the designers could study live animals up close to refine their designs.

Fig. #4-03 is a studio publicity still from the 1931 Tarzan the Ape Man, and the ape apparently challenging Tarzan is quite impressive in appearance, but we can't be certain this photo is a correct image of the ape costume. It's a studio publicity still, and they were famously and intensively retouched by expert photo retouchers. The reason we know this is a studio publicity still and not a frame from the movie is because Tarzan is wearing tennis shoes, in this picture, and as I recall, he didn't in the actual movie.

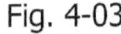

Fig. 4-03

The 30s would end with what stands even today as one of the most spectacular makeup achievements, the fantasy film, The Wizard of Oz. The Cowardly Lion, the Tin Man, the Scarecrow, the Wicked Witch, the Munchkins, the Flying Monkeys, and the Wizard himself collectively demanded that the craft of special makeup, prosthetics and full costume suits be pushed to new levels of sophistication. The workmanship was exemplary for its time.

During this era (pre WWII) when monster and fantasy films were popular, the film industry became unionized, and the union process invariably leads to specialized divisions of labor. One of those divisions was a formal policy that an actor could not do his or her makeup. A makeup artist must do the cosmetic application. So the era of the stage-trained actor doing his/her own makeup for a film, came to an end. But as makeup for film became increasingly technical, to look correct under intense lights and the newly introduced color films, even the simple makeup to make a person look natural became complex enough to warrant the specialized and trained talent of the makeup artist. So this division of labor was not only inevitable, but logical.

But another division occurred which bewilders many people. That is the division of who handles hair or fur.

Fur creature effects are generally either hair directly applied or affixed, or fur cloth (fake fur) or real fur pelts, sewn like cloth. So one would think the hairdresser would handle the real hair affixed, or a wardrobe costumer (who sews) would handle the fur cloth or pelt (which is sewn). Yet this became one of the domains of the makeup artist, whenever the hair or fur was intended to look like the natural body hair or fur of some live creature or animal.

The exact circumstances for this curious and seemingly illogical division of labor are probably lost in historical shadows, but the result are well documented.

For the costumer, the wardrobe specialist, this person would be in charge of the design and construction of any costume which is intended to actually appear to the movie viewer as a true costume, worn by a character in the story. The costume may be real or fake fur, but as long as it is intended to be a true costume worn on the body, the wardrobe department handled it. A fine example is Raquel Welsh's fur bikini in "One Million B.C."

If a costume is intended to look like the real and natural body of a "creature", it becomes the job of the makeup artist, even though it is in fact a suit or costume. So if you want a human performer to pretend to be a real bear, for example, the makeup artist usually makes the bear costume, and makes it look like a real creature.

Want a costume that looks like a reptilian aquatic creature, like "The Creature from the Black Lagoon", see the makeup artist, not the wardrobe department.

Want a costume that makes a man look like a real gorilla, see the makeup artist.

The second curiosity is that the hairdresser will do all traditional hair styling on an actor or actress' real hair, and will style and apply any traditional wigs worn by either. But applied facial hair (beards and mustaches) and other body hair are the task of the makeup artist, not the hairdresser, even though it is hair, and it is dressed or styled.

Out of these two curious divisions of labor, the makeup artist became the designated person to create "creature costumes" with any materials, and would oversee the hairwork for any fur or hair covered costume instead of the hairdresser.

Within the "creature costume" process, if done in a Hollywood studio, there were further divisions of labor, depending on what was being made. We can alter the human appearance by either prosthetic appliances (sculpted and molded pieces that are glued to the face or body, and makeup applied to blend them with the skin foundation color) or with masks and costumes.

If the work was a prosthetic appliance, the makeup artist could (and likely would) do the facial impression and plaster cast, sculpt the prosthetics in clay, make his molds, run his foam latex and then apply these appliances and do the final makeup and any hairwork needed. (Note: When I say "His" in this section, it isn't an oversight of mine. During this time, the Hollywood makeup union rules did not allow a woman to be a makeup artist. Only men were allowed, and women were allowed only to be hairdressers. So the use of the "His" at this point was a reflection of that unfortunate gender prejudice at the time. Thankfully, that discriminatory rule has long since been abandoned and women now work as makeup artists very successfully in Hollywood.)

But if it was a mask and costume, those same skills were given to other studio crafts workers. Rick Baker learned this on the 1976 King Kong, where he was given a waiver to work as a sculptor so he could sculpt the King Kong face, but that was all he could do, aside from his performance in the costume. The rest of the work was assigned along strict union rules as follows:

Only a sculptor would sculpt the clay into an anatomical or biological shape. Only the mold maker would make the plaster molds around the sculpture. Only a propmaker would run the foam that filled the mold to make the pieces. Only a painter would paint the cast parts. Only a costumer would sew any undergarments and padding accessories. Only a special effects technician would build and rig any mechanical apparatus that might function in the costume (like the crab-like pincers on the arms of the classic "Metaluna Mutant" from This Island Earth, one of the 50s classic science fiction films).

But for most of the classical creature costumes, a makeup artist was the supervisor, the designer and the person credited with the creation. For example, the creature costume for "The Creature from the Black Lagoon", and the Metaluna Mutant from "This Island Earth" were both made at Universal Studios, and Bud Westmore, the Makeup Department Head, was credited with designing and creating both those creatures, even though other union crafts did most of the work.

Off the studio lot, outside the jurisdiction of the union's rigid division of labor, an artist could do as he or she pleased, and so one artist could sculpt the facial and skin parts, make the molds, cast the pieces, sew the fur, and paint the skin tones and cut, dress and/style the hair or fur. The finished costume could be rented to the studio, usually with the stipulation that the person who owned the costume could be the one who dresses up in it and plays the part on camera. Thus were born the "Gorilla Men", costume performers who actually made their own gorilla costumes.

Gorillas were popular characters in films of the 30s and 40s. The 1933 King Kong, its sequels and knock-offs (like Mighty Joe Young) were immensely popular. Wildlife photographers who actually filmed wild gorillas and chimpanzees were famous. So the demand for gorilla costumes was common. Cheap costumes, the "costume party" variety, were easy enough to make, but laughably unrealistic. The realistic ones were quite technically and artistically challenging.

Unlike the chimpanzee costumes for the early Tarzan movies, where real chimps were available and used in the films as well, the "gorilla" movies of the 40s were hindered by the fact that real gorillas were not readily available for study up close. The real mature gorillas, even in zoos, were not approachable or handled by most animal trainers. The movies wanted the gorilla to be a scary monster. And there apparently was not any great concern for anatomical accuracy in the Gorilla costumes of the 40s, because they are rather wildly incorrect, as compared to real gorilla anatomy. So we actually saw a decline in realism for most of the gorilla examples from the films of the 40s.

Most makeup artists were more concerned with the more routine aspects of the makeup craft, the traditional character makeup with cosmetics, facial hairpieces, injury makeup effects and aging techniques (as well as reverse youthening for aging actresses who the studio wanted to keep young and beautiful).

So that era of the makeup profession saw few makeup artists actually willing to give enough time to master the intricacies of a sophisticated gorilla suit, and the rental of an existing costume was more cost effective than the total fabrication of a new one. So between the studio producers looking for the cheaper option, and most makeup artists actually shied away from the challenge of doing a realistic gorilla suit, the Gorilla Men satisfied both situations.

Fig. #4-04 shows Charlie Gamora's gorilla from "The Monster and the Girl" (1941). It is surely his finest creation.

Fig. 4-04

Charlie Gemora was the most famous within the profession, but Ray "Crash" Corrigan was probably better known to the general public. How could he not, with a catchy nickname of "Crash Corrigan", and a well publicized career as a stunt performer, who also set up his own tourist ranch, Corriganville.

These "Gorilla Men" took the time and invested the effort in developing some quite sophisticated gorilla costumes for their time, although by today's standard, they are quite crude.

The 50s saw the rise of Science Fiction, and so the creatures became more imaginative and original in design. For this, the studio Makeup Department was unrivaled in fabricating elaborate and original creature costumes. Classics from the era are the "Creature from the Black lagoon" (truly one of the creature business masterpieces), and the huge-brained alien insect-like creature, "The Metaluna Mutant" from "This Island Earth.

Going into the 60s we would see some major makeup/creature accomplishments. The Time Machine (Morlocks), The Seven Faces of Dr. Lao, List of Adrian Messenger.

In the 60s a new "gorilla man" emerged. He was Janos Prohaska, and he continued the legacy of Gemora and Corrigan, making his own creature costumes and renting them to studios, with the condition that he would be the performer inside. His career was tragically cut short by a filming crew plane crash in 1974, when all on board were lost returning from a filming of a TV program on human ancestors. But in 1967, he was alive and quite prominent in the film industry as the "go to" gorilla man, and thus he was consulted by producers of TV documentaries on the PGF. One of his interviews survives and can be studied as he describes his thoughts on the PGF creature.

Fig. #4-05 shows Janos Prohaska in costume interviewed in the 1971 "Bigfoot: Man or Beast" documentary, talking about the PGF footage.

Fib. 4-05

That's the way it was, up until the 60s when I got started, learning the basics of theatrical makeup in college in 1967, the same time the PGF was taken. So as I began my formative training as a makeup artist, the general presumption was well established that the Makeup Artist designs the fur and foam creature suits, and so I accepted the idea that this was something I needed to learn more about.

One interesting and remarkably influential magazine of the era helped foster the development of the makeup special effects and "creature work". It was "Famous Monsters of Filmland", published and written by an eccentric man named Forrest J. Ackerman. "Forry" (as he was called by all who knew him) was truly the most dedicated man alive in the matter of documenting and preserving the legacy of Hollywood's monster creations, and acknowledging the legacies of the people who made them and the performers who wore them. His monthly magazine was loved by kids who enjoyed the monster and science fiction movies of the 50s and watched the older monster/horror movies of the 30s and 40s on TV. I grew up reading the magazine faithfully in the late 50s and early 60s. Once entering the profession, I had the opportunity to meet Forry in the 70s and visit his home in the Hollywood Hills, which was a personal museum of all things monster; books, posters, photos, films and relic props donated by studios. His knowledge of the subject, and his collection of material, were unrivaled in his day. My earliest knowledge of how makeup prosthetic effects were molded, cast, and applied, came from the pages of his magazine. The classic creature body suits of the 50s were well documented in his pages.

Fig. #4-06 shows one issue of Famous monsters, which I used as a visual reference to sculpt a replica of the Boris Karloff Frankenstein face. The magazine was a treasure of vintage monster photos.

His magazine was unique in the depth and comprehensive way it took the reader behind the scenes of creature design and construction, long before such fine magazines as Cinefex, Cinefantastique, and Fangoria expanded on the idea and published the behind the scenes stories of how movie special effects work is done. Famous Monsters magazine may rightly be considered one of the inspirational sources for the "Golden Age of Makeup Effects" to come.

So I, like many of the mature makeup effects people today, was inspired to learn about creature suits and effects, from that single and enchanting publication, for so many years, the only one of its kind we could read.

Fig. 4-06

It has been argued that Roger Patterson was an artist, a saddle maker, and thus creative enough to fabricate his own costume to hoax the PGF, or modify a purchased costume. That brings up the question, back in the 60s, how did a person learn to be a "creature maker", a makeup effects artist? Today there are an abundance of professional schools and training programs one can enroll in, and they are quite comprehensive and effective. I myself was the director of the first such makeup artist school, from 1973-1979. But in the 60s it was a very different situation.

Aside from the above referenced "Famous Monsters of Filmland" magazine, which had some behind the scene photos of work in progress, there was essentially nothing published that would be considered comprehensive enough to teach a person.

Hollywood was strictly unionized and the makeup artist union, IATSE Local 706, had a policy that its member makeup artists could only train union apprentices, and could not offer any outside instruction in motion picture makeup technique. When Mike Westmore offered to teach a class at Los Angeles Valley College in prosthetic makeups, the union only approved that when the course was described as "Advance Theater Makeup" and not movie makeup. I was in that first class given by Mike (starting in January 1969), and near the end of that first course (around Mid-May), Mike announced there was an opening at Universal Studios for a makeup artist to do the Studio Tour Makeup Artist Show. Twelve of us from that class applied for the job, and I got it. A few others from the class would subsequently be hired as well, after me.

In so far as books on makeup technique goes, the two most authoritative texts of the era were Richard Curson's "Stage Makeup" (third edition) which I used as my textbook in my 1967 Stage makeup class in college, and Vincent Kehoe's "Technique of Motion Picture and Television Makeup". Both had minor sections about molded facial prosthetics, but neither had any information sufficient to teach anyone how to build "creature costumes". We would not see any textbooks describing special makeup effects and creature costumes until the 80s. And in the 60s the two makeup books described above were not authored by any makeup artists in the Hollywood Movie makeup group, because their union wanted to keep all their education restricted to their apprentice training. Kehoe was active in the New York film/TV scene, but not connected to the major "creature" work in Hollywood. Curson's background was classical theater.

So, in Roger Patterson's time, there was no school for makeup artists and creature effects, no books with a comprehensive description of creature suit making (on a practical "how to" level) and the Hollywood artists could only train studio apprentices. Occasionally there was private mentoring of a veteran artist, to help a promising newcomer, as Dick Smith mentored Rick Baker as a teenager. But for the most part, the craft was a closed endeavor, almost impossible for an outsider to learn.

This is why I personally find implausible the usual hoax suspicions based on Roger fabricating his own suit, or modifying one purchased. Having taught makeup, I have seen how people learn it, and seen how far even very talented people can go on their own. I have seen that real talent only develops with experience, practice and time combined. Yet there is no documentation of Roger ever practicing, developing his claimed creature suit-making skills, or having any documented contact with creature artists who could mentor him. There were no publications comprehensive enough in technique and process to teach a reader how to do the work, not even the above described "Famous Monsters of Filmland" (It could wet their appetite for wanting to learn, but it could not actually teach one the craft).

The myth of the self-taught "genius" whose first ever attempt is a magnificent and unique accomplishment, such a myth has no parallel in the history of makeup artistry, and the field has seen many true geniuses. So the myth of Roger Patterson being somehow qualified to make a creature suit, while enchanting for the skeptical crowd, has absolutely no merit as an option to explain the mystery of the PGF subject.

Going into the late 60s we arrive at the year the PGF itself was filmed, 1967. And if the film was a planned hoax, the film could not have been made at a more awkward and inappropriate time.

Perhaps it is just a strange coincidence that Roger Patterson filmed his famous footage in 1967, as two landmark "Ape" films were poised ready for release to theaters, and the idea of a "guy in an ape suit" would be impressed into the minds of the general public more powerfully than ever before. The Patterson Gimlin film was reportedly taken and then announced on October 20, 1967, and both "Planet of the Apes" (abbreviated POTA) and "2001 - A Space Odyssey" (abbreviated 2001) would be released early the following year. Each film, in its own curious way, may have strengthened the public mind to the idea the PGF's creature was just a hoax, a guy in an ape suit.

One thing that should be cleared up is the error that the ape makeups and suits of these films were not "the technology of the time" that the PGF was filmed. In David Daegling's book, "Bigfoot Exposed" he writes:

(on page 112): **"It is very important to remember the date of the film (the PGF). In 1967, we were still a few years away from Planet of the Apes and 2001: Space Odyssey.
Ape and Monster costumes were not very sophisticated and apparently hard to come by. . ."**

Daegling is incorrect.

Planet of the Apes was released on Feb. 8, 1968 (a mere 3-1/2 months after the PGF was filmed on October 20, 1967) and POTA was filmed from May 21, 1967 to August 10, 1967, before the PGF was taken. The makeup designs and tests would have been done even earlier in 1967.

2001: Space Odyssey was released on April 2, 1968 (mere 5-1/2 months after the PGF filming) and principle photography began December 1965 and was completed in September 1967, before the PGF was taken. And the ape suits would have been developed and fabricated well before they could be filmed.

So in both cases, the ape costumes and effects of both films can be considered the technology of the time. But how that technology actually compares with the PGF is a matter often misunderstood by people who are not knowledgeable in the actual processes and techniques of "creature work".

The work done for POTA actually has little resemblance to what we see in the PGF. The POTA apes were done with facial prosthetic appliances, glued to the actor's face, blended and then painted to finish, and lace front wigs were then applied to finish the look. The actors wore full costumes of cloth or leather, to appear dressed in armor, garments or similar civilized fashions. The PGF Creature is a full and apparently naked body of something humanistic in anatomy, but covered almost entirely by a hair or fur, as humans normally are not. If it is a hoax, the technique for making a full body costume is far different from the techniques of POTA.

But the Hollywood publicity machine was relentless in publicizing the transformation of the human actors into ape characters, and so the public was being reminded over and over that the magic of Hollywood makeup artistry could make a human look like an ape-like creature. Two months before Patterson made his famous film, Life magazine, in August 18, 1967, ran a four-page spread on the Planet of the Apes film, complete with detailed photos showing the transformation of actor Maurice Evens, into the old orangutan, Dr. Zaius. It named John Chambers as the makeup designer and mentioned a $1,000,000 budget for the film's ape makeup effort. We must wonder, if Roger Patterson were conspiring to create a hoax film, how he would view his potential success in this environment of Hollywood makeup magic educating the public in techniques to make humans look like apes.

John Chambers, who designed the makeups and was leading the team of 175 makeup artists doing the work on the film, was awarded a special Acadcmy Award for his effort (this was before the Academy had a regular makeup artist award category).

As Roger Patterson's filmed creature received its first national magazine article and publicity, in the February 1968 issue of Argosy magazine, and Roger was trying to convince the world he had actually filmed a real "Bigfoot" creature, the mental image of human actors impersonating apes with well crafted and believable ape faces and costumes, was ever present on the minds of the public, as Planet of the Apes was released that same February. The fact that the film went on to tremendous box office success is a good indicator of how much public awareness there was for the film and its human-as-ape characters. This certainly may have fostered or boosted the urban legend that the PGF most obviously was just a guy in an ape suit.

Fig. #4-07 shows the cover of that Argosy magazine issue.

But now, more than 40 years after these events, we can approach the two films, the 16mm PGF and the 35mm Hollywood produced POTA, with a more reasoned perspective. If we want to get to the truth of the Patterson Gimlin Film's mysterious subject, the POTA films are not where we must look for clues or solutions. There is actually very little the POTA apes have in common with the PGF subject.

Fig. 4-07

The 2001 ape characters wore full body suits and masks, depicting ape characters that were apparently naked and real. In that respect, they have a far greater comparative value in the mystery of the PGF subject. They were designed by a brilliant innovator in the field of makeup artistry, Stuart Freeborn, of the UK (where 2001 was filmed). His designs of facial masks with soft lips and a hard dental structure underneath, so the lips could snarl and move independent of the teeth, was a major advance in makeup creature effects, and would in years to come evolve into the industry standard for such character masks. The Chambers POTA appliance techniques, where the teeth were a simple plastic insert glued to the foam rubber muzzle prosthetics, fell into disuse after the POTA films and TV show passed their prime.

Fig. #4-08 is a publicity still showing Stuart with some of his "ape" actors in costume.

Stuart Freeborn and some of his "apes" from the movie 2001: A Space Odyssey.

So if we are going to look for the highest technology ape suit effects of the era to compare with what we see in the PGF, the 2001 apes are the ones to compare with.

The release of 2001, and its impact on the aspiring makeup artists who were just starting to learn the craft, was profound. On the other hand, the "old guard", Hollywood's well-established veteran makeup artists, saw 2001's ape designs as a nuisance to be ignored, if possible, and belittled, if mentioned. The Hollywood establishment rallied behind John Chambers, and continued to describe those makeup techniques as the finest of the land. In truth, the "old guard" veterans did not want to go back to being students, and learn how to employ mechanisms in the masks to animate the lips and other expressive facial areas. It was the kids starting out, who were already in a learning mode, who embraced Stuart Freeborn's designs and decided they wanted to learn that as well. But it would be a decade or so before these new kids moved up the ladder to positions where they could be the lead designer on a film, and incorporate the concepts of facial animation mechanics Freeborn had introduced to them.

Once that occurred (in 1980 with the release of "The Howling"), the entire makeup artist profession was literally turned upside down, with special makeup effects being boldly featured in films as if they were the star, the makeup artists becoming household names, and every new job demanded that these new makeup wizards push the envelope with some new and innovative technique or process. For the makeup profession, it was a blessing, and became quite literally "The Golden Age" of makeup effects. For the PGF, it may have created an even larger cloud over the curious little "Bigfoot" film, as more and more realistic and dynamic fur suited characters were created with stunning realism. What was lost to the public was that these fascinating new innovations were done with materials and techniques that did not exist in 1967, and thus were a useless comparison against the PGF creature.

I was part of that "Golden Age of Makeup Effects", and it was an exhilarating time. Material and process innovations were happening year by year, sometimes month by month, it seemed. A new film would come out, the makeup effects would be dazzling, and on the next job you bid, you were invariably asked, by the producer or director considering hiring you, "how was that done?" And if you answered "I don't know", you didn't get the job. You had to answer with some authoritative reply. Of course, if you actually did know how the effect was done, you could easily and confidently explain the process. If you didn't know how it was done, and you wanted the new job, you had to give an answer, even if it was a fabricated bluff. You had to give the impression that nobody could fool you or create something you couldn't analyze and explain. That became the professional practice of the era, the "Nobody can fool me. I can figure out how it was done every time." attitude. And that "nobody can fool me" attitude would damage the entire PGF mystery, because makeup artists would be hesitant to say the film might be real, fearing that it might be proven a hoax after they said that, and then they'd look like the fool who couldn't figure it out. Better to say with cold confidence, "it's a fake. I could tell that right away. It didn't fool me for a second."

That was the downside of this Golden Age of Makeup Effects, in so far as the PGF mystery is concerned. The arrogant "Of course it's a fake" attitude was deeply ingrained in so many people in the makeup profession, that it raised the urban legend (that the film's been debunked, proven to be a hoax) to new heights. The real mystery became harder to solve, with this false bravado passing for analysis polluting the discussion.

I suppose it would be fair to ask, at this point in time, that if this was the prevailing attitude for makeup artists in the "Golden Age" era, and I was working in that era, why wasn't I affected by this attitude as I describe others were. Why would I say with absolute sincerity that the film mystified me, and that I just don't see it as a fake? A part of it was that, I suppose, I was never comfortable with the "I know everything, Nobody can fool me" bravado. All my life, I strived for real knowledge, more so than the pretentious appearance of knowing everything. I loved to learn new things, and so if something new was developed and introduced, I strived to really know and understand how it was done.

Unlike any other makeup artist drawn into the PGF mystery and commenting on the film, I was a teacher of makeup in the 70s (and returned to teach again in the late 80s and early 90s) and you cannot be a responsible teacher and bluff your students. If they ask a question, like "how was that done?", you are duty bound to answer truthfully, if you respect your teaching craft, and I did then and continue to do now. So that was my personal higher priority, a factual and truthful answer. And when I went back to the makeup work as a freelance artist, and the Golden Age dawned, I approached each job bid as an opportunity to teach the producer or director why what I was proposing was a fine way to give them the effects they wanted. So that teaching concept guided me more that the usual PR bravado that pretentious artists embrace so often.

Later in the 80s, I had also branched off from the movie work to include museum exhibit models, and wildlife art, I had moved toward a practice where scientific accuracy and factual precision were assets. Those priorities, along with my teaching philosophy, guided me when I finally entered the PGF controversy with full force.

One other factor of the Golden Age of Makeup Effects which was common, and which has personally served me well in my own investigations, was the practice of "reverse engineering" an effect. Reverse Engineering has a long standing tradition in the business/technology world, where one company introduces a product which becomes successful, and rival companies reverse engineer it to know exactly how it was designed so they can plan how to compete with it.

In the makeup effects business, as we witnessed this constant barrage of bold innovations, we learned to look at a film and reverse engineer the makeup effect we were watching, so we could explain it to others and duplicate it (or even improve upon it) if we got a job requiring a similar effect. So this practice of reverse engineering an effect, by studying the film showing the effect, had the curious result of better preparing me to embark upon this quest to resolve the PGF creature mystery. Exactly as I would study a film with innovative makeup effects, to know what was done and how, I applied that same methodology to the PGF and its creature, to know what I am looking at.

So, in retrospect, the Golden Age of Makeup Effects had a most intriguing impact on the PGF film. The downside was that it encouraged a lot of "Nobody can fool me. I can spot a fake every time" bravado among the makeup community, and this has confused a lot of people who are trying to figure out if the PGF is real or a hoax, and so they look to the opinions of makeup artists to help them decide. This bravado, passing for analysis, has certainly hurt the effort to get to the truth.

The upside is that the reverse engineering which we needed to master, is exactly the type of factual analysis method that may ultimately solve the film's mystery. That may be one of the greatest benefits of the Golden Age of Makeup Effects on the PGF story. The methodology is ideal for the task.

The Golden Age of Makeup Effects would come to an abrupt end, in 1993, a mere 13 years after it blossomed. The cause of its demise was the new king of the hill, Computer Graphics Imagery (called CGI), This is not to say makeup effects ceased, because they continue today, but nearly all the more spectacular effects which makeup artists did in the 80s were taken over by the CGI processes. So the star (makeup effects) now became a supporting player in the special effects process. Each year after that, there was less work, creature shops started closing, and makeup effects people started transitioning to other fields, often to computer graphics related work. I made the transition to computer graphics myself, in 1997.

What happened in 1993? The film, Jurassic Park was released, and it featured a truly spectacular full-scale Tyrannosaurus Rex animatronic dinosaur figure crafted by Stan Winston's fine team of creature effects artists, but it also featured a CGI T-Rex created by Industrial Light & Magic (ILM), the leading CGI producer of the time. The CGI dinosaur was a perfect visual match for the physical one, with scenes intercut between the two versions of the dinosaur, and the CGI version could do something the real animatronic could never do. It could walk.

I was actually involved with a theme park robotics company in 1990 when Steven Spielberg first acquired the film rights to Michael Crichton's novel, Jurassic Park. At that time, Spielberg was involved with the development of the E.T. attraction at the Universal Studios Orlando facility. Creative Presentations, the company I was with, was the builder of the E.T. robotics for that attraction, so they had a a relationship already with Spielberg and Universal Studios.

Fig. #4-09 shows in the top image a collection of various ET and a larger ET-like character, because the intended ride would have these many individual figures scattered all along the ride path in different scenes. The image below is my sculpture of an ET for an intended stroller costume a little person would wear to actually meet park visitors for PR purposes. I was the project manager (and sculptor) on that job. But the smallest we could make the figure to be worn by a little person 3'10" tall was deemed too big, and so the concept was revised to use a robotic ET that was smaller.

Spielberg was heavily involved with the Orlando Theme Park's development, and thus was familiar with the name Bob Gurr, as Bob was the finest large-scale theme park hydraulics effects designer of the time. So they turned to him to design the walking T-Rex for the film, assuming then it must be a large scale hydraulics physical effect, and that CGI simply wasn't good enough yet to handle it. And Bob, knowing the capabilities of Creative Presentations, Inc. recommended they be allowed to bid on the actual fabrication of his design for this walking dinosaur robot.

Above, the various ET robots for Universal Studios Orlando 1988, made at Creative Presentations, Inc.

Below, my ET sculpture for a "stroller costume" (worn by a performer) for the same park attraction. The size was eventually judged too big and a smaller one was made robotically.

I had actually read in the Hollywood trade papers about Spielberg's acquisition of the film rights to the novel, and brought that to the attention of the CPI owners, around May of

1990, and encouraged them to plan a campaign to be allowed to bid on the dinosaur contracts, with no expectation CGI could be competitive as an alternative process. And I had actually begun working through some design concepts myself for a walking theropod dinosaur (which I had assumed would be a large Allosaurus, because the T-Rex was not a Jurassic dinosaur, but rather a Cretaceous one. I took the story title literally and thought all the featured dinosaurs would be Jurassic era species. They were not, as it turned out. But both Allosaurus and T-Rex were bipedal king predators of their time, so the design was similar.)

In anticipation of our being able to bid on the job, I expected their immediate question would be "Have you ever made realistic dinosaurs?" We were in the process of making a half-scale T-Rex for Knott's Berry Farm and I was supervising the visual design and sculpture, so I took our molds for the head, neck, teeth, and tongue, and cast a head/neck figure of that T-Rex, painted it, and took it to a nearby natural location for a series of promotional photos we could show Spielberg and Universal if they did invite us to bid on the job.

Fig. #4-10 shows, on top, the T-Rex I made for photo portfolio purposes, photographed outdoors at a nearby landscaped location. The bottom photo is the head sculpture in progress, which we were doing for Knott's Berry Farm's "Kingdom of the Dinosaurs" attraction that wanted to upgrade. I was project manager on the overall job, lead sculptor on the total figure, and did the figure-finishing as well.

Fig. 4-10

I approached the walking dinosaur design with a concept of a figure supported by a boom arm projecting out rearward from under the creature's tail, and that boom arm extending to a heavy-duty vehicle behind the tail, properly counter-balanced. A Green Screen panel between the vehicle and the dinosaur would allow the vehicle to be easily removed from the footage by chromakey matte systems. This design would allow both legs to actually walk with a natural leg cycle of bends and swings, a true walk motion pattern.

A few months later, Bob Gurr's design was faxed over to the offices and CPI was asked to bid on building it. Bob's design had a figure supported by one rigid leg that merely swung forward and backward, and that support leg was secured to a track system which would be covered by ground debris (hopefully). The non-supporting leg would have full walking range of motion. There was talk of building a prototype and testing how quickly it could get up to a running speed, and how far it could run before having to slow down to a stop. How long that track would need to be to run the course was debated.

I sat in on one meeting where this design was discussed, and I voiced my objection to the design. The support leg would look completely unnatural when compared to the free moving natural motion leg, like a creature limping from an injury. My candid prediction was this design was destined to fail, if built, and it was not wise to fail Spielberg on a high profile job. He was far too powerful in the industry.

After that meeting, my advice was no longer welcomed, and so I was phased out of the bidding preparation (despite being the most knowledgeable person on staff about dinosaurs, and the only person on staff with a real movie creature effects background, compared to the team who knew only about theme park robotics). So I resigned myself to watching them crash and burn if they did get the contract, and left the company to resume my freelance career. But Stan Winston was also very aggressively bidding on the film job, and he did also have strong working history with Spielberg and Universal Studios. So he eventually got the job.

But it was ILM's aggressive push to improve CGI creatures that caught everybody by surprise. Nobody saw it coming. Nobody could deny, after seeing it, that it was the way of things to come. Stan Winston himself was famously quoted as saying, after he saw the CGI T-Rex footage ILM did, "I have seen the future, and I'm not in it."

Stan's team did in fact produce some spectacular animatronic dinosaurs, and his makeup effects shop still continues today, after his sad passing, as one of the primary makeup effect facilities serving the film industry. So the reality was not as devastating as his prediction. But the makeup effects business would never again have the power and prestige that it had during its Golden Age.

It is vital, when we connect any analysis of the PGF to the craft and profession of makeup special effects and the fabrication of creature costumes, to have some understanding as to what the work entails, who does it, how that person develops his skills, and also to understand that the craft changed radically around 1980 as new materials and techniques were introduced that transformed the potential for realistic results. But the PGF and its Hominid must be judged by 1967 era materials and technology. The "Golden Age" as I call it, only has relevance in that it inspired greater reliance on reverse engineering as a technique of analysis. So when I take you through the design and fabrication of an ape suit or creature costume, in Chapter Seven, I will be frequently separating what was available in 1967 as compared to things available in the decades to follow and today.

Fig. 5-01

Bluff Creek, California

Chapter Five - Bluff Creek, Where It Happened.

Bluff Creek, California is an obscure woodland area in Northwestern California, made famous as the "Bigfoot Film" location. Understanding where the site is, and its specific structure and layout is helpful in understanding the film, by providing a context for the filming event, and it also allows us more data of the filming event that can be examined to address suspicions of hoax.

One of the most powerful tools to try and expose a hoax, especially a historical one, is new technology. A clever hoaxer tries to mislead investigators by not just insuring there's no evidence to lead to the hoaxer, but goes one step further and strives to create "false positives", false evidence that seems truthful, and leads away from the hoaxer. If one studies how investigations are conducted, a deceptive person can create these false positives to derail investigations. The problem with false positives is that you can only create such for investigation techniques known to exist and be used when the hoax is pulled off. You can't look into the future and see new technology which could be applied then, and design false positives to fool the future technology.

This is one of the reasons new technology for analysis is so vital to the solution to the PGF. If Roger Patterson really was a hoaxer and the PGF really was a fake with a guy in a costume, the new technology must trip him up. Die-hard fanatics believing in a hoax should seize this new technology and the film image data as their best opportunity of finding irrefutable proof of hoax, because of this factor. The hoaxer would almost assuredly fail to create a false positive of something he never envisioned that an investigator would examine. And that would be the perfect method to expose the hoax.

New technology is telling us remarkable things about the film site and the event as it occurred. These are things no hoaxer in 1967 could imagine we are looking at and studying. Computer graphics simply did not exist as a technology in 1967. I worked in the computer lab at Los Angeles Valley College in 1967, as a part time student worker, and back then, we input data on 80 bit punch cards, and we output data onto an 80 place line printer. The only computer graphics back then were paper printouts where the clever programmer made row after row of alpha-numeric characters and spaces to produce crude shaped printouts like signs or trees, or simplistic faces. The college had an IBM 1620 computer, and occasionally I got to run my applications on a IBM 360, the 'big dog" of that time.

If you were to tell me back then what I'd be doing today with computer graphics, I would seriously think you'd been reading too many science fiction novels, and lost touch with reality. So I look at the analysis technology today and embrace it as an excellent tool for unlocking the truth of the PGF precisely because nothing a hoaxer in 1967 could have done will fool me with the tools I have today.

So this consideration of studying the Bluff Creek film site has the value of giving an excellent channel for trying to find "false positives" a hoaxing filmmaker might try to pass off, and also find real evidence that a hoaxer could never anticipate would be studied, which may expose a hoax or validate a real experience.

In movie making, we routinely "cheat" shots, meaning we take simple shortcuts to get a shot when the most precise option is simply far too impractical to accomplish. And those "cheats" usually fooled people then. They wouldn't now. One of the most common was filming "day for night". Shoot in the daytime but make it look like dark night. Properly done, it generally fooled the movie audience. It might not fool a veteran filmmaker. It definitely will not fool the analysis technology of today. A second and very common "cheat" was

112

to cheat the location, use one area and make your viewers think it's a different location. A hoaxer in 1967 would almost invariably have done some "cheats" with confidence nobody could expose them then. But if he did, we can expose them now.

And this is where the understanding of the film site has great value. It represents an excellent way to look for "cheats" that a hoaxer may do, but a truthful filmmaker shooting an unexpected and spontaneous encounter would not do. And if you find the cheats, you can say conclusively that the footage was staged with deliberation of intent.

So a study of the Bluff Creek setting provides an excellent opportunity to look for evidence of hoaxing. Knowing the site and its landscape does also provide a necessary foundation for understanding what occurred and was filmed that day (detailed in the next chapter). So allow me to take you on a brief tour of the Bluff Creek location, where it all happened.

Through the miracle of Google Earth, here we see a map of Northern California, and Bluff Creek is the white circled pin labeled PGF.

Fig. #5-02   The upper image shows the PGF site in relation to the Northern California border with Oregon, and how far inland it is from the Pacific Ocean. The lower image was a Google earth measure from Yakima, WA (Roger's home town) to the site, in "as the crow flies" miles. Suffice to say, driving surface roads would make the trip somewhat longer.

Fig. 5-02
Where is Bluff Creek, the PGF filming site?

The location is in the white circle above.
Below shows the distance from Bluff Creek to Roger
Patterson's home in Yakima, WA. It measured 408.59 miles

People claiming a hoax like to embrace the story as told by
Bob Heironimous, the man who says he wore the costume. As
he describes, they drove to Bluff Creek, he put on the suit, did

one take of the complete walk, and then they packed it in and wrapped the shoot. That's about fifteen minutes to half an hour, getting him suited up, performing, getting him out of the suit. If Roger drove over 400 miles to that remote wooded area, with Bob and the horses, camping gear, etc. for about 15 minutes of work, one minute of filming, why didn't he do it somewhere closer to home?

No filmmaker in his right mind, and intent on staging something, that I ever knew would do such deliberately.

Are we sure it actually was filmed at Bluff Creek, CA? I ask this because people making movies always consider when location work is planned, do we really need to go there? Can't we cheat the location? (In these terms, to cheat a location is to use a more convenient location that sort of looks like the intended location, and is easier to get to and film at.) So if Roger wanted to claim the encounter was filmed at Bluff Creek, but he wanted to make it easy on himself, why not just pick some woods near home and call it Bluff Creek? After all, one forest looks like another. If he's going to cheat his creature, why not cheat the location too?

Fig. #5-03 shows two woodland scenes Roger had filmed earlier in the year for his documentary. Neither location has anything in it so distinctive that we can positively identify the location. Roger could claim this was Bluff Creek, and we could not prove otherwise. Realistically, many images of woodlands and forest could never be positively identified from examining film images.

Fig. 5-03

Where were thes scenes filmed? Can you identify the location?

(source: Roger Patterson's documentary footage reel.)

So if we want to play the devil's advocate and question everything, why not question the claim the film was shot at Bluff Creek. So researchers have tried to examine and verify the location seen in the film, and those examinations have verified with 100% certainty that, yes, beyond any doubt the film was shot at Bluff Creek, California. Many researchers have visited the site and the photographic evidence verifies that the marked Bluff Creek location is truly where the filming occurred, beyond any doubt, and the evidence is as empirical and reliable as evidence can be.

So, we know that's the spot, where it occurred. I've even been to the site, in July 2012, and I was astonished at how many trees and other landscape objects were still there and positively identifiable. It allowed us to meticulously map out the setting and verify the way it was, and the ways it's changed.

Fig. #5-04 shows what we found in the 2012 site visit and survey. The top image is a frame image composite, middle shows the individual trees, and the bottom image shows those same trees overlaid on a faded scene, so they can be easily located.

The fact that the event was so quickly and so often researched by people visiting the sight, plus its unique configuration of trees, negates any possibility that the location was cheated. I simply felt that it should be verified by factual review, because people less well versed in this research material might raise the question. So, for the record, the Bluff Creek site we study is verified, beyond any doubt, as the true location where the filming occurred. And it can still be found today and 100% verified as the same location.

So, we know that's the place. We know it's over 400 miles from Roger's home. We know it's a long way to go for an investigation into something real, and a long way to go for staging a hoax. So now we must ask the question: Why Bluff Creek?

Fig. 5-04

What is still there in 2012 - Trees, stumps and logs we found.

Above, a composite image of the site

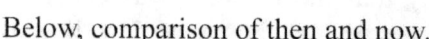

Middle, the trees and objects we found

Below, comparison of then and now.

Let's approach it both ways.

If the film's a hoax, why go to Bluff Creek?

Look at a map Roger himself put in his 1966 book, "Do Abominable Snowmen of America Really Exist?"

Fig. #5-05 A year before his famous encounter, Roger published his own book about the Bigfoot phenomenon (Reprinted intact in Chris Murphy's fine book, "The Bigfoot Film Controversy"). This diagram by Roger shows the locations of many reported events, tracks and incidents. It should be noted there are more in Washington State, near Mt. St. Helen, than in Northern California.

Roger had documented in his book the locations of tracks and apparent sightings in California, Oregon and Washington states. So he knew of a lot of places, most closer to home, where he could plan an expedition and stage a filming with his costumed performer. He had been exploring the Mt. St. Helen area closer to home in late August. All he had to do was fake something there, like footprints, and then say, "wow. We have to go back and keep looking there. This is a a good place to find evidence."

What skeptics seem totally oblivious to is the fact that a real encounter, spontaneously filmed, just requires you be in the right place at the right time, have a camera ready to go, turn it on, and chase your subject. It doesn't require any elaborate planning. A hoax, on the other hand, requires planning, a lot more preparation, and a lot of deliberate consideration about how to accomplish it. And the more deliberate things you want to attribute to Roger as doing to trick us into thinking it's not a fake, the more complicated that deliberation and planning becomes.

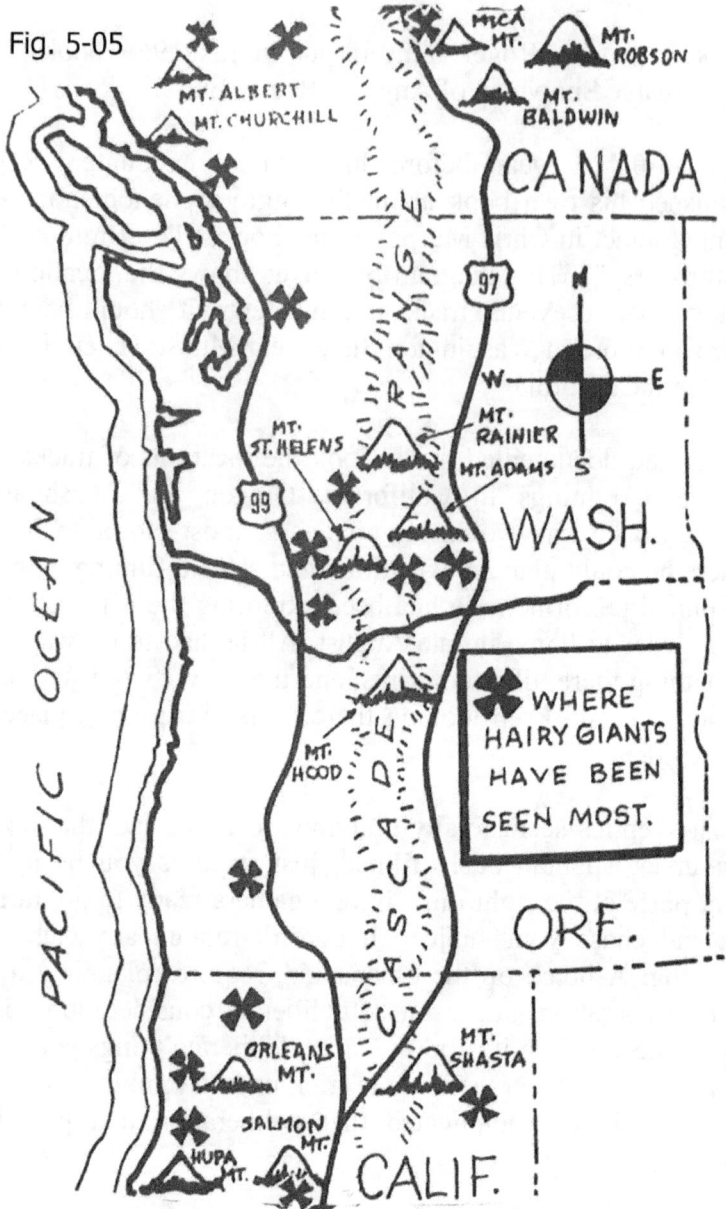

Fig. 5-05

So, if you want to argue for a well-planned deliberate hoax, you must address these questions about when and how all this deliberation and clever planning occurred. And if that's so, you must address the question of why Roger would choose Bluff Creek as a site, despite its many inconveniences, when there were plenty of closer and easier locations he could have justified as well as a place to search for Bigfoot evidence, and have an encounter.

And then you're stuck with the fact that the only justification a hoaxer would have to go to Bluff Creek is because he's deliberately making it look complicated so it doesn't look like a convenient hoax effort. And that's a lousy justification. Frankly, as a filmmaker myself, I can tell you we DO NOT (my emphasis) look for ways to make things more complicated. We are forever looking for ways to make things easier.

Now, let's explore the inverse. If the film's a hoax, why should Roger avoid Bluff Creek?

Certainly if the film were hoaxed, Roger would reasonably choose a location with some history of activity as his rationale for going there, but with his prior research on Bigfoot sightings (as documented in his 1966 book), Roger had many locations to choose from which qualified as having history of activity. If Roger were staging a hoax, we must assume he had other logistical issues to consider, because a hoax requiring a prepared fur costume and a person arranged to wear it (who fits the costume), and thus entails considerable planning and preparation. Anyone who has done location filming (and I have done an abundance of it over the years of my career) knows that a distant location filming requires a lot of both planning and an excess of precautionary supplies and equipment to insure a successful filming. The further we go from "home", the more things we take "just in case". We look at a multitude of "what if's" (what could go wrong?) and try to anticipate having solutions arranged or solution supplies brought with us.

So given there was both an abundance of woodlands closer to home and many sites with historical Bigfoot suspected activity (which Roger knew about), Roger had many options that would be justifiable for a cover story of "why here?" that were more convenient for the work of staging a hoax. The Bluff Creek site, however, would rate a poor choice for staging a hoax, as it presents numerous logistical challenges that sites closer to home would not.

Also, the Mount St. Helen area Roger was investigating in late August/early September was much closer to home and we must assume Roger was already planning the hoax event because he would have needed at least that much lead time (likely even more) to get the fur costume ready. So if there were a hoax, it should have been already well planned while Roger explored the Mount St. Helen area. So there would likely have been a better choice for filming, and Roger would have only needed to fabricate some "interesting indications" of Bigfoot activity there to justify going back for his encounter.

He would have also had the advantage of getting familiar with the location before coming back to stage the encounter, and anyone who has done location filming readily appreciates the advantages of familiarity with the location before you finally pack up to go there and film. The Mount St. Helen area could be evaluated for a hoax filming while he was there, while his last described Bluff Creek trip was 3 years before. It is not likely that on that trip, Roger was choosing sites for a film hoax to be done 3 years later.

So in this regard to site choice, his actions do not seem like those of a hoaxer. A hoaxer/filmmaker chooses conditions convenient to accomplish, and conducive to success. The Bluff Creek site was neither, if a hoax is being planned. And so this lends some persuasive support to the film's authenticity, by not showing any indications of a hoax, when we should see them in this microscopic analysis of his every move and action.

So now we'll try the alternative. If the encounter and filming are real, why was Roger in the "right place at the right time?"

The explanation is actually quite straightforward and unsuspicious.

It has been reported and reasonably documented that Roger chose that location because two months before, John Green and Rene Dahinden had investigated some footprint evidence in the vicinity, at Blue Creek Mountain in August 1967, and news of that activity was passed on through people having interest in the subject to Patricia Patterson, while Roger was away on another search expedition, reportedly to be around the Mt. St. Helen area. When he returned home, his wife told him of the activity in Northern California, and so that became his destination for his next search trip, in October, 1967. Mrs. Patterson verified this account to me personally, so I take it as a reliable account. And as much as I have determined, there is no dispute or conflicting account.

If we are to look at this account in the reference frame of an authentic encounter, it seems like a plausible motivation for Roger to go to this specific site.

So, in this question of "Why Bluff Creek?" which is the more logical explanation? Granted, any answer is argumentative, and so I simply leave it as an open question.

Understanding other reasons why the site tends to support an authentic filming encounter needs a proper foundation of the site before specifics are discussed, and no researcher before me has given as much effort to map out and cross reference the site and the subject path, the camera operator's path, and the unification of the landscape from the various segments, as I have. So I'll begin this section with a primer on the site itself, the specifics of its landscape and arrangement of objects that are used in the film analysis.

Identifying Elements of Bluff Creek

For clarity in evaluating the setting, I named or numbered various trees and objects. Now I'd like to explain that naming system.

The center cluster of trees in the Segment Five lookback are quite distinctive, so they were grouped with the initials TC (for Trees in the Center group) and numbered 1-5

TC-1 is the first tree Patty walks behind after her lookback.
TC-2 is the second tree she walks behind.
TC-3 is the "Shadow Tree", so called because it has a black shadow on the right side of its base area, important as both an identifier for Segment Five and Six, and for shadow studies and elapsed time during filming.
TC-4 is the "Fork Tree", with a trunk that splits into two upright sections like a fork.
TC-5 is the "Leaning Tree" for obvious reasons

The north face background trees all were numbered with a TN prefix. Labeled in the illustration are the Big Tree, and the "Q-Stick" tree, as they are the most important, but we also have the "Red Tree" (surrounded by red leaves), the "Ladder Tree" with short branches like the rungs of a ladder, and "Laurel and Hardy" a skinny tree leaning against a much fatter one.

The "Big Log" (aka the "main log") is that large horizontal one that crosses the setting.

Then we have several stumps. The "Smiley Stump" is usually seen directly below the "Q-Stick" tree, and it has a funny pattern on it sort of resembling a person smiling.

The "Leaning Stump" has a very noticeable rightward lean. The "Big Stump" is short and fat.

In the early part of the filming, Segments One and Two, important objects are the "S Branch" and the "White Post" (both illustrated).

124

Fig. #5-06 shows the individual trees and the identifier code we use for site analysis.

The Main trees we study at Bluff Creek
A. TC-1 The first tree Patty pases behind.
B. TC-2 The second tree patty passes behind.
C. TC-3 The "Shadow Tree" for the black shadow on its side.
D. TC-4 The "Fork Tree" because the trunk splits in two like a fork
E. TC-5 The "Leaning Tree" sself explanatory.
F. The "Big Tree", the biggest one at the site.
G. The "Q-Stick", like a pool q-stick, two colors and no branches.
H. The "Sentries", small trees standing in front of the Big Tree.
I. The "Big Log" so prominent in the scene.
J The "Smiley Stump" because it has markings like a smile.
K. The "Leaning Stump" that obviously leans to the right.
L. The "Fat Stump" as compared to the others.

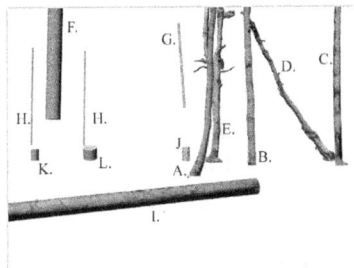

Below is the initial landscape Roger filmed, with some features we use for reference.

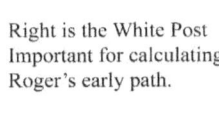

Left is the "S" Branch, which has the "S" curve in it.

Right is the White Post Important for calculating Roger's early path.

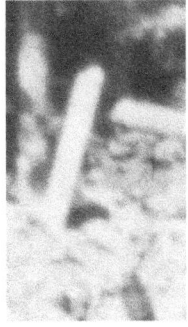

As described earlier, filmmakers often "cheat" locations when they think they can get away with the cheat, and the PGF has two distinct landscape sections that seem to have no connection. The segments One, Two, and Three are all one fairly distinctive landscape, and segments Four, Five and Six are another fairly distinctive landscape. There is no obvious connection between the two. A filmmaker hoaxing the PGF could have staged the first three segments in one spot and then found a different location nearby for the other three segments, thinking no one would ever reverse engineer the landscape from the film footage as we can today. If a filmmaker did cheat the locations, our analysis technology today could find the cheat, and it would be irrefutable proof of a hoax.

But through careful analysis of every frame of a full-frame contact print, I was able to find evidence unifying the two seemingly separate landscapes into one site map, and everything about Patty's path and Roger's pursuit path are perfectly consistent with this unified landscape model. There is no "cheat". Time and again, I will look at things that could be absolute and irrefutable proof of a hoax, things a filmmaker back then would likely have done thinking nobody would ever find the evidence to expose the trick. And every time, the PGF holds up as authentic and without cheats, tricks, discontinuity, or evidence of deliberate staging.

When Roger Met Patty

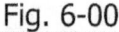

Fig. 6-00

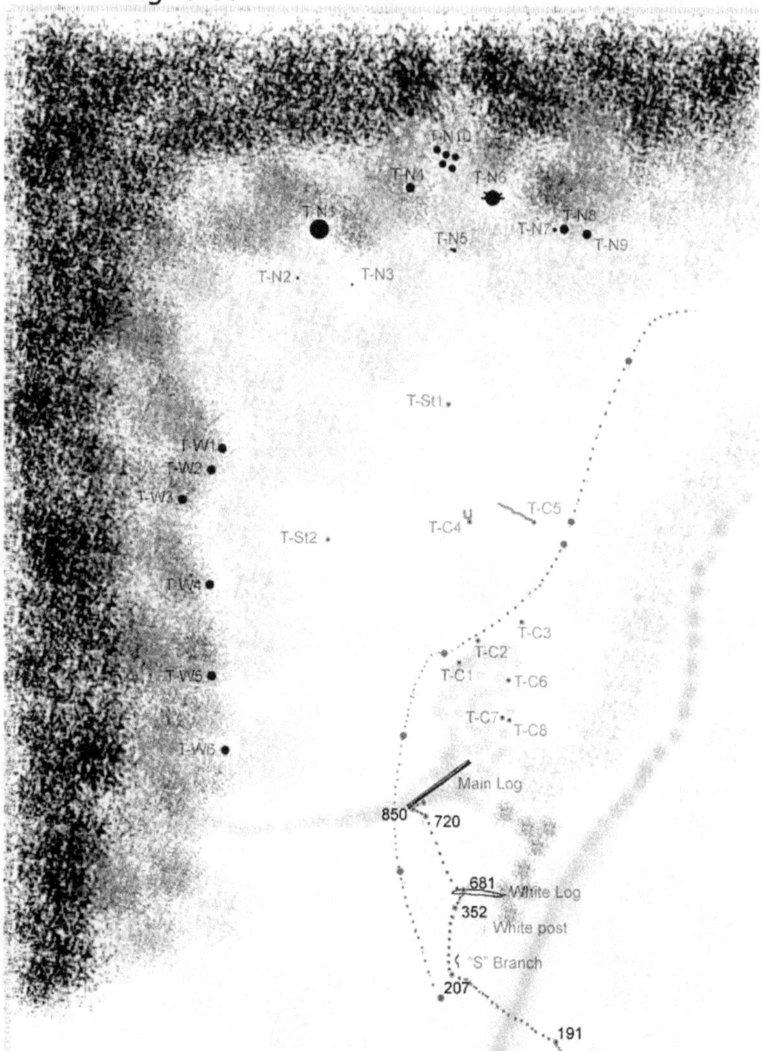

T-N10

T-N4   T-N6

T-N1   T-N8

T-N5   T-N7   •T-N9

T-N2 •   T-N3

T-St1 •

T-W1•
T-W2•
T-W3•   T-C5

T-St2 •   T-C4

T-W4•

T-C3

T-C2

T-W5•   T-C1   • T-C6

T-C7•• T-C8

T-W6•

Main Log

850 •720

681 White Log

•352

White post

"S" Branch

207

191

184

Bluff Creek Layout - Chart #2
Camera and Subject Paths

090

001

May 30, 2011   Chart by Bill Munns

128

Chapter Six - How the Event Occurred

Most people who want to research how the event occurred go to various interviews and comments by either Roger Patterson or Bob Gimlin when they described what happened. And these men's accounts are good in familiarizing us with the basics of what happened that day. But these recollections aren't precise, and occasionally there are conflicting descriptions. Skeptics seize these conflicting accounts as proof of some kind of deception. They aren't. They are just the simple fact of human frailty, and the fact that our recollections are not perfect and can change over time. We may also describe an event differently in different contexts of discussion. Most of the PGF's prior researchers took these recollections as their primary source of information as to how the event occurred. I did not.

Because I understand the film medium, and have made films myself, I had a better understanding than previous researchers about the relationship between when we see in the film and what goes on with and behind the camera. The film actually tells us as much about what the camera and camera operator are doing as the film tells us what the filmed subject is doing. And if the camera is moving about, as Roger's camera was, then that movement through the landscape tells us a tremendous amount about the camera movement in relation to the landscape. You simply need to understand the process and the film will tell you remarkable things about what happened that day.

So I evaluated the descriptive accounts of the events of that day, by Roger and Bob, and then considered the story the film itself can tell about what happened that day, and the film evidence is by far the more empirical, and Roger and Bob's accounts are more testimonial, anecdotal. And as I stated in Chapter Two, the empirical evidence is the best, far superior to testimonial or anecdotal evidence, for clarity, precision, and scientific reliability. So in my presentation to you about how the event occurred, I will rely upon the film to tell you what happened.

But I do need to begin with an outline of the technical processes, those foundation principles, which are so vital to understanding how this part of the analysis was achieved, and why I was prepared to accomplish this type of analysis that eluded previous researchers.

Learning and reasoning skills have been divided up into specialized categories over the years and when I was in school, almost 50 years ago, it was common to require middle school and high school students to take a battery of "I.Q." tests, to try and measure the student's learning capabilities. I recall there were well over a dozen categories for the tests I took back then. Of that bunch of tests, the ones I excelled in (scored in the top 1%) were Abstract Reasoning and Space Relations. Curiously, both skills are highly valuable and very relevant to this aspect of the PGF Mystery and analysis. The Abstract Reasoning is a form of fundamental logical deduction that tries to remove social and cultural influences from the reasoning process, to reduce the reasoning to its more fundamental core principles. The Space Relations category is perhaps poorly chosen as a title, because it has nothing to do with "Space" (as in "outer space", the sun, planets, stars, etc.) Space Relations has to do with understanding basic objects in pure X,Y,Z, 3 dimensional space, their relationships of size, position or orientation, and changes of either size, shape or position.

The relevance of Space Relations to photography is vital, because the photographic process takes a three dimensional space and array of objects and reduces them to a 2 dimensional representation of same. We usually refer to this as a loss of depth perception, when that third dimension, depth, is removed from the image. Understanding the Space Relation between the camera and what it photographs is one of the fundamental assets for successful special effects photography and cinematography, because we use the loss of depth perception to advantage in creating optical illusions. Having grown up in the era of "trick photography" as a prime special effects technique, the era before Computer Graphics Imagery (CGI), I was fascinated by these trick photography techniques and studied their underlying process. And my aptitude for understanding Space Relations made it very easy for me to fully comprehend the techniques and processes. And this same aptitude gave me the core understanding to excel in 3D Computer Graphics once I began my professional computer work in 1997.

Reconstructing a three dimensional environment from photographs is simply a sort of "reverse engineering" of the photographic process of reducing 3 dimensional scenes to 2 dimensional images. It has become a formal process called "stereo-photogrammetry" and has many technical and professional applications today. It has been very effectively adapted in CGI movie work for two specialized processes in the Special Effects pipeline, Photo-Modeling and Matchmoving.

Photo-Modeling allows for the use of multiple photos of some object to reconstruct the three dimensional shape of that object. MatchMoving is the process of being able to determine the path of a moving camera as it films a scene, so the camera path can be replicated and computer generated image elements can be composited into a filmed scene for some effect, with those computer generated image elements having the same relationship to their moving CGI camera as the real objects have to their real filming camera.

As applied to the PGF and related films and photos of the Bluff Creek site, this collective set of films and photos allow for an excellent reconstruction of the Bluff Creek setting itself, and as well an excellent determination of where the creature subject walked through the scene, and finally, where the camera operator was at various filming points moving through the scene chasing the filmed creature. Oddly, my apparent high score in the mental process of understanding "Space Relations" was an ideal qualification for doing an analysis of this process and result.

This analysis does not rely on any testimony of Roger Patterson (the camera operator) and it does not attempt to calculate any position, path or activity of Bob Gimlin, because he is not seen in the film nor is he holding the camera. I also deliberately am not describing anything about what happened before Roger turned on his camera, because the image data does not tell us anything. All accounts of what happened before the camera started are anecdotal, and thus imprecise. The entire analysis at this point is based on film image data only. It describes the location site, the movement of the filmed subject, Patty, and the movement of the camera operator in relation to both.

The scientific paper on the image data integrity (described with link in Chapter Two) has a well developed section on this concept of reconstructing a three dimensional view of what a camera sees, explained in laymen's terms, and so if you want to understand more of the basic concept of the process, I highly recommend that material. Because it's readily available, I choose not to repeat it here, but rather simply refer to it.

Suffice to say, understanding the concept and having all the image data to work with, we can confidently develop a factual description of what occurred as Roger's camera was running.

As earlier described in Chapter Two, there are six filming segments for the PGF. Now I'd like to explain each in greater detail, and diagram about where Roger and Patty were when the start of each segment occurred.

Segment One is frame 001 to 094. Patty is seen full body, and from a back view as she walks away.

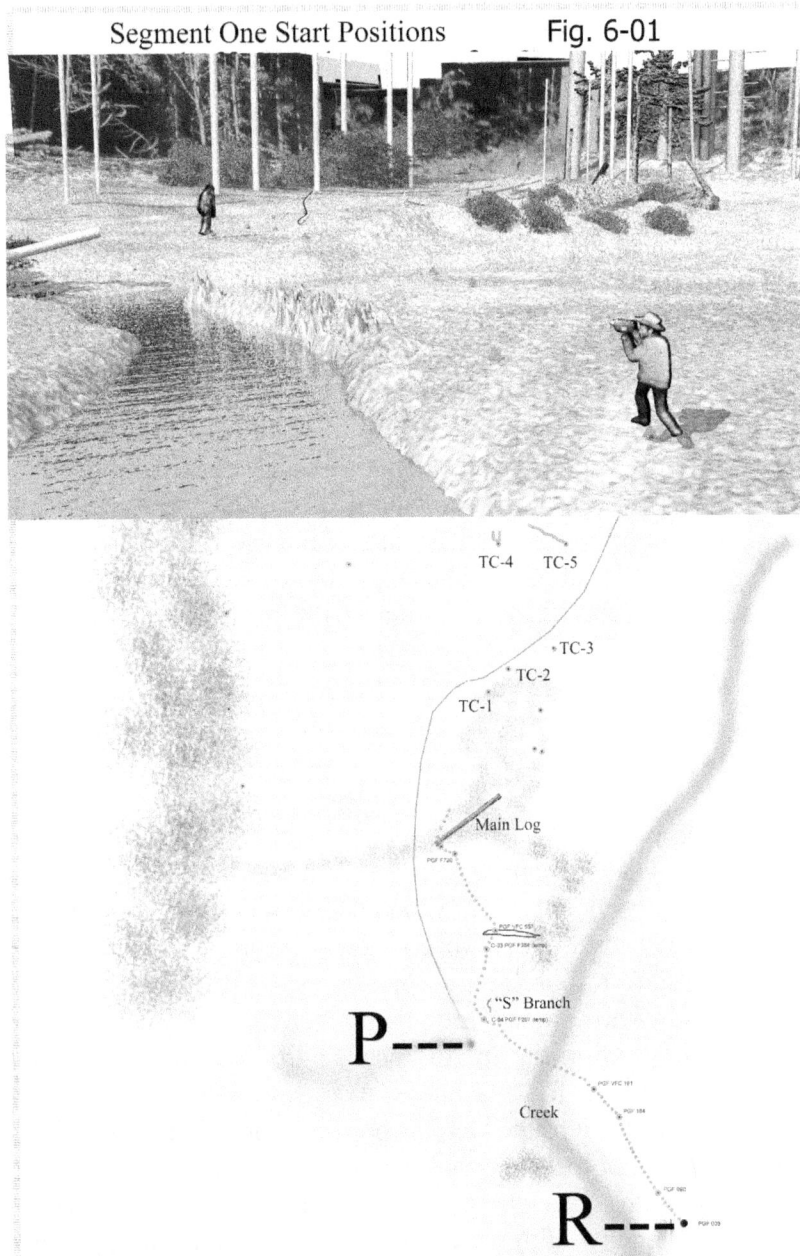

Fig. #6-01 Segment One Map. The "P" shows where Patty's start position is, and the "R" shows Roger's start position relative to her and the landscape. A 3D computer graphic visualization shows Roger and Patty in a reconstructed scene.

We will start at film frame 001, and the diagram shows the Bluff Creek layout, and the proximate start positions of both Roger and Patty. They are about 100-120 feet apart, and on opposite sides of the creek. Patty's path is marked in a set of solid lines, while Roger's path is marked in a string of small dots.

The most curious thing about segment one is that the camera starts while the operator is moving forward. We determine this by examining film frames 003 and 006 and a comparison when the distant background is anchored in both at the same point, the foreground terrain shifts larger and downward in frame 006. We use frame 003 because frames 001 and 002 have motion blur negating a comparison with frame 006.

Fig. #6-02 The upper image is frame 003, the middle one is frame 006, and the lower one is a comparison by a technique I call a tonal inversion.

Both frames have been cropped to the area they have in common. The technique of creating the bottom image is described in detail in Appendix Five, Image Analysis Techniques. In essence, what it shows us here is that middle grey tones are things the two pictures show the same, and the black and white tones are changes of something, usually because of motion, as in this case. Patty's white foot is her forward stepping foot as she walks. And on the lower part showing the foreground landscape, the black and white shifts of the ground objects shows the white are frame 006 positions, lower than the blacks, frame 003 positions. Ground that a photographer is walking over will go under the photographer has he/she walks, and the under translates to lower in the 2D photographic image.

Fig. #6-03 Illustration "Segment One Motion 2" enlarges the Patty figure and the close foreground shifts to better see the black and white evidence of the things which have changed position. The fact that foreground objects (bottom) shift more than distant ones is evidence of a forward motion on the site.

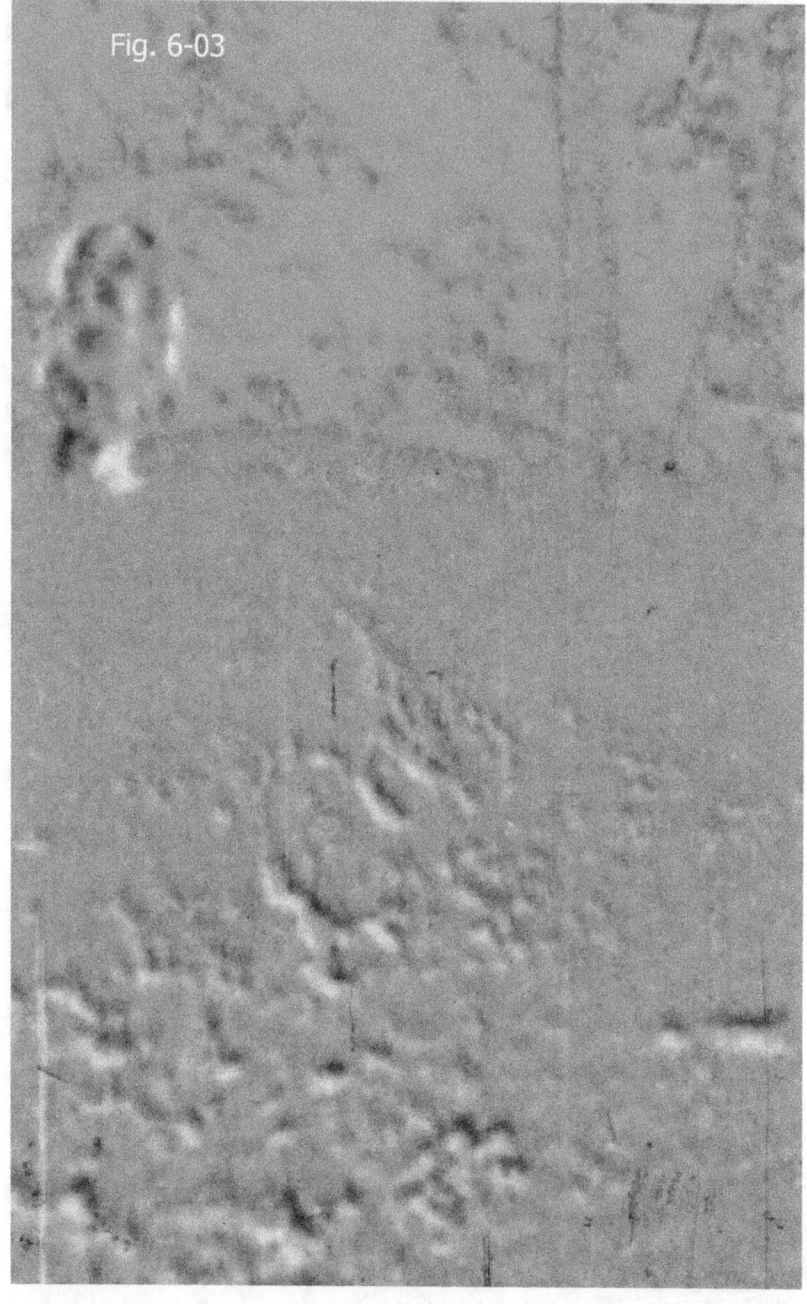

Fig. 6-03

The camera will continue to move forward for the rest of this sequence, which ends on frame 094

So Roger turned on the camera while he was walking or running forward. This is irrefutable fact, based on empirical image data.

Why Roger did this is argumentative. My reasoning about the "why" is as follows:

If it's a hoax, Roger has a lot to do before he can start his camera. He had to make his final decision as to where his mime performer (wearing his costume) would go, and where he (Roger) would run to chase after his mime. Then he had to go with his mime performer and his assistant over to cross the creek and get the mime suited up in his costume and brush it out to look good. He would then position his mine at the starting point, and have him wait while Roger went back across the creek to get the camera ready. Once Roger picked up his camera and checked to see he had it wound up, and the focus and F-Stop on the lens set correctly, he could then look to his mime and give the person some kind of visual cue for "action" (the standard movie command to begin performing). Typically, you start the camera and then call or signal "action" but if we assume Roger did not want to edit his film, maybe thinking it would look more authentic, he'd call or signal "action" first and then start his camera. So he has given the signal to his mime (and it would have to be a visual signal because the distance, the noise of the creek water between them, and the head mask reducing the mime's hearing, all together would make a spoken cue hard or unlikely to be heard), starts his camera, and begins to chase the mime while filming.

Except, that's not how it happened. Roger was not standing still and started his camera and then ran. He started his camera while running. Would a hoaxer have given any thought to the difference? Why should he? If he starts his camera and starts running, who will notice it, with the analysis technology of the

137

time? If he deliberately did run first and then start the camera, and nobody in his time noticed it, and he did it to make the hoax more convincing, sooner or later he'd have to find a way to bring it to the attention of researchers in hopes they'd find his story more compelling, more believable. But nobody noticed this when he was alive, and he never brought it to anybody's attention. Indeed, no researcher ever fully understood this until I began my image analysis 7 years ago. So it certainly doesn't seem like a false positive created by an ingenious hoaxer.

A hoaxer would have stood there, checking his camera, stood there giving his mime the action cue, and stood there to turn on the camera, and then start running when he was sure the camera was on and running smoothly.

But a real person encountering something very strange, something he had hoped one day he might find, frantic to get off his horse, get his camera out and start filming as the thing walked away, would start chasing the thing as soon as he was off the horse and had his camera. And he would likely start his camera while running, as Roger actually did, because he was likely never standing still.

So, on this first fact, that the camera was started while the operator was in motion forward, I find implausible for a hoaxer doing, because the hoaxer would be acting with deliberation in all his actions, and standing still while he got everything ready. He would start the camera with deliberation while standing still, and then start his chase. He would not deliberately start the camera while running as a false positive to make the hoax more believable, because analysis technology of the time couldn't find the proof of that action.

I find this fact supportive of a real spontaneous event.

Segment One ends at frame 094, and if the camera was running at 16 frames per second, that's just shy of 6 seconds of film. If the camera was running at 24 fps, that's just shy of 4 seconds. I mention both because the camera speed is still unresolved as to which setting Roger had the camera on. More on this in Chapter Twelve, Loose Ends.

So, Roger has been filming for 4-6 seconds, moving forward constantly as Patty moves away, and his camera stops. At the last frame of this segment, frame 094, we do see Patty clearly as she approaches a structure we call the "S" Branch from the left. She's actually a ways beyond it, but in the camera image, her forward foot is about at the left base of it in the ground. So we know where she is when the camera stops. Then it restarts for Segment Two.

There is no splice between frame 094 and 095, the end of Segment One and the start of Segment Two, and the start frame of Segment Two, frame 095, has the over-exposure consistent with a camera start. So these sequences were not edited together. They are the actual segment order of filming, and nothing has been edited out or re-arranged.

Segment Two is frames 095 to 190. Patty is seen sometimes blocked by foreground objects, and her lower legs obscured by terrain, and in a back view.

Fig. #6-04 Segment Two Map, again marks positions for Roger and Patty where we calculate they are at frame 095.

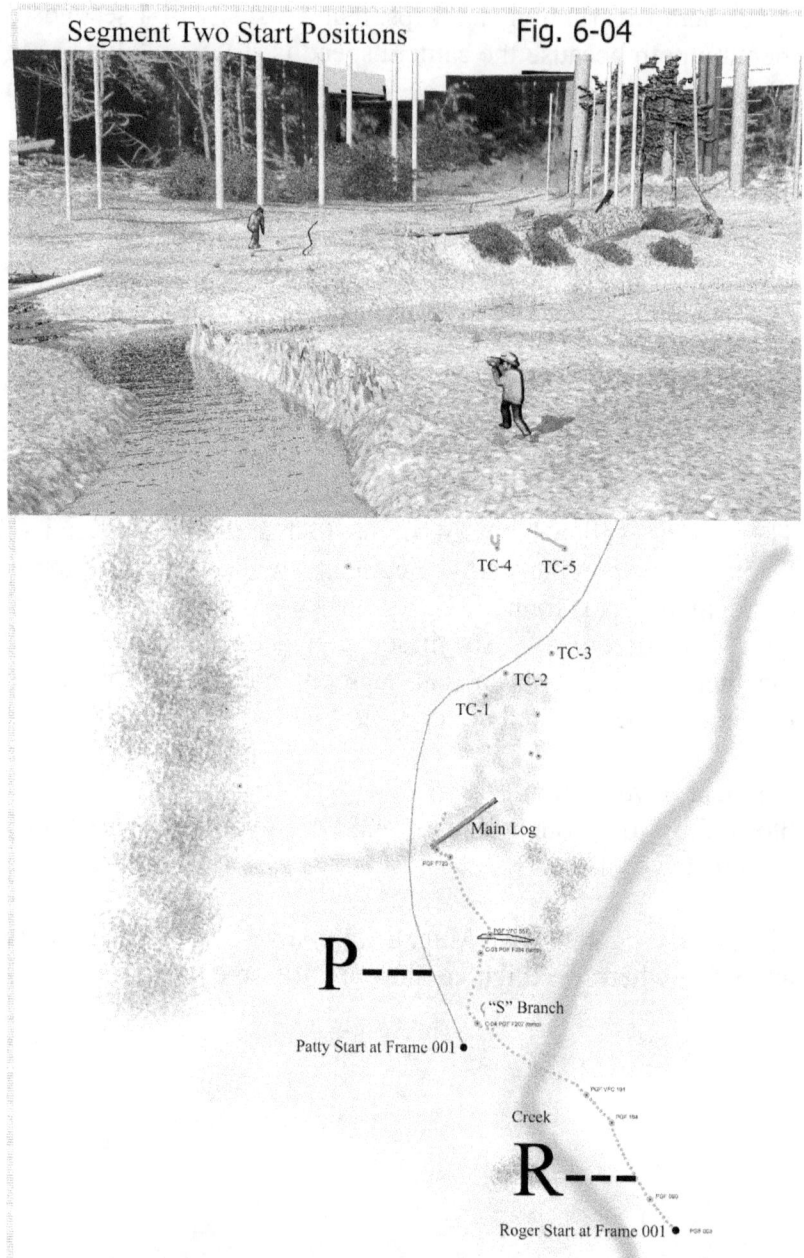

Segment Two Start Positions   Fig. 6-04

When the camera restarts for Segment Two, there is more shaking of the camera and Patty is behind the "S" Branch and in front of some dark background, so we don't see her clearly until frame 108. We see the landscape has changed, the close foreground is much closer, meaning Roger has continued to chase Patty while the camera was briefly stopped. The "S" Branch has shifted its position relative to two marker trees in the background, and in frame 108, Patty is finally visible to the right of the "S" Branch. Using her own body as a measure of approximate proportion, she's taken three or four steps since frame 094. At her pace, seen in the film, she takes a step about every 11 frames or so (such as her step from frame 063 to 074, with alternating feet). So three steps would be about 33 frames, and 4 steps would be about 44 frames. But only 14 frames separate the two images, frame 094 and 108. So we are estimating about 20-30 frames of time in the interruption between Segment One and Two. Less than two seconds elapsed when Roger's camera went off and he switched it back on, and he was running the whole time, because the image data shows a significant changed forward position.

So, once again, we have the irrefutable fact, as shown by the empirical film image data, that the camera was stopped, and restarted, the restart position is well forward of the stop position, and Patty has taken three or four steps in the time elapsed.

Why did Roger's camera stop and restart? Once again, the "why" is argumentative. In my opinion, this is why.

If it's a hoax, with a hoaxer acting with deliberation, there's no reason to stop and start the camera. Nobody noticed it and Roger never brought it to anyone's attention. Analysis technology of the time couldn't determine it even happened. So doing so with deliberation makes no sense as creating a "false positive" event in the film with hope it would make the film seem more authentic. So I do not see any rational purpose for Roger deliberately doing this. And when a camera operator is filming with deliberation, his finger is usually quite secure on the start trigger or button. It doesn't slip off when a cameraman is calmly doing his job.

A man spontaneously trying to film an unexpected event, where he has to watch the subject figure run away, he has to chase it on terrain that's uneven and unfamiliar, he may sincerely wonder about his own safety this close to the thing, and he has to run the camera and try to keep it steady because his running is making it wobble around, that man's got a lot on his mind, and keeping his finger on the trigger isn't his top priority. All he needs to do is hit a patch of loose ground to momentarily throw him a bit off balance, and he could release the trigger, causing the camera to stop. And once he realizes he doesn't hear the camera spring-driven motor running, he quickly restarts it while still running.

So once again, I find no plausible reason a hoaxer acting with deliberation would stop and restart the camera in this fashion, but I find it entirely plausible for a person really experiencing a spontaneous and unexpected event to do so.

Roger will continue to move forward, nearing the creek, and his ground is lowering slightly. The far side of the creek embankment is rising up a bit as Patty moves left to right and away from the creek, and so this closer terrain of the creek embankment is starting to partially block Roger's view of Patty. The last clear frame where we see Patty is frame 187, and the creek embankment now blocks all of her legs. Only her upper torso is visible. So Roger is losing sight of her.

This second segment is 4-6 seconds long, and the camera stops at frame 190. Roger is near the creek but has not crossed it yet.

Segment Three is frames 191 and 192. This is likely a camera trigger slip, and Patty is suggested but not verifiably seen, but the creek Roger must cross is seen in part.

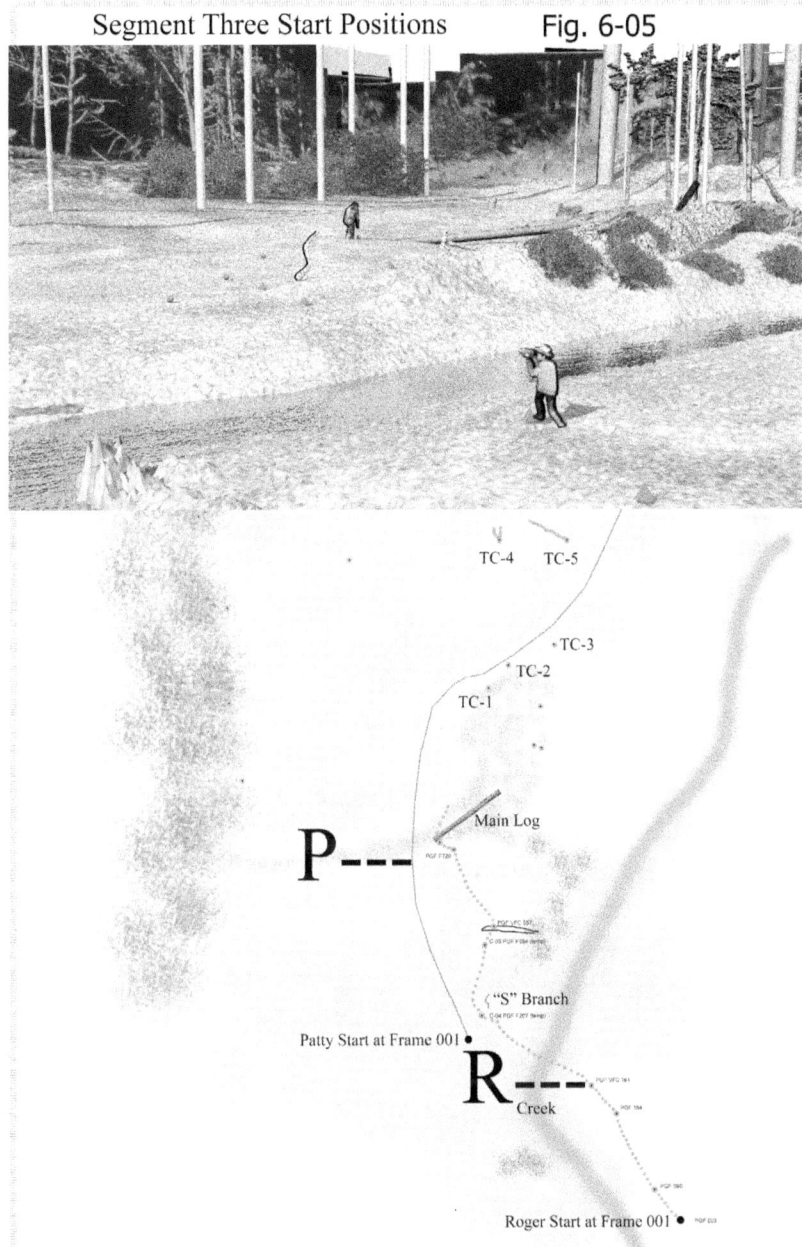

Segment Three Start Positions   Fig. 6-05

Fig. #6-05 Segment Three Map shows Roger almost at the west bank of the creek, and Patty continues on her path.

Segment Three is especially curious because it's only two frames long. To accomplish this with a Kodak K-100 camera (the one verified as being used by Roger), you must pull down the trigger lever so the spring wind motor engages, and then IMMEDIATELY (my emphasis) let go, in a 15th of a second or less, so the shutter brakes to a stop.

There is no edit before or after this segment. It has not been shortened by editing. A dark mass that may be Patty is well forward of the last position, maybe 6-8 steps ahead but this is not conclusive. But she is definitely not close to her location in Segment Two, frame 187 or earlier. So a few seconds have elapsed from Segment Two Stop to Segment Three Start.

Segment Three shows the creek ahead of Roger, so he still has not crossed it yet. There is another odd thing about the frame 191, the start frame of Segment Three. It appears to have a motion blur which is rotational, not linear, and the blur is not global, as motion blur tends to be. This is quite rare. Chapter Twelve, Loose Ends, discusses this further.

Fig. #6-06 shows a filmstrip of segment three, with the last frame of Segment Two, the two frames of Segment Three, and the first frame of Segment Four.

So the two-frame trigger slip did occur as the film shows it. Fact. Once again, the "why" is argumentative.

Fig. 6-06

Frame 190
Last of Segment
Two

Frame 191
First of Segment
Three

Frame 192
Second (and last)
of Segment Three

Frame 193
Start of Segment
Four

If the hoaxer planned to do it, why should he? No one can find it with the analysis technology of the time. Roger didn't plan it and then draw attention to it when it went unnoticed. But if he were hoaxing this film, it simply must have been a deliberate action, and a deliberate action with no goal is irrational. Could a hoaxer accidentally have done it? Not likely. When you film with deliberation and calm planning, the camera operator keeps his finger off the trigger or shutter release until he's ready to shoot. It's like a person carrying a firearm, in a calm deliberate situation. You don't point the gun ahead, rather up or down, you have your hand off the trigger, until you are prepared to fire. A camera operator does the same, in general, in a calm and deliberate situation, keeping the hand off the trigger or start switch until it's time to film.

So from the hoaxer position, there's no rational reason to do it deliberately, and not a likelihood of it being done accidentally by a hoaxer, acting with calm deliberation. It's just not what you should see on a hoaxed film.

But going back to the firearm analogy, a person carrying a firearm in a spontaneous and uncertain situation, chasing something which has potential to attack, you hold the firearm forward, pointing where you look ahead, you have the safety off, and your finger on the trigger, ready to fire in an instant, if you suddenly see a reason or cause. A cameraman in a spontaneous and unexpected situation, intent on capturing the target on film, behaves similarly. The camera is pointed forward, and the finger is on the trigger or shutter at all times, ready to fire up the camera on a moment's notice and take any footage he can get.

So, realistically, the person acting with deliberation tends to keep the finger off the trigger or shutter until he is definitely intending to film, while a person holding a camera in a spontaneous and uncontrolled situation holds the finger on the trigger or shutter release at all times to hopefully capture that unexpected event on film. And walking on the uneven slope of a creek bank, walking down toward the creek, as the film verifies Roger was, any slight loss of balance can cause the finger to slip as the arm tries to balance or steady the man. So this two frame trigger slip can reasonably occur if the filming is a spontaneous and unplanned encounter, but is unlikely in a deliberately planned hoaxing action.

This argues for a spontaneous and unexpected event.

If we continue to try and reason for a hoax, let us imagine Roger deliberately wanted that shot of the creek, so he could then embellish his account of the event by telling how he ran to the creek, then struggled to frantically cross it, and then resume his pursuit of Patty. Then, why only two frames of footage, and why was the creek only in a sliver of the lowest part of frame. You'd never see the creek in a projection of the film at any speed, and it might even get cropped out by the projector aperture, which is smaller than the full image frame. If he wanted to have a creek on film so he could add that to his story, he should have shot about a second of footage, and aimed the camera to see the creek more clearly. So even approaching the hoax idea this way, it still doesn't hold water. He didn't get enough footage or show the creek clearly enough to impress anyone and he never brought it to anyone's attention as a suggestion of his frantic and spontaneous filming behavior.

It's not a false positive, not deliberate, and not likely to occur by accident in a deliberate filming situation.

It's real and spontaneous, an accident of the finger by a man trying to carry a camera down an embankment to cross a creek with rapid flowing water, to get to the other side where something was walking away and he could hardly see where it was anymore.

So Roger has this two frame slip of the finger, and the camera is off, as he finishes going down the creek bank, and crosses the creek. The film does not tell us how slowly and carefully, or how hastily and briskly Roger goes across it, but he is wearing cowboy boots, and they tend to have smooth soles and a smooth sole isn't the best thing for traction on slippery rounded rocks on a creekbed.

This brings up another issue with the event. If it was a deliberate filmed hoax, Roger chose his location. He chose his staging plan, and chose where his costumed performer would start and where he would start to chase that performer. So why would he choose to start his performer on one side of the creek, and then he had to cross it during the chase?

One, if he drops his camera because he slips on some slippery creek rocks and the current of water, the camera and film are ruined. Pack it in and go home. All is wasted, all is lost. So why choose a plan that required him to cross that creek and risk that disaster? Makes no sense to me, the risk of dropping the camera in the water while crossing the creek.

Also, starting on opposite sides of the creek, all the water noise is between them, and when you add the mask reducing the mime performer's hearing ability, a hoaxer has now chosen a situation that essentially guarantees he has no verbal communication with his mime, and the way he has staged it, the mime has his back to Roger, so once Roger gives the "action" cue, the mime sees nothing of Roger until the lookback. Why would you plan a deliberate hoax and stage it so you have no control and no communication with your performer being filmed?

Is Roger the incredibly brilliant hoaxer, so cunning that he planned to start filming while running and devised that incredibly clever two frame trigger slip, yet so dumb and disorganized that he didn't anticipate the hassle of crossing the creek and didn't think maybe he needs some communication and control over his mime performer? Well, he can't be both? Is he really brilliant, and planned everything, or really dumb and just drove over 400 miles to blithely "give it a shot" and see if he gets anything usable?

This is one of the more profound contradictions that advocates for a hoax cannot address, because there is no logical explanation.

Yet the film image data tells us on no uncertain terms that it occurred that way, Patty starting on one side of the creek, Roger starting on the other, the hasty starts while running, the two frame slip of the finger, etc.

That brings us to Segment Four.

Segment Four is frames 193 to 233. This segment has very few sharp frames because Roger is running around a branch structure and charging forward toward Patty as he films. A lot of this is just the camera pointed down at the ground.

Fig. #6-07 Segment Four Map shows where Roger is after he crossed the creek, and where Patty is, continuing on her path.

Roger has crossed the creek, and is moving up the creek bank, toward the "S" Branch we saw Patty passing behind earlier. But when the camera starts for Segment Four, Roger is still moving up the bank. We know this because we see the "S" Branch with its distinctive curvature on the left of frame, the distant trees behind on the right, and a few frames later, we see the "S" Branch on the right, and lower in frame relative to the height of the background trees. When the distant background trees are anchored in a frame comparison, and we see the foreground object move left to right and lower in the frame, what is happening is the camera is actually moving right to left and rising higher. Roger is still climbing up the bank when he starts the camera for Segment Four. He sees Patty is still in view, walking away, and so he takes a shot at her, with his camera. The motion and his frenzied pace make almost every frame of Segment Four too blurred for any good footage of Patty (except a few frames here and there), so from a hoax standpoint, he wasted his film. Ah, but the clever hoaxer that he was, Roger just wants us to think he wasted his film, so his footage seems more sincere.

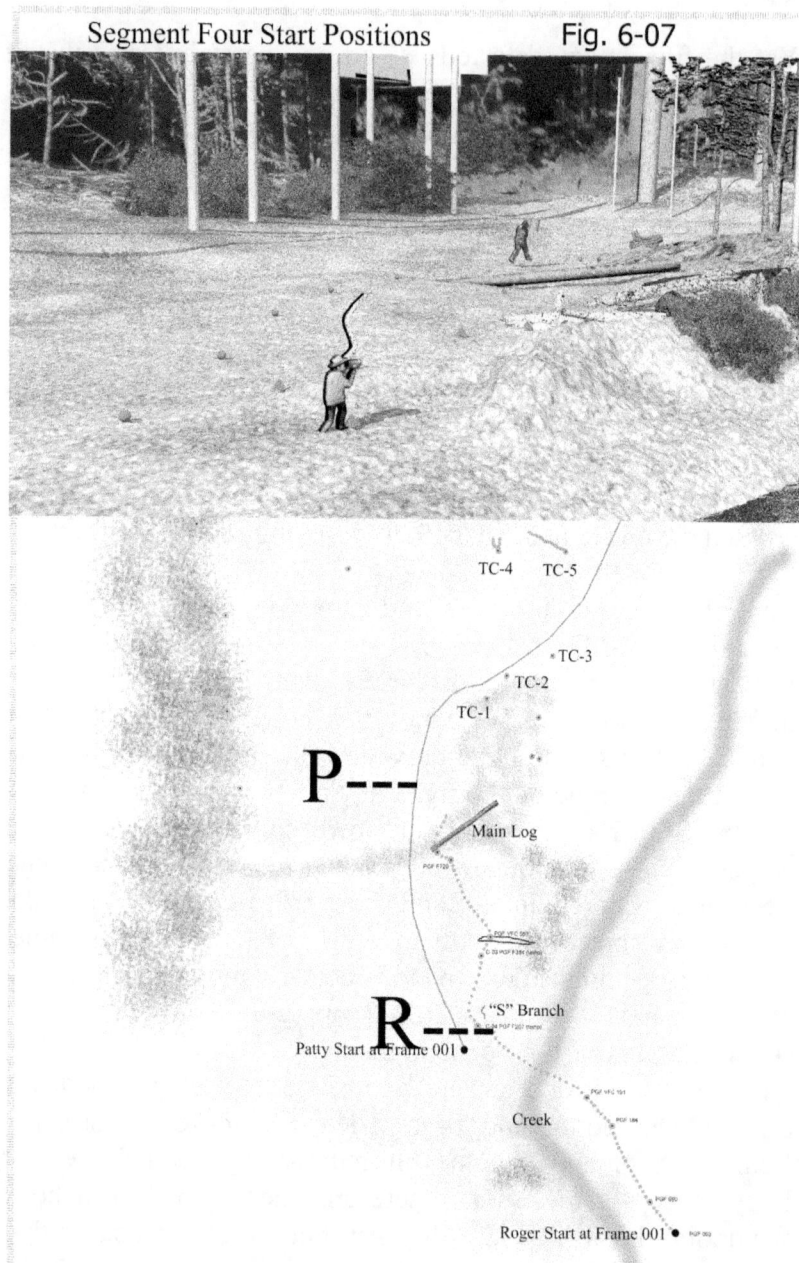

Segment Four Start Positions Fig. 6-07

Seriously, this again breaks down into fact and argument.

Fact is, the camera was started while the operator was still hurriedly climbing up a creek embankment, moving past a large branch structure, and then charged forward so hurriedly that the film got more shots of blurred ground than shots of Patty.

Argumentative is the "Why?" once again. Is this the action of the brilliant hoaxer, the inept hoaxer, or the man responding to a spontaneous event he has no control over, and is just trying to capture some kind of footage? When you consider that before Roger climbed up the creekbed bank, Patty was out of view as as he crossed the creek. So he had no idea what he'd see when he got to the top of the bank. Maybe she'd veered left into the woods? Maybe she'd doubled back to confront him, and maybe attack? Maybe she was hiding behind some debris, and he couldn't locate her? He didn't know what to expect.

What Roger saw when he climbed the creek bank was that Patty is still on her path walking away, but she's in the open, nothing to hide behind, and Roger is quite close. He has a great clear shot at her. And that, in my opinion, is why he started his camera as he was still climbing up the creek bank. He was desperate to get good footage of her, and now was the perfect chance, once he's crossed the creek. As soon as he realized this, while still climbing up the bank, he impulsively started his camera.

My opinion is spontaneous event. I'm sure some will disagree.

This sequence is a mere 40 frames, 2 1/2 seconds at 16 fps and less than 2 seconds if 24 fps.

Why did he even bother filming it, if it was deliberate and calm, a staged event? On projection, it's just a jumbled mess of images, and impossible to appreciate he started the camera while still climbing up the creek bank. With each Segment, the concept of a hoax simply becomes more and more irrational, in my opinion.

Segment Five is frames 234 to 686. This is the most important segment, because it includes the lookback by Patty and also because Roger has planted himself to hold his camera steady as he pans to follow Patty walking across the terrain. Most of the good body images come from this segment.

Fig. #6-08 Segment Five Map with Roger in his fixed position for the lookback, and Patty continuing on her path.

Segment Five is the most important part of the PGF, because Roger has his subject in the open, he's closer to her than any other time his camera is on (based on the image data showing her physical size in frame is the largest, which comes from being the closest). So Roger plants himself where he stands, and holds his camera steady, to film as Patty veers right to almost a profile path briefly, looks back at him, and that turns away to resume an angular path away from him. This is the clearest footage of Patty.

Finally, she passes behind some trees and the film has only hints of her through that cluster of trees. Roger shifts his position about 5 feet forward trying to get a better shot, but Patty is still behind the trees. For a few seconds she partially emerges from the trees, but it's now obvious Roger won't get any great footage in this position.

Segment Five "Lookback " Positions   Fig. 6-08

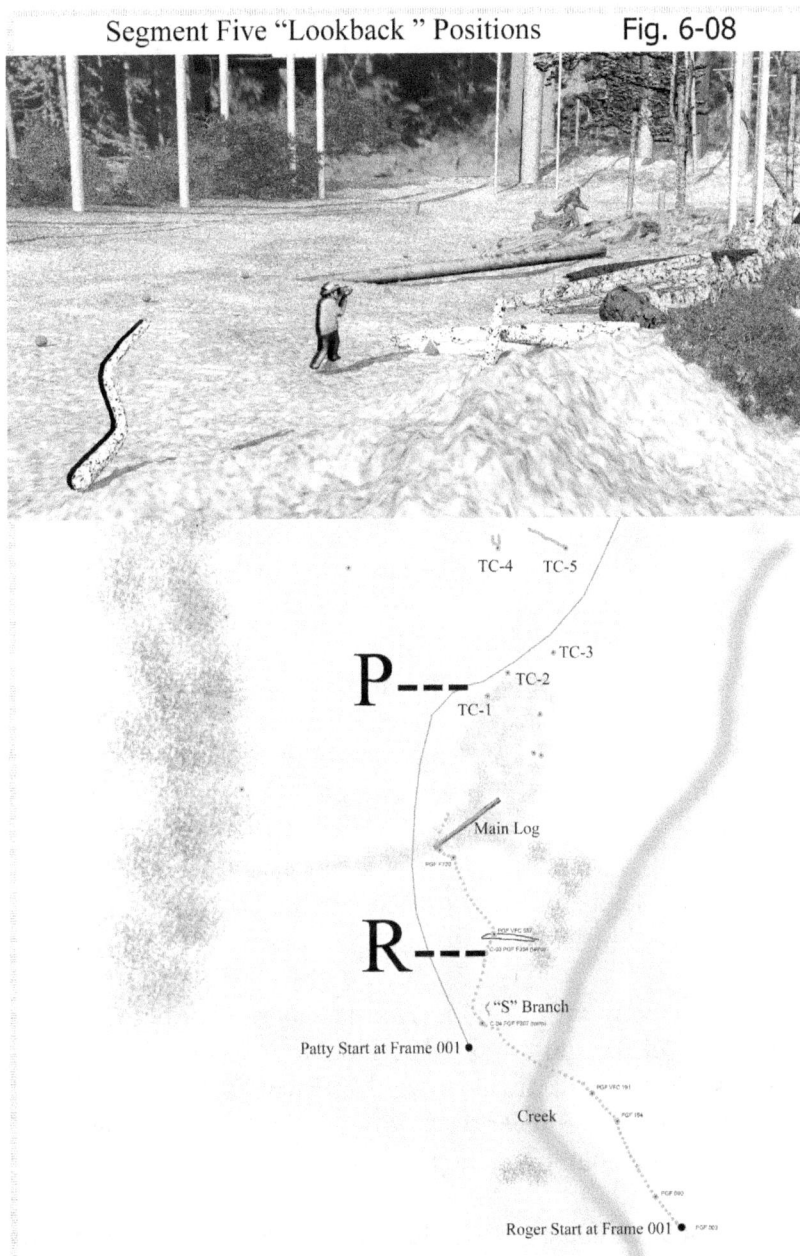

In this segment, what Roger does is what a good cameraman would do, plants himself, holds the camera as steady as he can, and tries to get a great shot. In doing so, the film evidence contradicts a common hoax suspicion. Advocates of a hoax, especially those who've never really studied the film, love to claim Roger deliberately shook the camera and made it full of motion blur to hide the flaws of his fur costume. This has been a long-standing hoax talking point. Segment Five refutes that on no uncertain terms. In Segment Five, Roger starts closer to Patty than any other time in the film. He plants himself and holds the camera as solid and steady as he can. Patty goes from a back view walking away to a side view and the lookback, so we have the best variety of different body views. All our best study frames for Patty's body, frames that could expose a bad costume, come from these remarkably clear images. Roger is not trying to hide artificial flaws in the appearance. In Segment Five, he's doing a remarkable job of giving us excellent evidence to appraise the body of Patty. If she's a poor costume, he's trying to give us as much opportunity as possible to find its flaws.

So the claim that the camera was deliberately shaken to hide the flaws of a costume is defeated by the truth of what the film tells us. In Segment Five, Roger is doing everything possible a cameraman could do to give us excellent image evidence for analysis.

That ends Segment Five at Frame 686. This is the longest segment at 452 frames.

Segment Six is 687 to 954. This is where Roger has moved again to get around some trees blocking his view, and films Patty's distant walk away toward the woods, and then Roger runs out of film. But Patty is very far away and so images of her have little anatomical detail for analysis.

Fig. #6-09 Segment Six Map shows Roger has now run forward close to the main log, where he gets a better view of Patty as she does the final walk away.

Segment Six Start Positions   Fig. 6-09

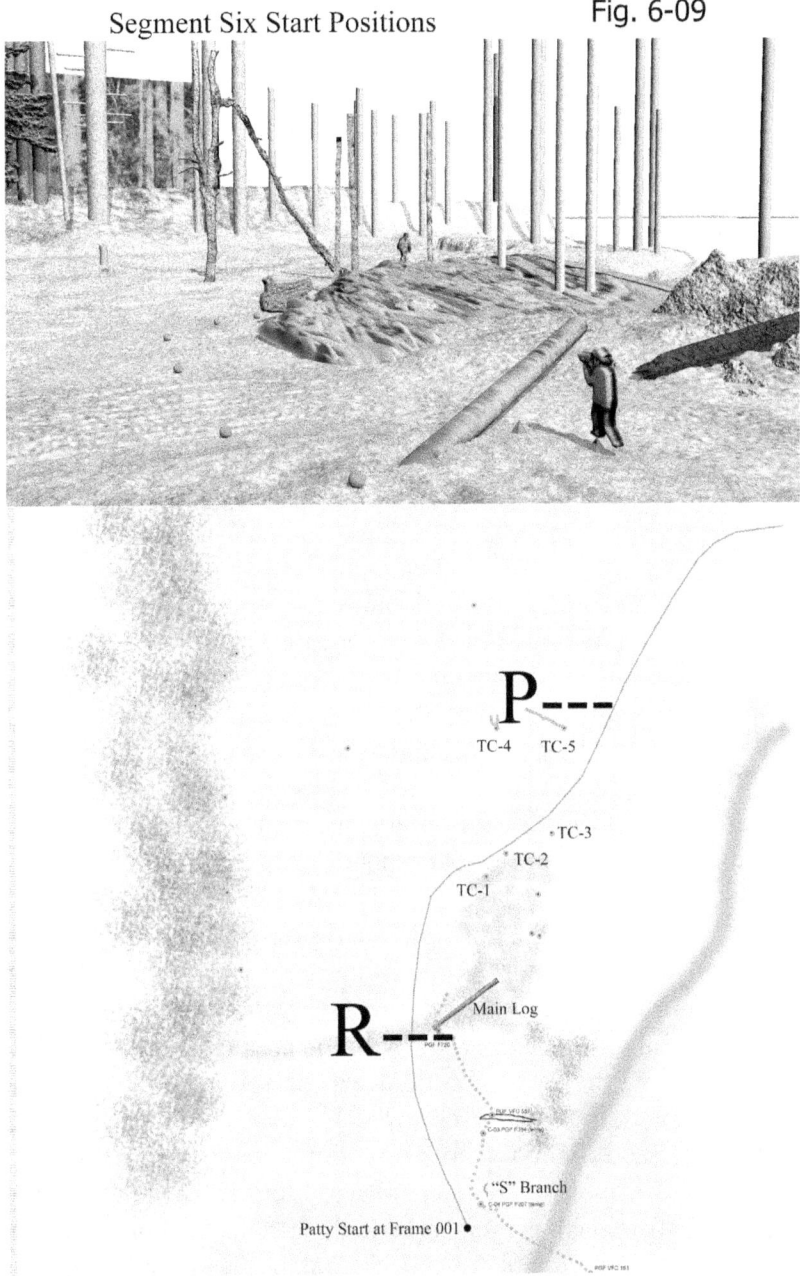

In this final Segment, Roger turns on the camera and runs forward to get a better position for seeing Patty walk away.

She's now further than ever before, and still walking away, so the best he can hope for is just her final walk away, unless she turns and does something unexpected. But he's still got film in his camera so he keeps filming.

He will run forward, while filming, and then shift his position constantly through this segment, and so it has a lot of motion blur with occasional good images in between. We rarely use this portion for analysis because Patty's very small, and has little detail, and there are few continuously good frames to make a good sequence.

Finally, Patty veers off to her right and toward the woods, and Roger's camera runs out of film.

How do we know this? When this type of camera runs out, with a 100-foot daylight load, the camera operator opens the camera door and pulls out the exposed roll and loads in a new roll. If the camera operator has a true darkroom, or a true pitch black camera changing bag, there may not be any exposure of the roll being removed, but the whole concept of a "daylight reel" is that you can unload it in a low daylight situation and only the outer layer of film on the take-up reel gets washed out. If the film is loose on the take-up side, you might lose a second layer of film, but one layer is only about 10 inches of footage on a 100' roll, so the loss isn't considered important. So in common practice, people unload the camera in low light and allow the outer 1-2 feet of film to be washed out by the light. The rest of the film is fine. We gain the convenience of loading and unloading the camera in light we can see by (so we can see what we are doing) and lose maybe 1-2 feet of film to washout. It's a fair trade off for people who don't want to learn to load and unload a camera in pitch black, literally blind, by touch alone (professional camera operators learn this, but amateurs rarely do. I learned both ways in my college filmmaking years).

When this washout occurs the total washout section fades into a partial washout section, where you still see picture but the frames have an overexposed orangish tint and some kind of light pattern based on how the film was wound around the take-up reel and how quickly you tighten up the reel to remove it. People usually cut off the washout part of a roll when they show it, but it's not unusual to keep some part of the washout where there is still some picture. And on the last few frames of the PGF, we see the washout pattern starting, and indications of print-through of light passing through the sprocket holes onto lower layers of film.

Unfortunately, for this book, the black and white only images prevents me from showing the study of the washout (color is needed), but it is printed in one of the Munns Report PDF documents available on the website www.themunnsreport.com

You generally only see this on the last shot filmed before the roll is taken out of a camera in a common low light situation. It is empirical evidence of such a removal of the roll, after the segment was shot, so Roger ran out of film. He would then reload the camera for what we call "Reel Two" footage, and that is a separate controversy dealt with later in this text.

So we know for a fact that Roger is again trying hastily to get some good shots, and that he will run out of film with this segment.

Argumentatively, nothing here is powerful in either the hoax or spontaneous scenario.

Behavioral Analysis

There is a legitimate division of overall science called "Behavioral Sciences", so I don't think I'm out of line here looking at the encounter from a behavioral standpoint. But any behavioral determination supporting validity of the event is vulnerable to the counter-claim that it's just a good performance intended to look real and spontaneous. So perhaps the better way to approach this is to look at the behaviors of

Roger and Patty, in relation to each other, and see if any of those behaviors appear to be an obvious indication of a staged event, a hoax.

If we first look at where Patty started and where she went to, a more direct path would have been to follow the creek shoreline. She would have gotten to the same end point, but more quickly.

But going to the creek shore would mean going closer to Roger. Is there any reason a performer hired by Roger would be afraid of him? I can't think of any. Is there any reason a real entity might be afraid of him? Yes, I can see that as reasonable.

So taking the long way would be a way to avoid Roger and get to the destination. It will force Roger to cross a creek, as noted before, not a wise choice for staging a filming, and it will limit communications between a performer and Roger as he films, which doesn't seem like a wise choice. But it seems a plausible choice for something afraid of Roger, and not concerned with his hardships filming.

Now we look at Patty's posture at the start. She's hunched over, her arms dangling loosely, and her steps not terribly long or aggressive. In the first two segments, we never see her taking aggressive steps or pumping the arms for a power walk.

Does that make sense for a performer when his director has just given the cue for "Action!" Slump over and take easy steps, and let your arms dangle. It doesn't for me.

Fig. #6-10 shows a sampling of good frames of Patty's posture when she's first seen starting to walk away. These images would be right after any "ACTION" cue for a hoaxer.

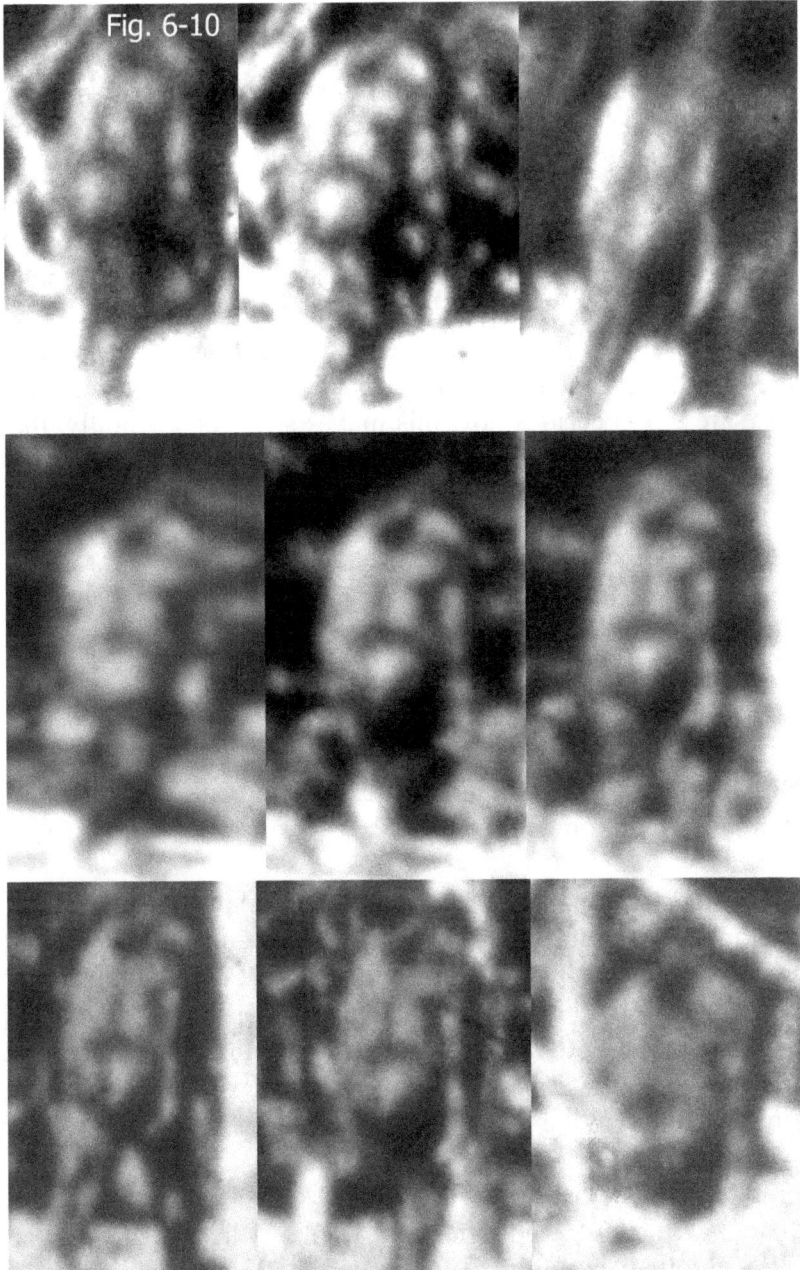

Fig. 6-10

When does Patty shift to a more alert posture, pump her arms more aggressively, and take longer, more powerful strides? When Roger is on the same side of the creek, at the closest he will be to her, when there's nothing in the form of a barrier between Patty and Roger. Is the mime afraid of Roger? If he wants to act mad or courageous or confrontational, why doesn't he turn and do some kind of threat gesture. Why just speed up to a more powerful walk?

But if Patty is real, she walked casually when Roger was over 100 feet away and there was a creek as a barrier between them. That seems like a "safe place". But in Segment Five, Roger is much closer to her, there's no creek to serve as a barrier, and the open landscape doesn't give much protective cover. It isn't as "safe". It is curious how this makes sense behaviorally, that when she felt safer, she walked more casually, and when she felt less safe, she walked more aggressively, especially after looking back and seeing how close Roger was in the lookback part.

Once she was much further away, the walk relaxed to a more casual stride again. She was further and further is safer. Behaviorally, it does make sense.

Shall we say the brilliant hoaxers planed all this? If so, I commend them on their brilliance.

But when you consider that a mime would have poor visibility looking out the mask, very poor hearing, and thus incredibly poor communication potential with Roger, coordinating their two actions, the fleeing creature and the chasing cameraman, is problematic. Roger would be giving attention to his camera and his shots, and the mime would be giving attention to not falling down walking in the irregular terrain path.

But somehow, the two actions seem to be well coordinated, like a real cause and effect is at play, the pursued figure adjusting its walk to the degree of threat posed by the pursuing camera operator.

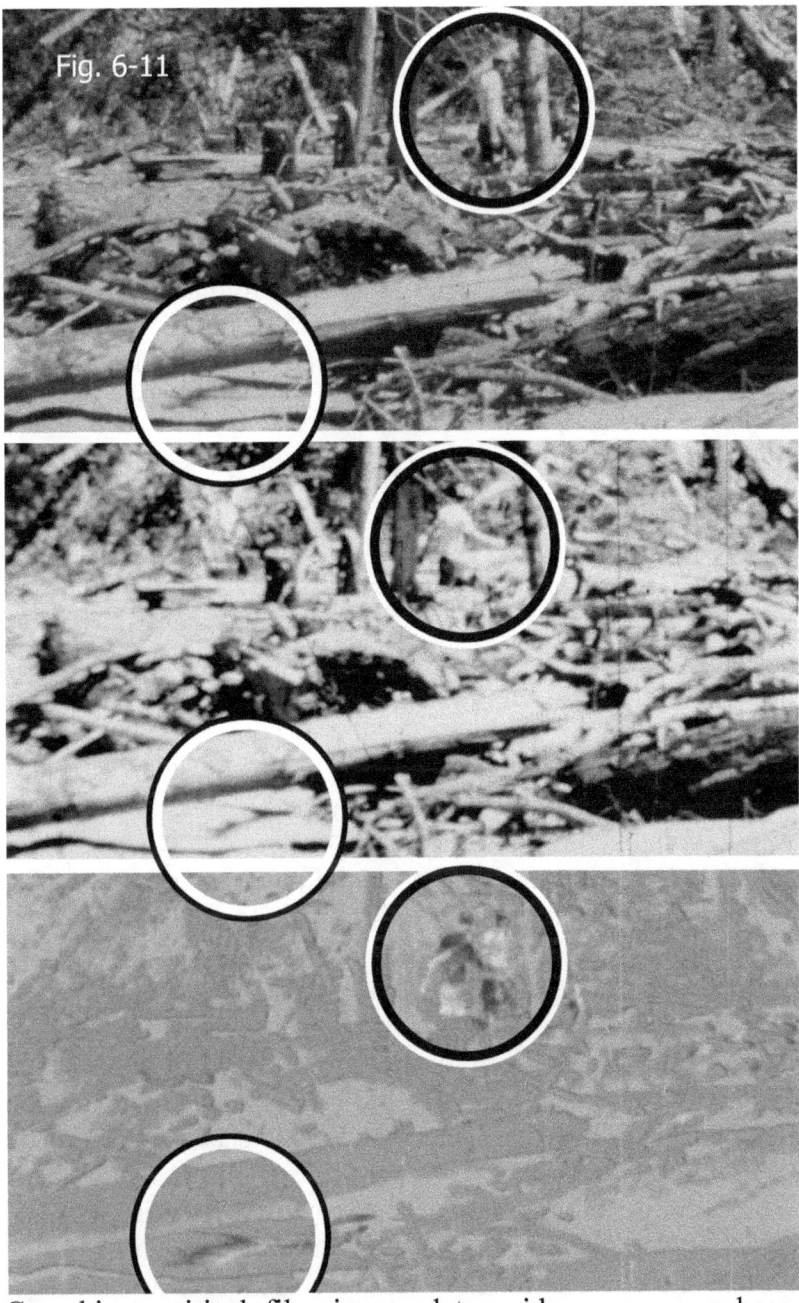

Fig. 6-11

Can this empirical film image data evidence prove a hoax (assuming it actually is one)? Yes, it could, quite splendidly and unequivocally. All you need to do is show some passage of time from one segment to the next, more than a minute or so, and you have proof of a hoax. Because there's no way Roger is going to tell a real Patty "Hang on and chill out for a few minutes while I line up this next segment shot." He could do that with a guy in a costume. So all we need to do is look for some evidence between segments of a passage of time.

To do so, first we need to study related film footage from Bluff Creek to see examples, and the two perfect study examples are Jim McClarin's first and second walk, and Rene Dahinden's posing with the measure bar at different points in the path.

Fig. #6-11 McClarin Walk Shadows

In this illustration figure, there are three panels. Top panel is Jim McClarin's walk #1, filmed by John Green. Center panel is walk #2 filmed by a second cameraman and camera, but we currently do not know the cameraman's name. Lower panel is the overlay comparison of the two images.

The three circles on the right, with white outside and black inside, show Jim McClarin in each panel. In the lower panel, we see both his positions, one with black pants and white short, and the other inverted tonality so the pants are white and the shirt black.

The left three circles, with black outside and white inside, show a tree shadow on the sandy creekbed ground. This tree shadow has shifted its position from the top panel to the center panel, and the lower panel compares the two positions with the top shadow position in black and the center shadow position in white. In the time it took for Jim to finish walk #1 and go back for a second walk, and the second camera set up in the same position, that tree branch shadow shifted that much along the ground. So the shadow is a factual evidence of a passage of time.

Also, the speed with which the shadow will change is in inverse proportion to the distance from shadow-causing object and the shadow itself on another surface. So a shadow falling on ground one foot away from the object casting the shadow will move slowly, but a shadow 25, 50, or 100 feet away from the shadow-causing object will move more quickly. Also, self shadows, like a tree trunk in sunlight with a sunny side and a shadow side, the shadow side will change slowly. So in these two scenes with McClarin walking, the fastest moving shadow is the one cast by a tree branch perhaps 30, 40, maybe more feet away, and most of the other shadows, either self shadows or very close shadows (close to the shadow-casting source) don't move as fast and thus are not revealed as distinctly in the black/white position shifts. In this case, the tree branch was located to the right of the scene, in a cluster of trees in the debris pile, and from that, we estimate its distance.

The Dahinden 1972 Site Survey

In 1972, Rene Dahinden went to Bluff Creek with other people and he did a site survey and made a 16mm film as part of that work. In this film, with some other person operating the camera. Rene Dahinden himself stands in the picture, holding an eight-foot white measure bar vertically. He does this in about eight or nine different positions. But each time, he takes a position, the camera starts, he just stands there for a few seconds, the camera stops, and then he moves to the next position to do it again. The top panel is one such segment. The middle panel is the next segment on the film, and Rene is now about two or three feet further to the image right. In the lower panel, where one image was copied and color inverted and set at 50% transparency and overlaid on the other, you can see the two positions for Rene easily. In one his measure bar is white, the other his measure bar is black. Most everything around him is grey, because nothing moved.

Fig. #6-12 Dahinden shadows

But in the bottom area of the lower panel, we see large masses of almost white and very dark, almost black patches. These are the shifting shadows from the trees and their branches and leaves from a cluster of trees off camera to the right, casting shadows on the ground. In the brief time that it took to stop the camera, let Rene move two or three feet forward, stand still, and start the camera again, these shadows have shifted. But most of the more distant landscape is just grey, so there is no significant movement of other shadows.

So we know that some shadows can shift in fairly short passages of time. I would not expect to see any noticeable shift in the time a real spontaneous filming would take (less than two minutes), but I would expect to see some shadow movement if a hoaxer paused for 10-15 minutes from one segment to the next to coordinate with his performer, plan the next step, choose his next camera position and action, etc. And I would expect a hoaxer filming would never anticipate 47 years later, somebody is studying tree shadows for evidence of elapsed time, so the hoaxer could confidently pause for 10-15 minutes to get the next segment planned before he shot it, confident no one would perceive that passage of time in the film. A hoaxer would never anticipate such. But if it were done, it would trip him up today and prove he hoaxed the film.

To test this idea further, I set up a HD camcorder outside my home, and put a 4x4 wood stud along a concrete wall, where the shadow of the house roof would fall on the wall in the afternoon. The 4x4 was to create a shadow on the wall less than 1 foot away, while the shadow from the house roof was 50 feet from the wall. This allowed me to study how fast a shadow can move. The results were quite impressive.

Fig. #6-13 Sample shadow shift timed.

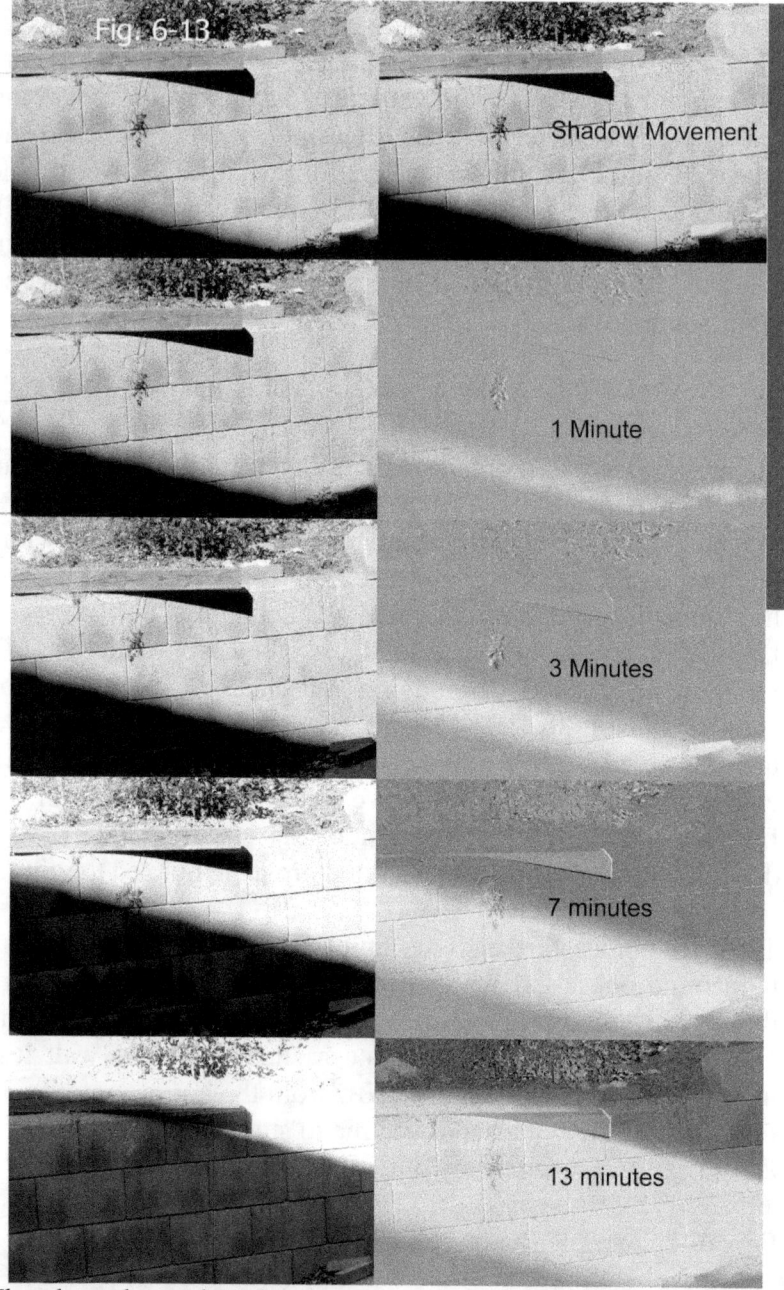

Fig. 6-13

Shadow Movement

1 Minute

3 Minutes

7 minutes

13 minutes

The chart shown has the shadows at the camera start, top image pair, then the shadow at one minute (left the frame grab, right the two compared by tonal inversion so the white diagonal line in the lower image is the shift of the shadow), third down is three minutes after the start, with a wider shadow, the fourth image is after seven minutes, and the final one is after thirteen minutes.

So while the failure to find a shadow shift doesn't prove the event is authentic, if we were to find a noticeable shadow shift from one segment to the next, that would represent an elapsed time that has no rational or justifiable explanation for a real event and would almost certainly be empirical proof of a hoax. So for that reason, I am frankly surprised that advocates of a hoax have never seen this line of analysis as important to pursue. And if I were trying to hide a hoax, I certainly would not bring this issue up for you, the reader to consider. But I truly want to get to the truth, and this line of analysis could potentially reveal a truth about the PGF as being real or false. That is why I pursue it.

A different method is required to compare various PGF segment frames because whereas in both the McClarin example and the Dahinden example, the camera was in the same place, and the lens focal length was the same, in the PGF segments, each camera position is different as Roger chases Patty, and so color inversion and overlay method no longer works. A side-by-side visual inspection is the alternative, looking for shadow discrepancies.

First I will compare frame 090, near the end of Segment One, with frame 110, early in Segment two. Both are relatively sharp and show shadows well. The camera is clearly closer in 110 and Roger's path is not true straight to the center of the picture, but instead straight to the White Post on the right, because its relation to the distant tree to its left remains near constant and this indicates a line of sight. Because the line of sight is constant, Roger's path forward is on that line of sight.

Fig. 6-14

Frame 090 above, end, Segment One

Frame 110 in Segment Two

Fig. #6-14 Segments 1 & 2 compared.

The best we can do comparing these two frames for shadows is look and see if there are any distinctly different positioned shadows, and none are apparent. So there's no evidence of a time passage more than a few seconds as Roger moves forward.

There isn't much to compare with Segment Three, the two frame segment, so I will next compare Segment Two with Segment Five.

Frame 184, almost the end of Segment Two, shows some common landscape with frame 290 in Segment Five. In the illustration, both are shown and the common landscape part is circled. Frankly, it's not much, and in color it's easier to analyze. But in the center of the circles is a tree trunk with some shadow patterns on it, and they are consistent between Segment two and Segment Five. So we don't have any indication there of a passage of time more than a minute or two.

Fig. #6-15 Segments 2 & 5 compared

Next I compare frame 207 from Segment Four with frame 275 from Segment Five. There's a shadow on the Big Tree, in the center of the white circle, and that shadow would be a great candidate for shifting in a short amount of time. But it doesn't. And no other shadows seem to have shifted either. So nothing in this comparison supports a lapsed time between Segment Four and Segment Five beyond the seconds it would take in a spontaneous filming.

Fig. #6-16 Segments 4 & 5 compared

Segment Two frame 184 and

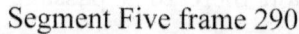

Segment Five frame 290

Fig. 6-16

Segment Four F 207 above and Seg. Five F 275 below

Segment Five frame 400 and

Segment Six frame 729

Finally, I come to the comparison of Segment Five, using frame 400 and Segment Six, using frame 729. Both show the "Shadow tree" TC-3 well, and that shadow on the tree would be the perfect candidate for shifting or even vanishing from the tree in as short a time as 10-15 minutes. So if there was any hoaxers pause between Segment Five and Segment Six even 10-15 minutes, the shadow shift would expose a hoax.

Fig. #6-17 Segments 5 & 6 comparing the shadow tree, TC-3

First, I am showing the two frames with a black rectangle around the shadow part of the Shadow tree. The next illustration (Fig. #6-18) shows the shadow tree (TC-3) from both PGF frames compared to a photo Rene Dahinden took on site at a later date. The three above images, each with a white dot at the base of the tree, show that TC-3 doesn't have any permanent black coloration on it, as in Dahinden's photo. The Lower left three images show the tree, in each case enlarged to see the trunk color more clearly. And finally, on the right side, you can see the source of the shadow that is found in the PGF footage, the tree which is casting the shadow, and a white dotted line for the sunlight path which made the shadow.

Fig. #6-18 Segments 5 & 6b

This kind of shadow would be the perfect fast-moving shadow that would shift noticeably in 10-15 minutes. But it doesn't. It's still on the tree, in both PGF Segments Five and Six.

So we cannot find any evidence of any elapsed time greater that the real time filming that occurs in total in a minute or two.

Fig. 6-18

In the shadow test noted earlier, in a mere 7 minutes, the roof shadow shifted enough (as scaled by the 4x4 piece of wood) to approximate the estimated thickness of the PGF site tree TC-3 (the Shadow tree). So based on that comparison, that shadow on tree TC-3 could conceivably shift enough to no longer fall on the tree trunk in a mere 7 minutes. So if there were any staged activity and a pause between segment five and segment six, a pause of even 7 minutes, we would see irrefutable proof of the pause on that tree, by the disappearance of the shadow.

The shadow caused by the 4x4 wood stud shows how slowly a shadow moves if it is very close to the source object casting the shadow, thus verifying my analysis above describing how some shadows move fast and some move slowly.

So shadow analysis would be a perfect technique to expose a hoax, and the fact that it doesn't is compelling evidence that the whole filming of the PGF took place between one and two minutes in total (one minute of footage and some pauses between of several seconds each).

So we must conclude that a real encounter would reasonably occur in that brief time frame. Would a hoaxer really try the whole film in one take, running, starting his camera, stopping his camera, starting it while still running, pause, do a deliberate two frame trigger slip, cross a creek, start the camera while he's running up a creek bank incline, charge forward, finally hold still for a perfect Segment Five steady shot, and the go back to starting his camera and running for a wild Segment Six, all in about two minutes or less, for about 1 minute of film? I would love to see an advocate for a hoax, with any real filmmaking knowledge, argue for that to have occurred, and explain if the hoaxer is really brilliant or really careless and foolishly daring.

Another curious note is that the K-100 camera has several trigger switch positions. One runs the camera only as long as you hold the trigger down, and stops as soon as your finger lifts off the switch, but the next switch position down runs the camera continuously even when you take your finger off the trigger, and keeps running until either you deliberately push the switch up to stop, or the spring wind runs out of tension. So if the suspected hoaxer wanted to do the whole thing like a single spontaneous event, one try, start to end, he could have just switched the trigger switch to the lowest position, so the camera just runs on and on, and he could take his finger off the trigger and concentrate on running and composing his shots and such. That would have resulted in a single filming segment, not six, and a person familiar with the camera (as Roger was) and intent on a bold one take try for the whole encounter staging, would likely have done that so the trigger was one less concern, for all the things on his mind.

A man trying to get real footage and make every shot count would not run the camera continuously that way for fear of wasting the little film left in the camera, and he'd surely lose his subject if he stopped to reload film. So a man experiencing a spontaneous encounter would try to control the trigger to only get good shots, as best he could, while a hoaxer boldly trying a one-take "go for it" staging would likely lock the trigger down and just let the camera run. Roger did what the real person in a spontaneous situation would do, and not what the hoaxer might do.

As noted previously, new technology has a remarkable potential to trip up a historical hoax by subjecting the evidence to analysis that the hoaxer of olden days could never have imagined the analysis could reveal. If the event was hoaxed, something in this analysis should trip up Roger and ring false, some error or "cheat" should have been revealed, some detail never considered by him should have been exposed to prove the event was staged. Some lack of continuity between what Roger is doing and what his filmed subject is doing should have surfaced. None does.

Sooner or later, we need to find evidence of a hoax in this film evidence, to keep that option viable. But the more we look at the good evidence, the more the hoax concept loses credibility. So far, it hasn't scored a single point of consideration.

So now, we must look to Patty herself, for any evidence, indication or suspicion of a hoax.

Fig. 7-01

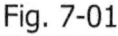

Chapter Seven - Making an Ape Suit

Basic Practical Considerations

The PGF discussion has spanned 47 years now, and sadly, that discussion has been confused by amateur appraisals of how a person in an "ape suit" can look convincing. Even many of the professional appraisals are very brief and contradictory. The goal of this chapter is to provide the reader with a concise outline of the practical design considerations that go into making any type of fur "creature costume" or "ape suit".

My philosophy is and always has been to try and teach what I know, as my presentation of my ideas and analysis, to educate people in the subject and (hopefully) allow others to understand why my presentation of analysis is correct and factual. So I will be taking this approach in this chapter, to lay out a fundamental course in creature fur suit making as a way of setting the foundation for analysis of the PGF creature subject.

A teacher should disclose his/her credentials up front as a foundation for any teaching exercise. So, aside from the knowledge I include herein, which should allow you to make some judgment of my knowledge, here's some photos of actual suits I have made in my career, plus some I performed in for the filming. A complete resume is also in the Appendix Section.

Fig. #7-02 is a sampling of various full body suits and creature costumes I have made over the years. The top photo is from a commercial for AST Computers, where the concept was a parody of the 2001 "Dawn of man" movie sequence. I did five full ape suits, with the hero ape having a 12 function servo-motor controlled animated face. Middle left and center is a

Birdwarrior from the movie "Beastmaster". Middle right is the Tar Man from "Return of the Living Dead".

Fig. 7-02

Chimp in gorilla suit and prosthetic nose by Bill Munns
Before, after, and suit understructure shown.

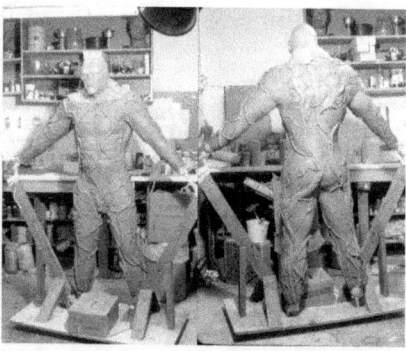

Despite the criticisms about my work on that movie, the Tar Man work stands as a fine effort. Bottom left are some examples of my work making gorilla suits for chimpanzees to wear, and bottom right is the full suit body sculpture for the title character in "Swamp Thing".

In this discussion, I will often use the word "mime" and/or "performer" to describe the person wearing a creature suit. In classical performing arts, a "mime" is a person who performs with no sound, no speech, using just body motion and gesture to create a performance. There are other terms for a creature suit performer. "Mime" is simply my preferred designation, in general.

Practical vs. Possible - Without a doubt, the most annoying and worthless ideas are offered by amateurs in the creature design and fabrication field who cannot tell the difference between what is possible and what is practical. To them, anything which is not categorically impossible becomes an option they can easily use whenever they are arguing an issue, despite their total failure to apply such knowledge in a practical way with consideration of other issues or factors. They are blind to the fact that impractical things which are technically "possible" come with a lot of baggage, stipulations as to when they may be reasonable and when they are not.

A professional who actually designs and builds suits and creature costumes will look at one material, one process and say "it's possible" but will look at all the related implications of choosing that material or process, how the choice impacts on other elements of the design. When those related considerations are factored in, the professional may conclude the material or process is not practical, even if possible, and will discount it as an option.

The amateur analyst in the PGF debate will simply look at the single material or process and if it can be said "it's possible", then they seize it as a solution to an argument they want to win, with no consideration for how this idea may ruin another "it's possible" idea they use for another specific argument. They fail to see that one "it's possible" contradicts the other. The professional designer, to the contrary, must look at all these "its possible" options and eliminate the contradictions because those contradictions can screw up the job. You only get the job done by making sure everything works, everything is practical in itself and in concert with the other things. The designer can not complete the job unless every material, every process, works together effectively. Similarly, one can only present a credible "proof of Hoax" by explaining how the hoax costume was designed, on the practical and specific level, and no descriptions contradict other descriptions. In other words, any such proposed hoax costume must be explained in a truly functional way.

Amateurs relying on these "it's possible" (but wildly impractical) things may win one argument but slam the door shut on another argument they rely upon, thus creating a contradiction of their own hoax theory. They, of course, ignore the contradiction, which is one reason Hoax believers have never been able to actually put together a single persuasive proof of hoax. They constantly rely on ideas that contradict other ideas they use, and so they can never get to a complete and unified explanation of the hoax (or even the costume they claim was used).

So you must be wary of the "it's possible" claim when a person can't demonstrate a full knowledge of all the conditions attached to that possibility.

Example: It is possible to make a costume split at the waist, and attach the head mask to the upper torso section, so when the upper section is lowered over the costume performer's head and shoulders, the mask eventually comes down to sit on the performer's head. But nobody does that, as a rule, because in the practical world, between filmed performances, the people handling the costume had to break the performer out of at least part of the costume to allow that poor person a chance to cool off. The inside of a costume gets very hot, and has been known to cause a performer to faint from heat exhaustion and dehydration. But if you've ever pulled a slip-over-the-head costume off of a sweaty performer, especially if there's any under padding, you know it's a very strenuous task, not easy or quick. If you can't get the performer opened up to cool off, the risk of heat exhaustion and fainting is increased. But pulling a head mask off is quick and easy, and from experience we know if we can get the head mask off, and blow a fan toward the face, the performer will be cooled sufficiently to rest up before the next session performing in full costume.

So while it's possible to attach a head mask securely to a costume upper torso piece, nobody does it in the practical world. It's possible, but not practical. Having a head that easily separates from the torso and easily is removed is practical, so we can easily cool off the performer. So that's how it was generally done.

Ignorant people with a bit of knowledge of creature suits and a determination to win the PGF debate and prove a hoax often cling to claims of "it's possible" when what they claim has no practical demonstration of use in the profession. I will be explaining what is practical, and why, in this section.

These are the fundamentals which form the basic suit design criteria:

1. How many pieces?

Realistically, making a full body suit in multiple pieces is simply a pragmatic reality for ease of fabrication, ease of use with the performer on set, and ease of repair if something is damaged. So every creature suit I know of, and every one I have personally built, is designed in multiple pieces. But there's no standard formula configuration. We design a suit to meet the needs of the job, and various jobs may have various requirements or criteria.

A basic design for an "ape suit" (having the appearance of a large primate) would be a facemask, a body suit, hand gloves and feet prosthetics. Six pieces total. Or we can split the body into two sections, a "shirt" and "pants" design (split at the waist), or a torso section and leggings design (full torso and snap crotch, and leg covers the hip area of the torso piece overlaps onto). That would be seven pieces.

You can attach the feet to the leggings if you want, and attach the hand gloves to the arm sleeves if you want, and in some instances, this might be desirable. I personally have never done either for an ape, but I did attach the hand to the arm sleeve for bear paws and lion paws on two different jobs.

For the grizzly bear suit, it was only from the waist up, and I designed it with a head mask (with animatronics), a torso that was a slip-over-the-head vest, and both arms attached to an undershirt and the bear paws fused to the arm sleeves. I did it that way because the chest/torso section needed to have heavy rigid shoulder braces with an aluminum frame inside to support the head, which not only stuck out in front of the mime's head, but had a lot of weight from the animatronics, so I needed a fairly rigid torso bracing system to secure that head to. The vest structure allowed the arms to be easily slipped on first, then raised up to slide into the vest armpits as the vest was lowered down onto the body.

Fig. #7-03 shows the three main components of the described bear costume.

So the job basically dictates the plan for how many pieces and sections, and why? There isn't any right way, no "industry standard", just the practical goal of achieving the right configuration to best get the job done.

Fig. 7-03

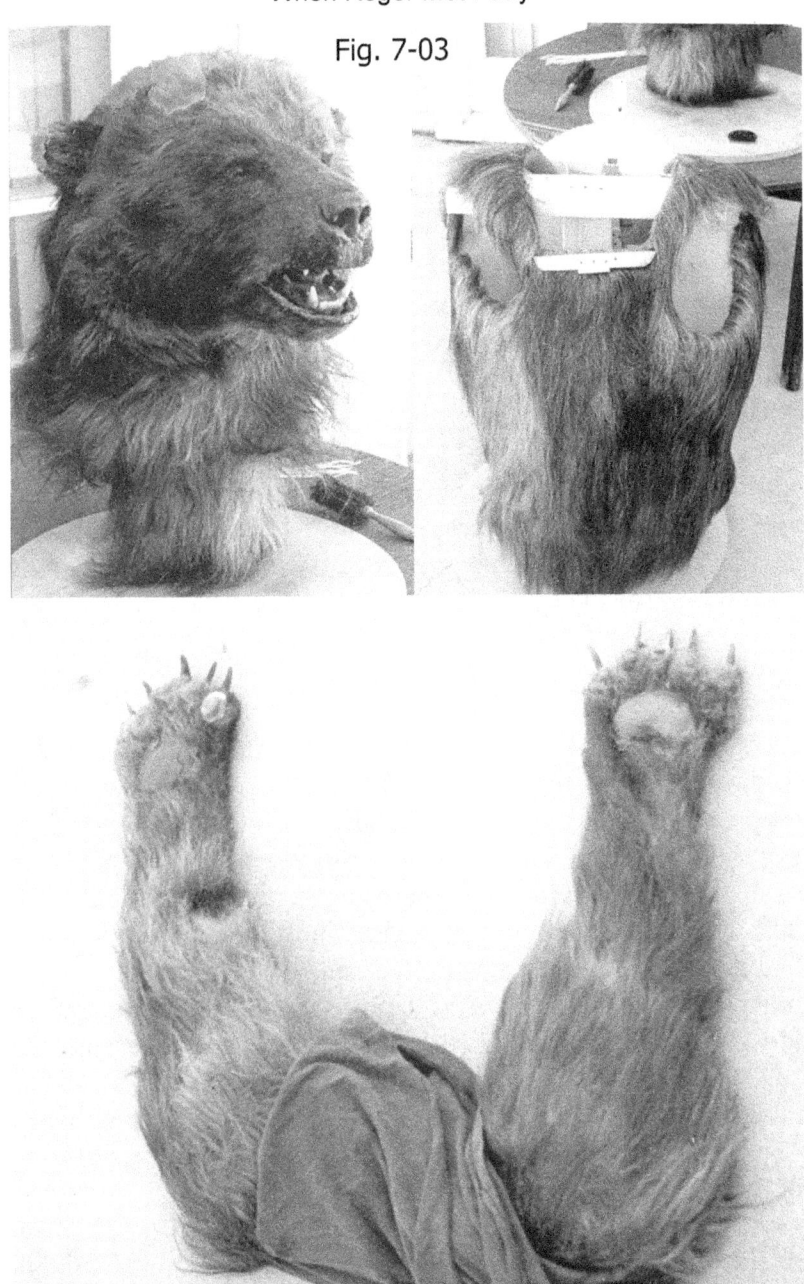

Skin and Fur

Now we consider that ape suits are divided as to material. Fur sections are usually tailored or laid up with some hair medium, while skin sections of the body are sculpted, molded and cast in some type of rubber (back then) or some skin-like plastic resins (now). The face on the head mask is the most obvious skin section, and the gloves over the hands is probably a close second as essential. A skin chest of some kind is common, and feet are common as well. Other options are knees, buttocks areas, or any special requests by the director.

So in the fabrication process, we divide skin and fur sections and fabricate them differently, but then we fuse the skin sections to fur sections, so a finished piece of the costume may have both skin and fur.

2. What kind of closures will be used?

Closures are the devices to close the suit and attach one piece to the next. They may be closed with zippers, Velcro, metal snaps, cord and eyelet ties (like shoelaces and corsets), hook & eye devices. But ideally, the closure should be mechanical in some way so the opening and closing is non-destructive to the suit. For the movie "Swamp Thing" however, my crew and I had to use glue to attach the foamed latex suit parts to each other, and then glue on model railroad landscape material as moss to hide the glue seams. This was necessary because the suit was worked extensively in swamp water, and so everything needed to be as securely attached as possible, and a carpet-seaming cement was the ideal glue to use. But for ape suits, the closures are generally mechanical of some form.

A closure seam is far more complicated than a simple furcloth tailoring seam (where two pieces of furcloth are joined to follow a contour and create the anatomical shape desired). The closure device usually adds a thickness to the body contour, and so an offsetting thickness must be added to the contour next to the closure to smooth it out. And the closure areas usually are particularly rigid, and so they should be positioned where the rigidity doesn't interfere with the mime's movement.

Fig. 7-04

Components of a sample closure seam.

If set up as diagrammed, the left fur will shingle above the right section, making the seam obvious.

If the left seam closure parts are lowered to the thickness of the underlying right parts, then the two fur sections will lay smoothly and blend better.

This adjusted depth of the closure is shown below.

But if this thick closure section sits on a flat under-structure, the seam section will bulge outward and be obvious.

So adjusting the under-structure or padding to have a recessed section the closure sits in will insure the fur lays flat and looks best.

Fig. #7-04 illustrates the considerations of making a closure seam. The general concept is shown here, but individual fabricators may have variations of this that they personally prefer.

The tailoring seam can join two pieces of fur pelt or furcloth by either sewing or gluing. Ideally, both sections of fur should be laying flat, one butted up against the other, so the hair lay is as smooth and consistent as possible. The shorter the fur, the more critical it is for the positioning to be precise. Sewing can work, but may sometimes bunch up and bend the fur base, and that can cause fur to stick out instead of laying smooth. The sewn seam works in taxidermy work because the hide is glued to a rigid body form, so once the two sections of hide are joined by sewing, the hide can be pressed flat and the glue holds it there. Also, being rigid, there is no subsequent movement to cause the sewing to cinch up anywhere. But for a fur costume worn by a mime, the mime's body movement can cause the sewing to cinch up, and might make the fur ruffle up to expose the seam. I'm sure some professionals do use this method, but in my personal work, I prefer a glued cloth gusset.

The cloth gusset is a patch across the underside of both pieces, and may be made from muslin cloth. Both sections lay flat and butt against each other, and once the glue sets, it is a very smooth and very strong permanent seam.

But a closure seam must continually open and close, so it needs more components to work. The most important component is the closure device, which could be Velcro, a snap strip, or a series of eyelets that can be laced up like a corset. But these devices have a thickness which must be accounted for, and there must be a way to attach the lower closure device to one piece while the upper closure device attaches to the other piece of fur. The upper closure in usually under the fur base, but the lower closure device is attached to a gusset which then must attach to a padding sheet to space the fur as high as the two closure device layers lift the fur on the other side.

And, ideally, there should be a furrow or similar indentation in the padding below the closure so the fur base lays flat across the closure. If it doesn't lay flat, then the closure will be very obvious, even when closed and brushed out.

So, as illustrated here, a closure seam is quite a piece of work, and all these components make it quite rigid. So a well-designed suit must put the closures where the rigidity of them doesn't restrict the mime performer's movement.

3. What, if any, inner structure will be needed?

The inner structure is usually padding to either bulk up the body to an apparent larger dimension, or define different and non-human anatomy. But inner structures can also be aluminum frames, animatronic control devices, cooling devices, air pumps to help the mime breathe in the mask, communication devices to make it easier to talk to the mime as he/she performs, etc.

For example, for some apes I did as a parody of the 2001 "Dawn of Man" sequence, for AST Computers, back in 1987, there were five suits, four simple ones and one "hero" suit with a 12 function radio-controlled face mask. All the suits had some back padding, an air pump to get fresh air into the mask, a two-way walkie-talkie headset for communication. The hero suit also had the full animatronic head, the RC receiver and battery set.

Every suit will vary in its inner structure needs, and you design according to the need.

4. How customized or generalized will the suit be?

This is a consideration of whether the suit will be custom fitted precisely to one person or be more generalized to fit a variety of people. But understand when we say it fits one person, precisely, that actually means it would also fit another person of the same body height, weight and proportions. But a simple face mask fitted to one person was likely sculpted on a head cast of that person, and a mask with animatronic movements may have an under skull structure vacuum-formed from a life cast of the performer's head and face, and that truly is customized to that specific person. A customized fit suit is the highest level of realism in most cases, because it is designed for that person's anatomy and potential for movement, and the anatomical inner structures are usually the best. Generic suits are more likely to have overly tight or overly loose sections and an imperfect fit and it simply won't move with the same positive adherence to the mime's anatomy.

But a custom fitted suit is more expensive and has more steps to the fabrication process, in general.

5. What kind of budget are we looking at?

Budget plays a significant part in the final quality of any suit. Some hair mediums are better, but more costly. Some skin materials are better but more costly also. And the design time for the highest quality sculpture and figure finishing, and the finest hairwork is more labor intensive, and that too compounds the cost. So budget affects the final quality.

6. What kind of hair or fur technique are we planning?

In 1967, we didn't have the spandex stretch furs we have now from National Fibre Technology. We just had plain old furcloth woven much like carpets are, with a base weave and a pile (the hair) and the base weave didn't have any real elasticity. We also had real animal pelts (like bear skins), we had wefted human hair (the stuff common wigs are made of), we had custom hand tied lace hair (called "ventilated" hairpieces) like the common beards and mustaches, and we had crepe wool, which was a common theatrical hair we hand laid and glued in place on the suit or the performer's body. Human hair and yak hair for hand-laid work was also occasionally done.

Mostly suits were either real animal pelts or furcloth. The other options were either very costly or very cheap looking.

Issues of hair color and color changing, hair length, hair density, options for styling, and cost dictated the decision of which hair material to go with on a given job.

So all these considerations do factor into the design of any job.

Now that an overview of the considerations has been given, I'd like to elaborate on the options for each consideration.

Design Concept

Any suit starts with the design specification. What do we want the finished product to look like? Do we want a very realistic gorilla or chimpanzee, or a sort of new species of ape, or a human/ape "missing link", a fictionalized monster ape, a pre-historic humanoid? The concept is our starting point. Also, we consider, is it bipedal, or a quadruped knuckle-walker? So our starting point is the final visual concept, what we want the finished "creature" to look like. Then we work inwards back to the person who will perform in it.

Custom or Generic

This basic consideration will influence our choice of the type of padding inner structure we will use, the number of suit pieces and where the closures are, and what type of fur or hair medium we will use. So once a concept is agreed upon, our first consideration is whether we must custom fit it to a specific performer, or make a more generic costume that many can wear, but it won't be a perfect fit. If we must custom fit it to a specific person, we need that person available to take measurements or make a body mold, so we have a fitting manikin to build things on.

It is possible to just build padding which simply shapes the outer anatomy and had a depth of thickness of 1" of reticulated foam to hold that shape, and stroller costumes for theme parks routinely use this method, allowing a shaped outside and a generic inside. But such costumes have their own style of movement, and I do not see that style in the PG film. When a costume has significant open air spaces or pockets between the foam under the costume surface and the actual person inside, the person sometimes moves within that space and the costume outside doesn't. It creates a disconnect between the real motion and the apparent motion of the outer suit. That disconnect between mime and outer costume can also set up unnatural oscillations of the costume, as it first lags behind the movement inside and then rushes to catch up, and then bounce back.

And when a movement of the mime inside forces the outer costume to move, it may sometimes buckle inward to the empty airspace, simply following the basic physics principle of taking the line of least resistance. I do not see either of those suit physics occurring in the film, so I would conclude the figure can only be replicated with a solid mass of fur, inner padding, and the performer within, but no significant air pockets or spaces a generic costume invariably has.

So I think it would be safe to say, if the PGF subject were a costume, she would be a custom fitted one, not a generic one.

Suit Sections and Closures

Now we move to the exterior design, and the decision of how many pieces. The decision on how many pieces is to a large extent dictated by the choice of where closure seams will be placed. And the two most important closure seams are the closure of the torso, once the mime gets inside it, and the closure of the head down on to the torso. By comparison, the closures of gloves or feet to the torso arms and legs is a relatively simple structure.

With the torso closure seam, three variations seem to be the most common in industry use, but, in theory, you can put a closure seam just about anywhere on the body. So I'll focus on the more common ones. Those more common ones would be a zipper up the back, a waist split between a shirt and pants, and a snap-crotch torso over leggings.

The first, the zipper up the back, makes it easy to let the mime step into the costume and then load the arms in one at a time, and then zip it up to close it. But the zipper can be a challenge to hide. I've done it successfully with a zig-zag set of fur flaps which alternate overlapping the zipper to the left, next one up overlaps to the right, next one up overlaps to the left, etc. each triangular patch shingles over the previous one, and when the fur is brushed out, the zipper is totally hidden.

Both the waist split and the snap-crotch torso have a flawless back because there's no closure there, so they would seem to be superior designs. But there are other considerations that affect a design decision, and a person who's designed and made suits will appreciate these lesser known concerns.

Both the waist split and snap-crotch designs require you to lower that torso piece down from over the mime's head, and the mime must raise both arms up and load the both arms in first, then the torso with the arms still up. Now when you are putting the suit on, it's relatively easy to hold the suit piece high and pull down on it over the mime. And the mime is likely dry and whatever underclothing he/she wears is likely also dry, and dry cloth of the suit tends to slide easily over dry underclothing (in general). So you are pulling down, ergonomically efficient for the people does the dressing, and you are pulling dry suit inner cloth against dry mime underclothing, which is not too strenuous.

But once a mime has worn the costume for any length of time and performed in it, the mime's body sweats profusely and heats up and the inner costume and undercloths are often soaked. Literally, you can wring water out of them, the sweat can be so intense. Now, to get a mime out of this design, you must lift the suit torso up over the mime's head, with the arms again straight up. But wet cloth clings more aggressively than dry cloth does, and a costume is not as easy to push off a mime as pull off a mime. But to pull the costume off, you must be standing so you are waist high to the mime's upreaching arms over his/her head. So you either stand on some type of platform, or make the mime sit down low so you are much taller standing, so you can pull from the arm sleeves at the wrist. If you grab the fur and pull up, you do risk tearing the fur away from any inner structure you've attached it to. So the lifting is awkward, and the soaked cloth inside from the mime's sweat both make pulling the torso off up over the head far more strenuous than lowering it on. You have to do it to appreciate how difficult it is. So that's the downside to a seamless torso section that must lower down the arms and over the head.

The alternative is for the mime to bend at the waist and hold his/her arms forward (up relative to the horizontal line of the torso at that point) and you stand in front of the mime and pull forward. But you may need somebody holding the mime in place, because the clinging of the wet cloth inside can hold so strongly that you might just pull the mime to fall over forward. Getting a mime out of such a suit is work, and if you choose this design, you'd better plan on the removal of the costume being a fight to tug and pull, and better reinforce the suit accordingly to withstand that struggle.

In terms of blending the hair to hide the closure either at the waist or along the hip arch line, hair which is longer will always hide such closures better than shorter fur or hair will. So if your hair length is arbitrary (meaning you, the designer can choose), your suit design will likely have a smoother look if you choose longer hair.

The second major closure seam decision is the neck seam that closes the head mask to the torso. As mentioned above, it isn't practical or wise to fuse the head mask to a torso piece permanently, because you must cool off the mime while he/she is wearing the suit, and the easiest way to cool off the performer is to break off the head mask and blow a fan on the face (plus a cold liquid the mime can drink). So we will assume you wisely do design the suit with a neck closure seam.

Before the era of spandex stretch fur materials, the hair options were essentially non-stretch, and that poses problems for the mobility of the head, and head-turning motions. A common old solution was a deep neck closure seam, meaning the closure went down a ways onto the torso, both front and back. This gave more fur material to the neck area from the very rigid closure area to the head, and allowed more head movement. The downside to this is that the neck fur would often wrinkle, buckle, or fold noticeably. Making the shoulders higher could hide some of those neck wrinkles and folds, so the higher shoulder padding was common as well. We see this on the costume Janos Prohaska wore for the "Bigfoot - Man or Beast" documentary where he was interviewed about the PGF. He first performs in his costume, and then pauses as a lady takes off the head mask and he does his interview with costume on but head mask off.

Fig. #7-05 shows Janos in full costume, then a lady helps him remove the head, and then he does the interview while still in costume. You can clearly see the deep line on the chest where the neck fur attaches.

Fig. 7-05

The common alternative was a straight neck split, so the mask bottom is hopefully lost under the chin and in between the shoulders. But the back of the neck was usually problematic, so these suits were done when the designer would recommend the director stage the scene to avoid showing the back of the neck as much as possible. If the neck is attached, it limits the mobility of the head, and if it is not attached, then tilting the head down can cause a noticeable gap in the fur.

The PG film figure does a rather dramatic head turn (the "look back" sequence) and there is a good section of footage of the back of the neck after that head turn. It represents an excellent point of inspection. If the film represents a human in a suit, that head turn, and the condition of the fur after the turn, presents one of the most challenging aspects of a suit for the artists who made it and are working it during filming to hide the neck seam.

The reason I focus on the neck seam is a threefold concern:

1. It is one seam that simply must exist on the figure (if the figure is a suited human), and it is located in an area that virtually guarantees it is on the film. And the sometimes argued poor quality of the image, which is occasionally stated as not sharp enough for us to resolve details on the figure anatomy, is in fact sufficiently sharp to show flaws in a neck seam if such flaws did exist in the filming of a human in a suit.

2. The head turn ("lookback") and the subsequent footage from the back showing the neck after the turn is quite literally a "worst case scenario" for trying to keep a neck seam from showing (for a hoax), while being a "best case scenario" for settling a "suit vs. real" debate.

3. The neck seam is an overlap seam, because of the direction of the lay of the hair, and its design is not only non-stretch, but even very poor at the simple bending or folding of furcloth alone. Thus the seam area, for all practical purposes, is a nearly rigid section.

Making the Suit Fit the Mime

I would need the mime chosen from the start, and I'd need that person available for several fitting sessions during the construction process. The initial session would be to either mold or pattern the mime's body so I can construct a tailoring manikin to build the foam padding and final suit. I wouldn't guarantee anything without such cooperation.

Without such a fitting manikin or form, I may risk one of the following:

If I make the padding too tight, for lack of proper fitting sessions, the tightness may restrict the mime's blood circulation somewhere inside the suit, inviting problems wearing it.

If I make the padding and suit too loose, the outer motion will have some of the unnatural physical dynamics I described above.

If I make the suit too short in the arms, legs, or torso (crotch to shoulders), the mime may not be able to close everything up.

If I make anything too long, it makes for potential baggy folds.

So I anticipate any formula for success will require me to have the mime come in to my fabrication facility at least three, more likely five or six times during the fabrication process, to get everything right.

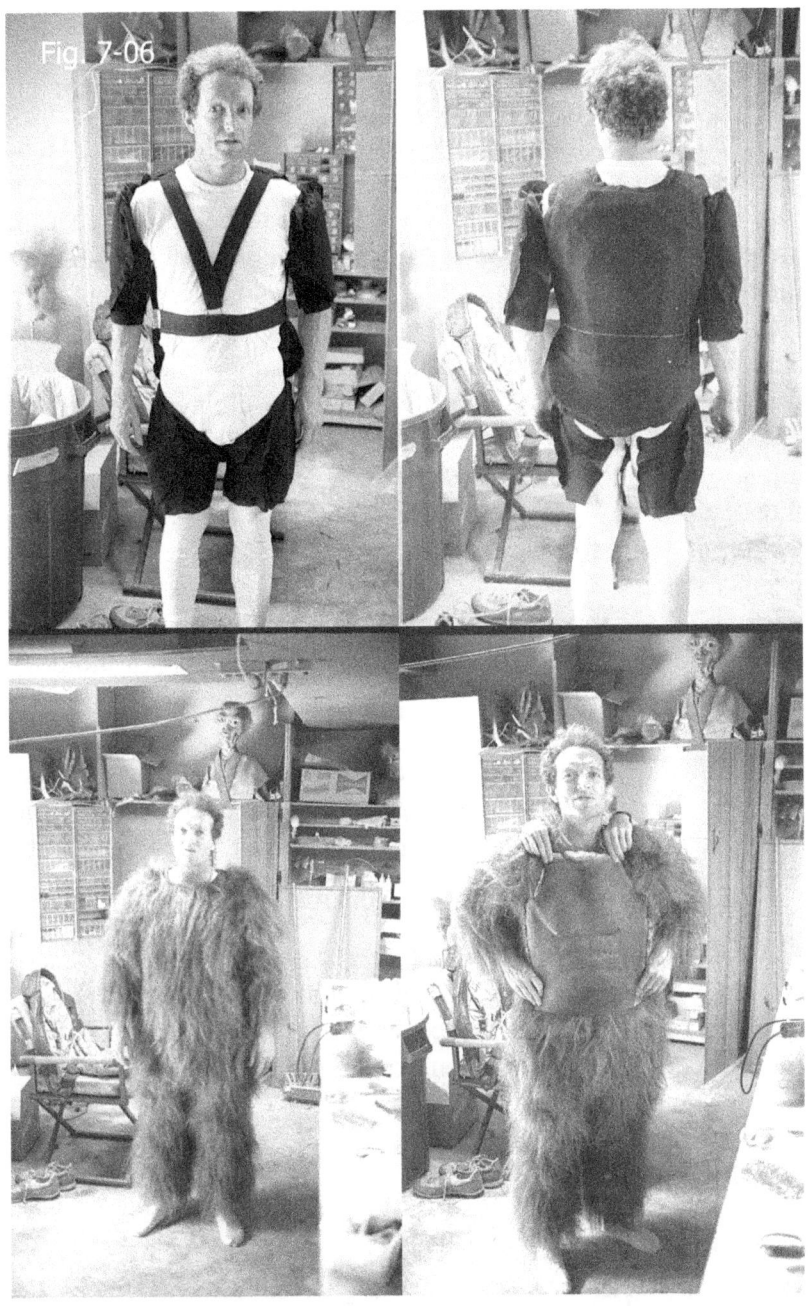

The Padding

Using the body cast or other tailoring manikin, I can place a base suit (long john's work quite well) on the manikin and build the muscle padding. While some artists like to sculpt the full body in clay and mold the sculpture to be able to cast the muscle padding, I personally find the irregular density of cast flexible polyfoam into such a mold an undesirable potential outcome. Shaping blocks of flexible sheet foam, with a truly consistent density, is my preferred technique. These blocks of foam can be glued to each other, but the gluing does impede the flex and compression of the foam, so glue is kept to a minimum, and some of the foam-shaped muscle areas may be sheathed in a light cloth pouch. T-Shirt cotton jersey is a common fabric for that process.

Fig. #7-06 shows padding made from reticulated foam sheathed in black cotton jersey material, and straps to hold the padding in place.

Once the muscle padding is assembled, with its closure devices (zippers, Velcro, snaps, or such), the mime comes in for a fitting session, to try it on, see how it fits, and test its mobility, how much freedom of movement it allows the mime. Tight areas (if any) with potential to restrict circulation are examined. Later, any adjustments are made based on the evaluations made during this fitting session. If substantial adjustments are made, another fitting session is needed to verify the modifications were successful and sufficient to allow the mime the necessary comfort and mobility.

The Fur

The tailoring of the final surface hair material then can begin. A critical decision is whether to glue or sew the fur to the padding suit, or glue or sew the fur into its own suit structure that can be dressed onto the mime separate from the padding suit. Both methods have been used successfully.

The nature of the fur, being short and dense, and the need for critically smooth and tight tailoring seams to allow the fur the smoothest blend from one section to another, or across a wedge or dart, all suggest my best blends of one fur section or piece to another will be done with cloth gussets and glue. That allows the fur cloth base of each section to butt up against the second piece with the hair lay maintaining the best flow and continuity. Sewing invariably entails some type of bending the base for a tight sewn join. A loosely sewn join allowing the two furcloth bases to butt against each other with no bend or overlap is a loose sewn seam, and one piece of the furcloth can easily fold, the sewing acting like a hinge.

So a glued seam with a cloth gusset panel under the furcloth bases will allow the smoothest, tightest seaming of furcloth sections and close tailoring seams best. But the sacrifice is some flexibility and mobility to the already stiff-backed fur. And the added stiffness of the glued sections also creates a discontinuity in the overall furcloth form flexibility, because certain lines or areas have more stiffness than others.

Closures are added to the fur suit structure to close up the segments needed to be opened for the mime's dressing and undressing.

So while this method will yield the finest fur blend, the closest approximating what the film shows, especially along the back area, it will further reduce any perception of muscle movement beneath the fur. The final perception, of how the figure's body mass will appear to move beneath the fur, this can only be determined by use, an actual test session with the mime fully suited up and observed walking, ideally photographed or filmed walking and turning the body.

Head, Hands, Feet

The body parts that represent bare skin are sculpted in

201

plastilina, molds are made, and the pieces are then cast in latex, usually.

The head mask can be either a slush latex face with no precise interior shape, just a wall thickness inward of the sculpted head, or foamed latex appliances molded with the precise inside shape of the mime's face. The first requires only a minute or two to apply in usage, while the foamed latex appliance generally is a 1-2 hour application process requiring a makeup artist on set.

But this facial segment must shingle over or attach to the back half of the head, so a mask is preferred to easily apply at the last minute when the mime is otherwise fully suited up. And the film has no apparent evidence of facial animation or motion, so the mask would suffice.

The feet should have a shoe or other secure footing for the wearer inside the cast feet, so a sports shoe, ideally a slip-on with elastic panels on the sides of the ankle, is preferred over a lace-up shoe. A shoe already tested to be a comfortable size for the mime is placed in the foot mold, and flexible polyfoam is mixed and poured into the foot cast to fill the space between the slip latex foot shape and the inner shoe, so the two become one unified structure, and the mime wearing this is assured the surest possible footing as he walks over the woodland terrain.

The Chest Section

Making a chest piece with the apparent breast shape isn't any challenge, but default industry process would normally be one of three materials: Slip latex alone, relatively thick, or a thin slip latex skin on both front and back mold pieces and a flexible polyurethane foam fill, or natural foamed latex. The slip latex prosthetic can be made from the mold of the sculpture alone, and slushed in or brushed up in layers to the desired thickness. The polyfoam and foamed latex materials require a two piece mold with the front being a mold of the sculpted breasts and chest wall, and the back being the inner contour of the suit or the contour of the mime's chest. The polyfoam is a room-temperature cure, while the foamed latex requires heat curing in a curing oven.

Fig. #7-07 shows a chest piece, with female breasts, being sculpted and set up for molding.

But none of these will impart any bounce or other fluidity to the breast mass, like we see in the PGF image analysis. These prosthetics will essentially hold their form and defy gravity. With a research grant, I was able to build test fur suits with breast prosthetic chest pieces made from the three common makeup artist materials of the era. We tested slip latex alone, a slip latex skin and a flexible polyurethane foam filling the breast mass, and a natural foamed latex prosthetic. These three were tested for any type of fluid dynamic motion, using a test station which subjected the prosthetics to a precise and repeatable force of motion and abrupt stop. The research also employed two female figure models who stood on the same test station and their natural breasts where filmed while being subjected to the same motion and abrupt stop.

The natural anatomical breasts demonstrated a fluid dynamic as we would expect, because the breast mass is a tissue structure high in water content, and doesn't have the potential firmness of muscle mass. The prosthetic breast pieces demonstrated absolutely no measurable fluid dynamic by comparison. The importance of this testing will be explained further in Appendix Six.

To accomplish a sense of fluid motion, with a costume breast prosthesis, skeptics look to one anecdotal story of "gorilla man" Charlie Gamora who reportedly put a water bag or pouch into one of his gorilla costumes to give the belly weight and some fluidity. But we don't have any documentation of the technical specifics of the design, we don't have any evidence of what motion, if any, it may have actually put into the suit, and we can't say with any certainty that the design could be adapted for breast prosthesis use. So this becomes one of those "It's Possible" (but not practical) arguments which skeptical people use to explain the PGF figure, but they can't actually prove in a practical way that what they argue could work. They can't even prove that Charlie Gamora's design worked, in terms of actually producing a fluid motion we can verify and study.

Now it is possible to design a breast prosthesis with some type of fluid inner pouch, but the design requires some sophisticated molding and casting techniques. Those techniques are more common today, but were rare in the 1960s, which I can verify because that's when I started doing this work. In the early 70s, I did a lot of core-molded puppet heads in foamed latex for puppeteer Tony Urbano, and so I had more experience with core molding that most makeup artists.

If we also consider that in the 70s the best makeup artist in America, Dick Smith, had to do breast prosthesis for Katherine Ross in the movie "Stepford Wives" and Dick was famous as one of the profession's greatest innovators, so if anyone in the business could do a fluid breast prosthesis, it certainly would have been Dick Smith. But still, he didn't do such, and simply made solid foamed latex prosthesis for Ms. Ross' makeup effect. The result was beautifully realistic in appearance, but these prosthesis pieces would not have any fluidity. So while one could argue that fluid-filled breast prosthetics could be done then, there is no documented history of any successful execution of the concept in the PGF era. And that does give us cause to wonder, if the best in Hollywood couldn't do it, why should we expect an untrained amateur to succeed in doing something which the best pros of Hollywood had not yet accomplished.

Fig. #7-08 shows Dick Smith's work on actress Katherine Ross for "The Stepford Wives".

Fig. 7-08

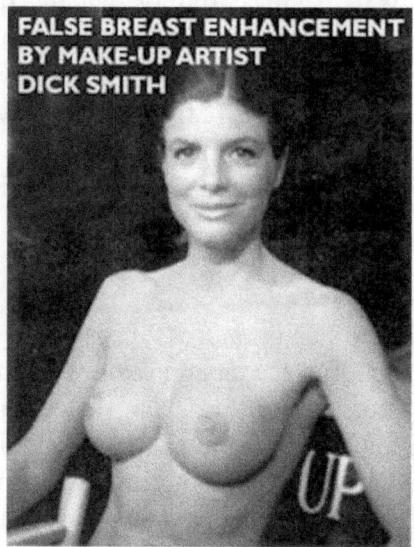

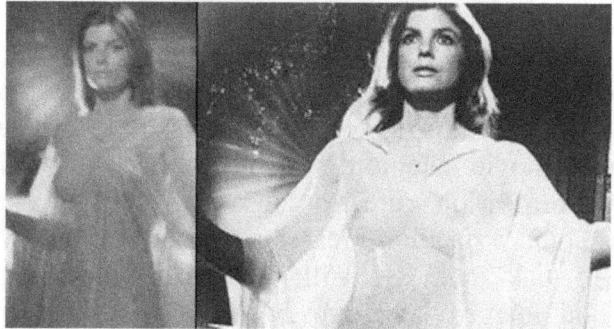

Final Assembly

The final assembly of the suit requires another fitting, to test everything, fit, mobility, smoothness of closures and fur blend over closures, and overall mime comfort. This final fitting also allows for a test run of the full dressing procedure and time to do so, plus confirming the number of assistants needed to go

206

with the mime during the filming day.

To assume the mime can dress himself fully, and groom the fur suit, without assistance is an invitation to failure, and most emphatically not recommended.

In closing this chapter, I wanted to address a common skeptical talking point, the idea Roger could make or revise an ape suit because "he was an artist". And to address that, I feel it's reasonable to look realistically at how people learn the "creature business", the craft of special makeup effects and prosthetics.

There are lively discussions of who made Patty, if one chooses to believe she is in fact a suit worn by a human performer. It seems the arguments of old centered around the prospects of some Hollywood makeup/creature people doing the fabrication of the suit, on the assumption it took at least a person or persons of demonstrated talent and experience to accomplish such a hoax (if indeed it was one). The "Patterson as lucky beginner/suit maker" idea has largely replaced the "Hollywood master secretly did it" theory.

So now, I'd like to review that "what if Roger did it" idea, but I'd like to review it in a larger, more generalized context. We need to consider two questions actually. One is how a person learns this craft, and the other is how one is qualified to appraise the skill of another person. Can a person's ability to make a suit be appraised by investigators today, and if so, how?

I've been reading over the last few months an increasingly optimistic appraisal of Patterson's "creature-building skills". It started with simple "maybe he could" suggestions and seems to have now blossomed into a full blown and absolute endorsement of his skills, with comments to the effect of "of course Roger could fix the suit. He was an artist, so he had the skills."

The usual argument states that because Roger has some demonstrated artistic skills (in drawing and sculpture, saddle-making), he must be artistic enough to make or remodel a creature suit. And this apparent conclusion, that he can make a suit simply because he's "artistic", I now must say is utter and laughable fantasy, a delusional monument to wishful thinking.

So the actual issue is, can people today, investigating the PG Film, determine if any non-professional person (whether it's Roger Patterson or any other person) was capable of making a full creature fur suit, or capable of modifying suit parts acquired from Hollywood professionals. How does an investigator determine who can or cannot make or modify a suit? How does anyone today appraise the potential "artistic" talents of another person?

So I felt it would be appropriate to describe some specifics examples of what qualifies a person to appraise the creature building skills of another person? First is the most obvious way.

1. Seeing actual creature suits that a person has built is a great start. I'm not aware of any other suits Roger ever made? (and of course we still do not know for certain about Patty, so she's excluded from Patterson's suit-building resume). If there's no evidence he made suits before the PG film was made, then it seems to me nobody can appraise Roger's suit making skills on this basis because there's no documented evidence he ever made one.

However, there are other possible ways the potential talent of another person to make a suit could conceivably be evaluated, under the following circumstances:

2. A professional artist building suits and having those suits used in films, frequently must hire assistants or crew personal, and that Pro will look at the applicant's portfolio of prior work

208

and artistic talent, and then the Pro make a judgment as to the applicant's potential skill to do various jobs. Then, if you (the Pro) hire the applicant, you see firsthand how good they actually are, as they work for you. This experience certainly qualifies a person to appraise the artistic skill potential of another person. I've done this many times over the years.

Finally, another process requiring the estimation of a person's talent potential may be through teaching people the craft.

3. Teaching people a skill or craft, such as training makeup artists, does require you to appraise the applicant's talent or artistic potential, based on any preliminary material they submit. Once accepted into the training program, a student 's skill, progress and potential for rising up to master the more challenging aspects of the makeup craft are constantly appraised by the teacher, and a teacher can reasonably be considered qualified to appraise the potential of an "artistic person" to learn the more complex skills of the craft, such as creature suits and prosthetics.

I was director of a makeup school for 6 years, and taught over 1000 students during that time, appraising them constantly as to who could do what, and I monitored their skill development. And once I returned to doing makeup work for movies, I hired many of my former students, and they assisted me in making suits and prosthetics. So I think I can reasonably say this process is a fine experience to qualify a person to appraise the suit-building potential of another person.

As far as I know, people saying "of course Roger could fix the suit", these people have none of the above qualifications sufficient to make any judgment on Roger's potential to make a suit, or even modify existing parts of a suit. They appear to be pulling this appraisal of Roger's ability out of thin air, an argument of pure wishful thinking, not reasoned analysis.

Another issue a true artist would understand, and a non-artist often misunderstands, is the matter of what specific artistic things an artist can actually do. Real artists have niches of activity where their talents seem to bloom, and other artistic things they simply cannot do. Some people are masterful at sculpting real fine living likenesses of actual people, to replicate a particular person, and other artists cannot. Some artists are great with wildlife subjects, and don't do people or costumed figures. Some artists work successfully in one medium, such as watercolors or pastels, while failing miserably working with oils or airbushes. Some artists are versatile while others are very narrowly focused on one medium or process only. A true artist would easily appreciate that the assumption any person can easily be assumed to do any creative task simply because that person is "artistic", this assumption is pure fallacy. Even true artists are in fact limited as to what ways they can be "artistic", and we never assume someone would be successful in an artistic medium until we actually see some solid evidence of the person's proficiency in that exact artistic endeavor.

There is simply no reasonable expectation that a person who has sculpted figurines or done some sketches is also automatically a natural to build complex creature suits.

In the interest of covering all possible options, there is the "beginner's luck" hypothesis. What if Roger was simply really talented and gave it a try and it just turned out good? Well, we can look to the makeup industry and see a fine example of "beginner's luck", with an artist who is undeniably talented. Rick Baker is indeed one of the industry's certifiable geniuses, and his extraordinary gift for creature work was evident to anyone who met him, right from the start. And his first fur creature suit, his "beginner's luck" suit is immortalized in the film "Schlock", where Rick made John Landis into the apelike "Schlocktopus". It's as good as anyone could expect from a first time try by a largely self-taught person, and a person who's sheer raw talent is truly extraordinary. And it's not as good as Patty.

A final consideration is that people with no true skill in doing this work fail miscrably to grasp the matter of how this skill develops, or even how artistic skills in general develop. The naive ones think you just get the instructions, or somebody just tells you how, and you just do it. No big deal, they think. How hard it is to operate a needle and thread, or a squeeze bottle of glue, or a single edged razor blade, they may ask?

Below I've put two photos from my own portfolio, two gorilla head sculptures. The one on the top I made around 1976, when I had been in the business for over 6 years, and having already recreated several classic makeup effects and suits to rival the originals in quality (including a complete Creature From the Black Lagoon suit). That gorilla head was my best effort, after 6 plus years full time doing makeup, and pushing myself to master the highest skills of the business.

Fig. #7-09 The two gorilla heads, and the two Sabre-Toothed Cat compared.

The photo at the middle is a gorilla head I did in 1983, seven years later, the best I could do at that time. See any difference? In terms of technique and material skills and fabrication processes, everything I knew doing the second head (at bottom) I also knew years before when I did the first one (at top). What changed, allowing my second gorilla head to be so much better than the first, was the maturing of skill, the slow but sure manner in which a physical skill is developed. It doesn't happen overnight. It doesn't happen reading a book or just talking to others who have done it. A second comparison is a figure of a Sabre-Toothed Tiger (Smilodon) at the bottom section. I made one in 1970, and thought it was wonderful then. 25 years later, in 1995, I had a job making one for a museum in Kobe, Japan. The Illustration shows my first one, 1970, and my second one, 1995, and I suspect you will easily see the difference that comes with skill and talent developed over time, and can appreciate that excellence in any artistic endeavor

211

develops slowly with intense practice and experience.

Fig. 7-09

Gorilla Head, 1976

Gorilla Head, 1983

Bill Munns

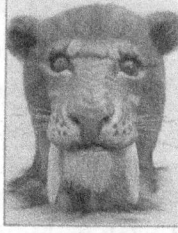

Is there any evidence that Roger devoted years of hands-on experience to developing sophisticated suit/creature building skills? If not, then any expectation that his skills simply bloomed overnight is a truly simplistic flight of imagination. Real fabrication skill develops from years of actually doing the task, real-world hands-on experience.

So now, each time I read somebody's remark about "of course Roger could, because he's an artist", as if it were well proven, I just laugh. His supposed "artistic talent" isn't even remotely enough to argue that we should believe he could make or modify a fur creature suit. You can say "what if" as you like with Roger or anybody else, as a simple hypothetical option. But you cannot rely on that "what if he could build the suit" to prove anything, or even substantiate other arguments. There simply is no basis to rely upon.

If anybody wants to "prove" Patty is a human performer in a suit, I would seriously recommend they focus on some explanation that includes a person or team of true professional suitmakers who built the suit and at least some of their team went to the location to assist while filming the footage. Then, their arguments would at least start with some credible foundation, which the "of course Roger could do it" theory completely lacks.

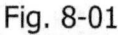

Fig. 8-01

Chapter Eight - If the Subject is a Creature Costume

Now that the foundation material has been presented about the design and construction of fur costumes, I feel it is appropriate to make the comparisons with the PGF subject figure herself, and discuss the elements of her body which challenge the costume making process.

For each anatomical feature, I'll explain what design considerations we evaluate to create the feature, and then explain why the PGF figure is problematic for a costume.

The Head Shape and Size

The fundamental rule for making a human head look inhuman is that we can only add to the real human head. We cannot subtract. And this fundamental rule has baffled and perplexed makeup artists for as long as they have been trying to make a perfect ape, gorilla, or chimpanzee head for a human to wear. Ape heads, especially chimps and gorillas, have a head that goes back once the brow ridge is defined above the eye sockets. The human skull, by comparison, goes up into the cranial vault instead of back. And that anatomical trait has doomed just about every attempt to make a perfect gorilla or chimpanzee mask for a human to wear.

Fig. 8-02 As an example, the illustration shows me and a full-scale Lowland Silverback (mature male) gorilla head I made in 1983. The gorilla head is bigger than mine, yet I cannot wear that head as a mask and perform in it, because the gorilla skull goes back above the brow ridge while my head goes up. If I put my eyes into the gorilla eye positions, my skull rises up above the true gorilla head. So as soon as I add enough head shape above the brow ridge to accommodate a human skull, the gorilla head is no longer a truthful shape.

Fig. 8-02

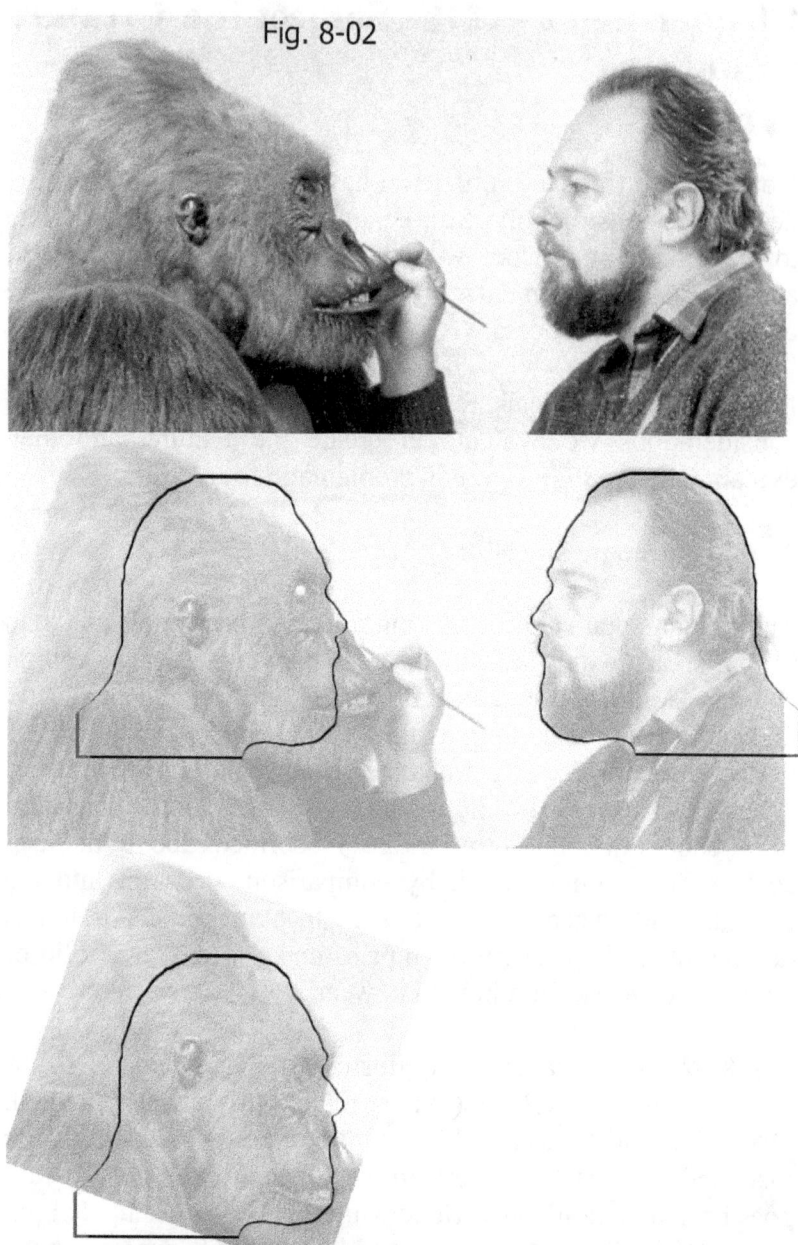

So that's how it's done with gorilla masks. Cheat the forehead higher than it actually should be, to accommodate the human head inside, but make it look small by enlarging other aspects of the head. This is a classical optical illusion to hide something enlarged out of necessity but is anatomically incorrect. By making other areas bigger, the enlarged cranium seems smaller by comparison. If you make the gorilla head longer front to back, it appears shorter top to bottom. But you get a bigger head. Now most people have never paid attention to this and so they looked at a good gorilla head mask and thought it was realistic and lifelike. It's not.

Now I'd like to compare several gorilla masks to the PGF figure, Patty. I've used three sample gorilla suits where there was a good comparative side view, and used one of the PGF still frames that shows Patty's head and torso in a good side view. First up is the 1986 movie "King Kong Lives" and the Kong character's costume and mask were made by fairly capable major studio professional artists and technicians. It's not a cheap suit made by an amateur. But when we scale the image of Patty beside the image of the Kong suit, both in side view the approximate equal distance from butt to the eyes, and then compare heads, Patty's head is quite noticeably smaller than the Kong mask. Yet it has the backward skull after the brow ridge which is not conducive to a human wearing a head mask. The Kong mask shows exactly how the above-described illusion is accomplished, raising up the forehead area to accommodate the human skull, but then elongating the front and back to make it appear as though the forehead is low. The end result of such work is a head much larger than a true gorilla head, and a head far larger than the PGF head, when the bodies are scaled similarly.

Fig. #8-03 compares Patty's head to the mask in this movie.

Fig. 8-03

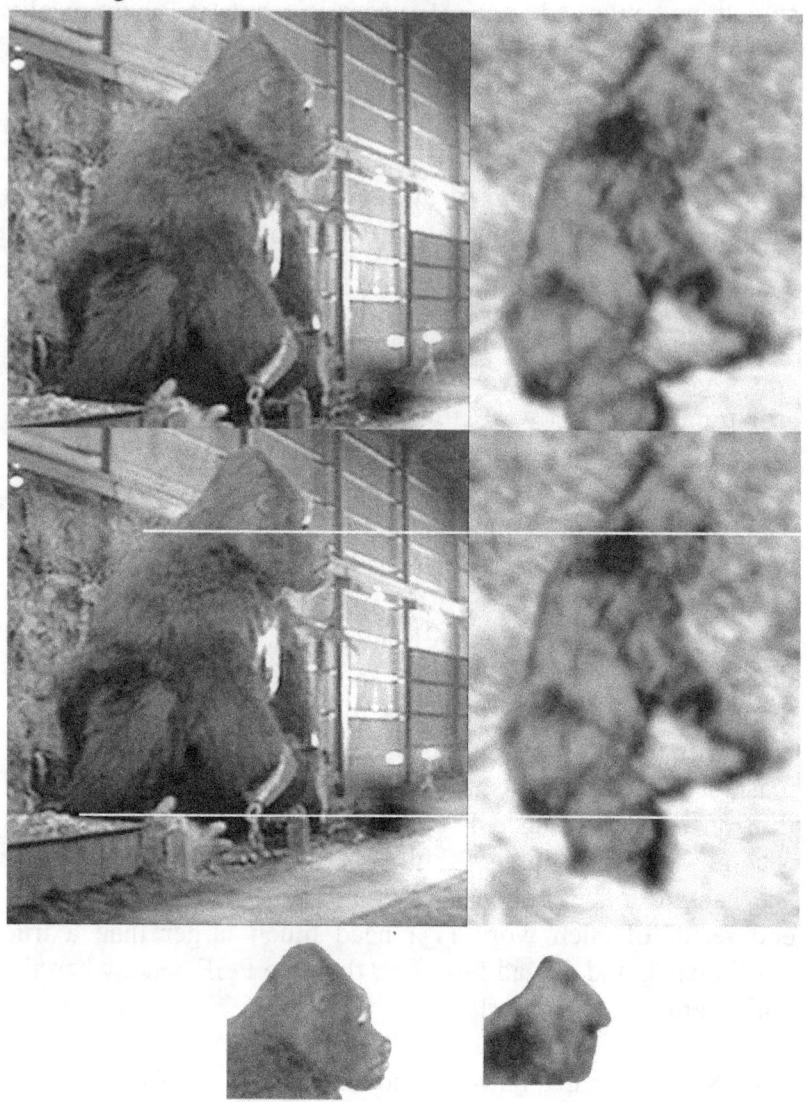

"King Kong Lives" 1986   PGF 1967
Comparison of head size.

Fig. 8-04

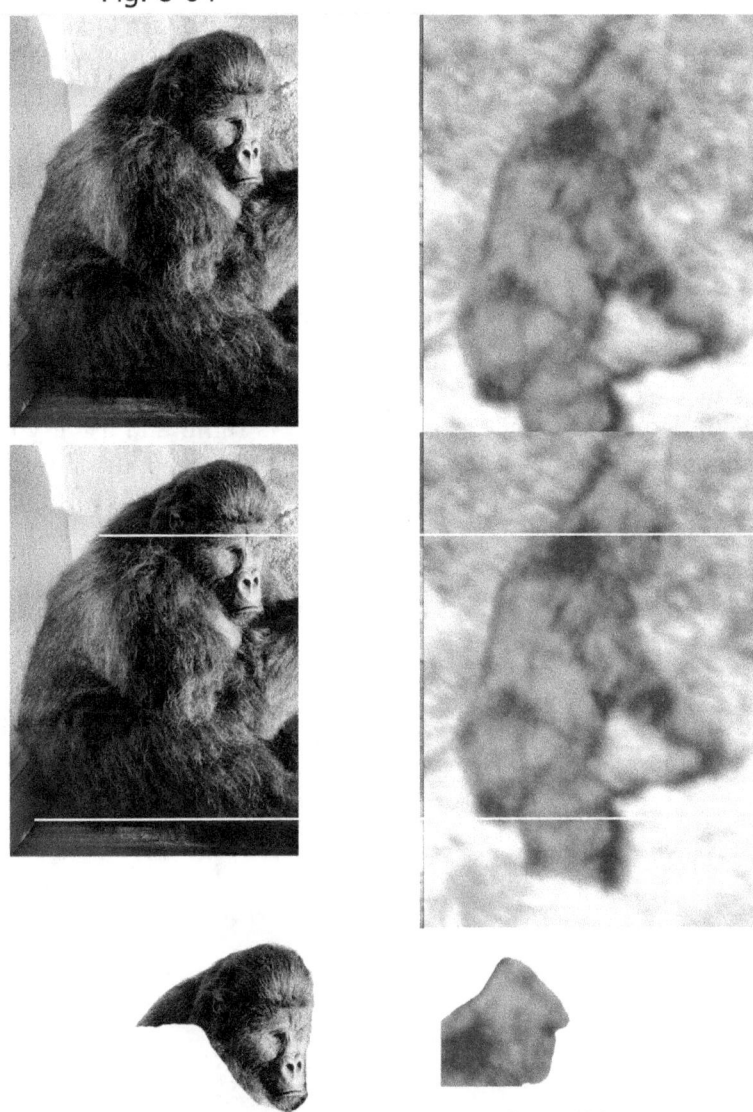

Charlie Gamora   PGF 1967
"Monster and the Girl"
Comparison of head size.

A second comparison is with the gorilla costume made by Charlie Gamora for a movie titled "The Monster and the Girl" in 1941? This is generally considered Gamora's finest suit, and he is generally considered the finest makeup designer of such suits in the 40s. And once again, when we scale the image to approximate the scale of Patty, butt to eyes, and then compare the two heads, this gorilla mask is quite a bit larger than Patty's head.

Fig. #8-04 compares Patty to Gamora's finest gorilla suit.

Third is the 1976 King Kong, and once again, the PGF head is smaller than the mask. And these masks were made by people who were making every effort to accomplish a true and realistic gorilla head shape. In the bottom row of images, the King Kong head is left, Patty's head is right, and a Patty head shape darkened is overlaid on the Kong head to show how Patty's head is smaller, especially in the top of the head above the brow ridge.

Fig. #8-05 compares Patty to the suit in the 1976 "King Kong"

So this represents the first major challenge to the idea that the PGF was simply staged with a person in a costume and mask or makeup.

Fig. 8-05

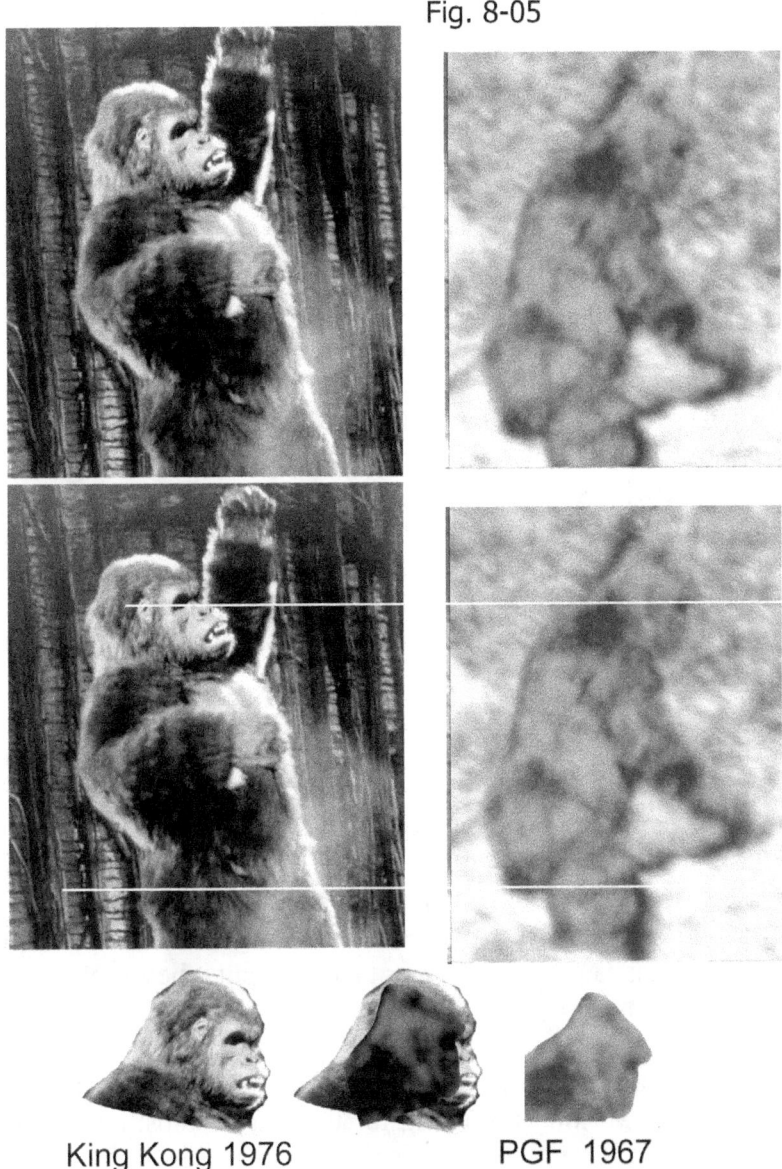

King Kong 1976   PGF 1967

Comparison of head size.

I have taken the PGF body and scaled it up to a size conducive
to a person of about 6'2" in height to wear, and scaled the head
accordingly. This was part of a research experiment to test the

description of a costume as specified by Bob Heironimous, who claims the PGF figure is him in a suit (as detailed in Greg Long's book, The Making of Bigfoot".) Once the head was made, so it matches the various shapes seen through the lookback sequence and shows the head from multiple angles front, side and rear, I tried on a mask of that dimension. I was barely able to slip it over my own head, and could only see out of it when I tilted it far forward and down, so it was on a severe down angle, and the mask nose and mouth were pressed firmly against my own, leaving nearly no space for breathing. This is not a good design for a mask, and even though one can drill or punch or cut holes in the mask for breathing, if the cut holes are not located in the mask nostrils or slit of the mouth, then the mask will photograph artificially, and it would show in the PGF. If the mask slips its position (as it constantly did), then even if you put breathing holes to align with your nose or mouth, the mask slipping misaligns those openings and makes it hard to breathe.

The issue of breathing in the mask is one of those "it's possible, but it's not practical" things I have already discussed. We don't design a mask with bad breathing passages and air flow, because there's enough physical hardship on a mime in a mask and suit with a well-designed breathing system, and we wisely don't try to make matters worse. If the mime passes out from lack of oxygen, we have to rush to get the mask off and try to resuscitate the performer, and that puts the mime at risk of harm or even death.

An amateur may think that because the PGF is only about a minute or two long (allowing for the starts, stops, and pauses before the next start), that you could say "the guy will only be wearing the mask for two minutes, so how much trouble can he get into in that short amount of time." The error of this thinking is that even for a two minute filming of the PGF, if it were a mime in a costume, he'd be in the head mask far longer than two minutes. You have to suit him up, and then when his mask is on, brush out the fur to blend well. Then you have to gather up all the bags or such the costume was carried in, gather up all your tools like brushes and such, and anything else you used to assist in dressing the man and grooming the suit, and take all that stuff back across the creek to hide it behind your camera start position so none of it is in the scene. Then you have to get the camera, check the spring wind tension, check the lens F-stop and focus, and maybe check the lens glass to see it's clean, and then hold the camera to line up your starting shot, and finally signal the mime with "Action". Then you chase him until he disappears about 150 to 200 feet away and you run out of film, and at that point, your mime in the mask is still trying to breathe. Then you have to go the 200 or so feet to his location before you can pull the mask off and let him breathe easily and freely. The mime may easily be in the mask for 15 or 20 minutes for that two-minute filming take. So you must factor the total time in the mask, counting dressing, grooming, cleaning up the scene, setting up the camera, doing the actual filming, and then getting to where the mime is far away before you can pull the head.

In designing a good working mask, you must pay attention to a design that gives the mime the best potential to breathe while wearing the mask, and the PGF head is very poor in that respect.

So we have a size that seems to run counter to all known technique for making an "ape" mask because it's strangely small when all evidence of professional work produces a larger head, we have an angle tilting the head mask down (so the wearer can see out the eye holes of such a mask) that is problematic to the head and neck posture of the PGF filmed subject, and we have a mask shape that obstructs breathing in a way which is potentially unsafe.

Anyone advocating a hoax can simply refute these concerns by making a mask that does have the shape and size of the head relative to the body, and letting a mime wear it and film that experiment and show us that the head shape and size, the head

thus more chance to be perfect, it is a disaster when viewed from the back and the performer is doing any aggressive activity (such as when he fights off the jet planes while standing atop the World trade Towers, the finale battle). The sequel movie, "King Kong Lives" (1986), was even worse in terms of being a lesson in all that can go wrong with the neck area.

Fig. #8-06 shows multiple screen frame grabs from "King Kong Lives, as a group in the chart upper half. Then the same images are repeated below, faded to half-tone, and only the neck area to study is full tonality in a circle. There are hairs flaring outward, gaps in the neck, odd lumps, and creases, all examples of the problems of a neck on a costume. Patty has none of these.

The usual deep neck closure seam down a few inches below the shoulder top gave enough fur material to allow good head movement, but you'd see a lot of wrinkling and folding in very cloth-like and non-anatomical ways. If you close the neck smoothly at the top of the shoulders and base of the neck, you'll likely restrict head movement, especially turning. Or you risk a severe head turn by the mime causing the mime's head to shift inside the mask so the mime's eyes no longer look out the eye openings, and the person's vision is blocked by solid sections of the mask. Or you can not attach the mask to the torso, and risk that if the head turns or tilts, a gap may appear which shows the separation between torso fur and head/neck fur.

Spandex stretch fur introduced to the industry in the 1980s allowed a new design option that actually does produce a very nice looking neck, even with head movement, but this was not an option in 1967.

Fig. 8-06

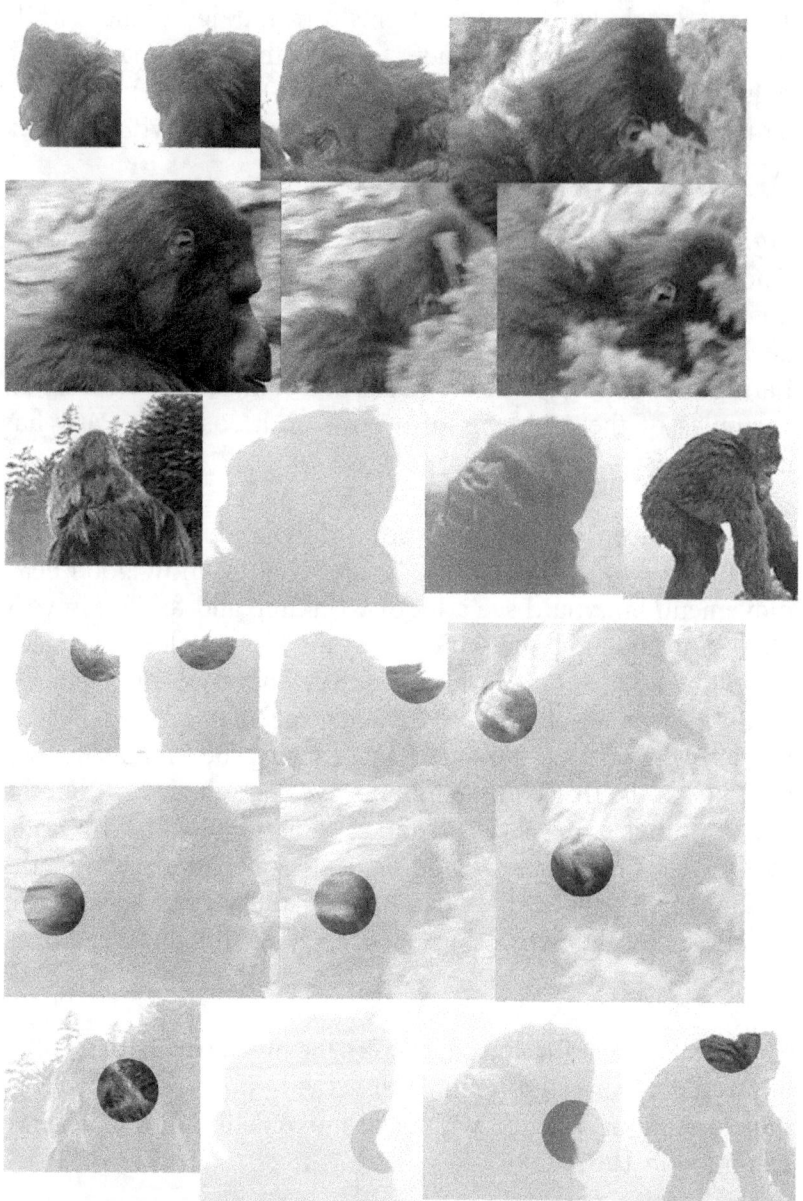

How does the PGF subject figure we call "Patty" manage to have a flawless neck before the lookback, an unwrinkled neck during the head turn, and a flawless neck after the lookback when she resumes looking forward and we see her perfectly from the back? She manages to do something that neither the 1976 "King Kong" or the sequel 1986 "King Kong Lives" could accomplish, despite both movies having a good number of talented Hollywood professionals and studio backing to produce excellence.

This has perplexed me for many years, and was the basis for my original appraisal of the PGF in the year 2000. The neck just doesn't do what costume necks did in that time, with the materials of that era.

Understanding the physics of how a furcloth neck rotates, and how the fabric can restrain the movement, isn't a complicated idea. If you have a disc of furcloth, with arbitrary measurements for an example, and the disk is 6" high, from the solid base to the top secure section, and the disk is 8" in diameter, then the actual circumference of that disk is 25.12". Now, we want to rotate the top section 90 degrees to one side, and that would shift the top furcloth edge 1/4th of that 25.21", which is 6.28". But the furcloth height is only 6" and it cannot elongate or stretch. So if it shifts 6" to the side, it essentially flattens down to almost zero height. The top ring or secure portion must lower 6" for that 90-degree head turn to be accomplished. But the neck is fitted to the torso and the mime's head cannot lower 6" to make that turn of 90 degrees.

A neck made of furcloth, with an 8" diameter of the ring, and 6" high rise from where it secures to the shoulders up to where it secures to the head, cannot turn 90 degrees to one side, unless the head can lower down 6" in elevation. The variables to this basic equation are: 1. Making the neck vertical measure longer will allow more head turn. 2. Making the neck disk diameter smaller relative to the disk height will allow for more turn of the neck with less compression of vertical height.

Fig. 8-07

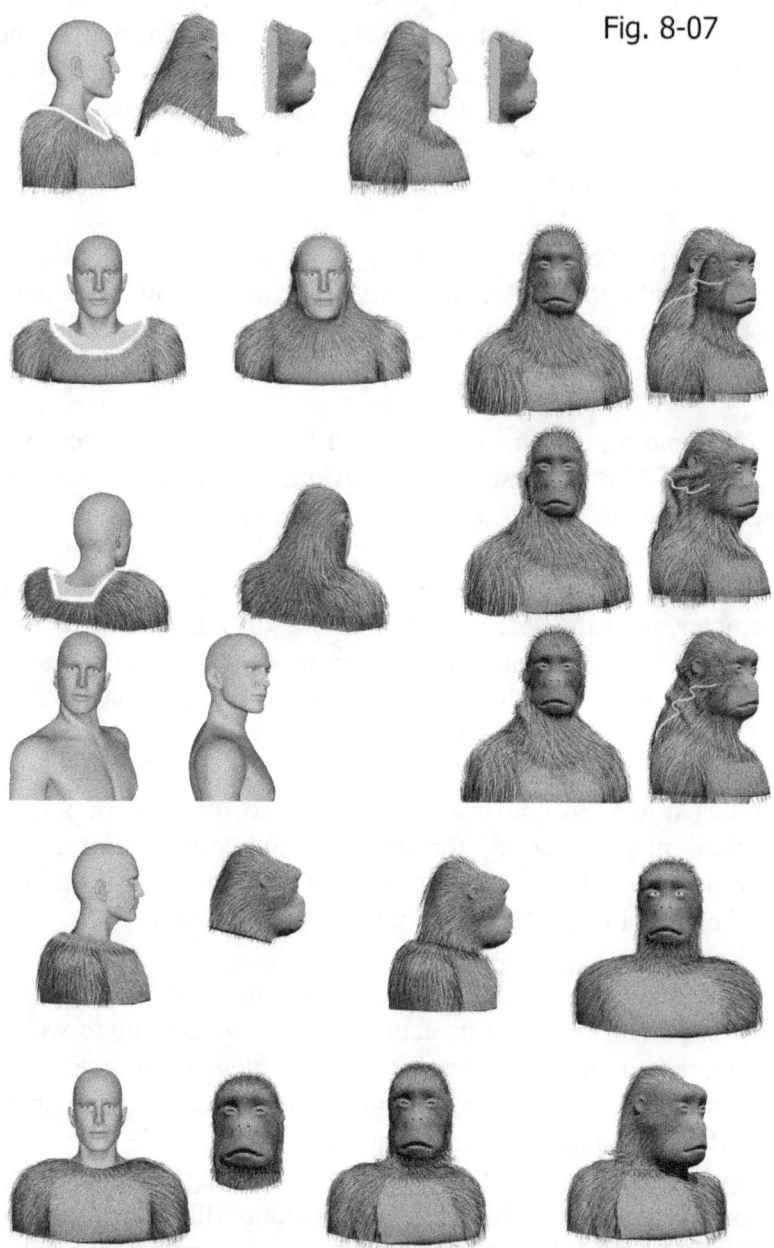

These are simple physical or mechanical realities, and anyone making a fur neck secured to the shoulder and secured to the head must deal with. And that's why sometimes designers don't connect the neck to the torso, or use the very deep neck seam well down into the torso, to try and find a solution to this mechanical problem.

Fig. #8-07 shows diagrams of various neck design options that are tried. None of them produces a perfect neck with the types of furcloth available in 1967.

Yet it would appear, if Patty is a fur costume, that somebody solved the problem better than any Hollywood makeup artist up to the early 1980s (when suit designs changes to work with new spandex materials not available in 1967). So considering 1967 fur suit technology, the idea that Patty is a costume, loses by a neck.

The Shoulder/Armpit

This concern is one that clearly is not "impossible" but none-the-less goes contrary to just about all fur costume design and logic, from the time. So a costume explanation can be made but the rational for someone choosing such a design defies all logic.

The chest section of an ape suit traditionally was a panel on the front of the chest, with the sides several inches inward from the armpit. The suit's fur torso piece would be made like a cloth-tailored vest, the arms would be tailored similarly to a shirt or jacket sleeve, and the chest section made with prosthetic materials was fitted into an opening cut in the chest fur, and the prosthetic glued in from behind. Then figure finishing artists would add hair to the chest prosthetic to give a good blend of the thick fur into the bare "skin" chest prosthetic.

Fig. 8-08

Example of a chest piece for an ape suit. Each designer may choose how close or far from the armpit to end the rubber piece, but it commonly does not go to the armpit.

Fig. #8-08 shows a common chest prosthetic shape for a fur costume, diagrammed at the far right image. The lower image is an example of the common tailoring seam for attaching an arm section to the chest section.

The traditional arm sleeve design would be a single piece of fur (cloth or hide) with one tailoring seam positioned up the inside region of the arm, so it's hidden as much as possible.

The primary challenges for working with fur are: 1. getting the "lay" of the fur (the direction the hair tends to lay down) so the lay from one piece flows into the next across tailoring seams, 2. getting the lay to realistically flow along the body naturally, and 3. getting the blend from thick fur to bare prosthetic skin to appear as a smooth and natural transition. So all our work with fur demands we give these three considerations priority in making design decisions.

With the torso section shaped like a vest, we generally aim for a fur lay to be downward. With the arm sleeves, we aim for a hair lay that is downward from shoulder to wrist, and when the arm is down beside the body, the hair lay looks good. But the shoulder is a bit tricky, and various suit designers may have their personal favorites for how they deal with the tailoring seam along the shoulder top from neck opening to arm opening, when the front hair goes down forward and the back hair goes down backward, and so these two sections have hair lay that are both going away from the seam, like center-parted hair on a person's head. One option is to force brush the fur along the sides of this seam toward the arm opening, so the hair on both sides of the seam runs parallel and shingles over the arm section of the costume. Delicate spots of glue on some of the hair fibers from each side of the seam will help keep the hair laying as brushed, as one technique.

The result of this is a torso costume section where the chest "skin" (the prosthetic) fades into the chest fur and arm several inches inward of the actual armpit and arm sleeve. The armpit is solid fur. And the cut of the arm sleeve joining the torso vest is a tailoring seam that goes vertical across the armpit line, dividing chest and arm.

But we don't see that in the PGF. The standard costume technique is clearly not being employed.

Fig. 8-09

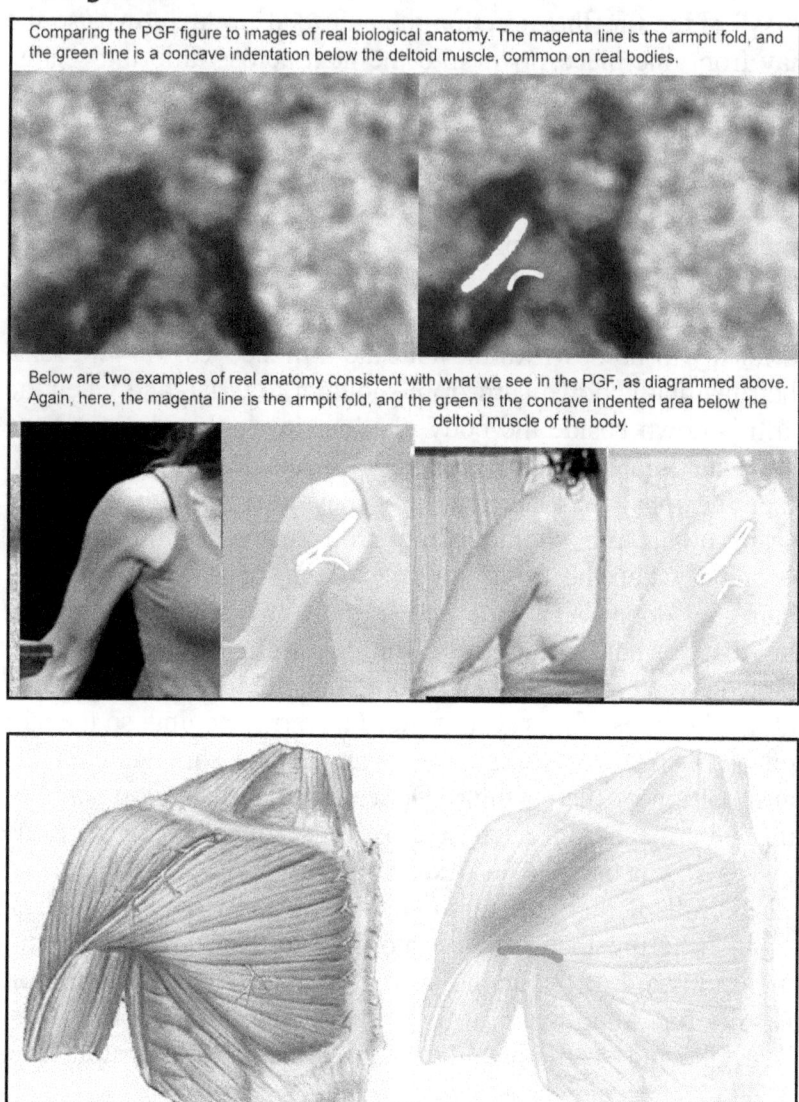

Comparing the PGF figure to images of real biological anatomy. The magenta line is the armpit fold, and the green line is a concave indentation below the deltoid muscle, common on real bodies.

Below are two examples of real anatomy consistent with what we see in the PGF, as diagrammed above. Again, here, the magenta line is the armpit fold, and the green is the concave indented area below the deltoid muscle of the body.

Fig. #8-09 shows the armpit fold on Patty, and diagrams of the reality of this fold on real anatomy.

What we see in the PGF differs from standard costume design in two respects. First, the chest section doesn't have a conspicuous fur perimeter, as just about any ape suit you'll ever see does. And second, the armpit has a curious flowing contour from chest to arm like a real anatomical arm does, and fur costumes never normally do.

The Chest Fur Perimeter

Generally, fur material has a certain degree of flexibility to bend, and the prosthetic skin materials have a certain flexibility to bend. But to join them together, we need to glue the prosthetic to the underside of the fur material, and then add more hair to blend the fur line to thin more and more until it is just bare skin. That joining section has both the fur material, the prosthetic material, and glue, and the result it that this junction of the two materials is more stiff or rigid than either material alone. So as a rule, we try to position that rigid area on a part of the chest which is usually not subject to much motion, shifting, bending, twisting, etc. A small panel on the front of the chest, ending high above the waist and ending well inward of the armpit, is a common and good design.

We must also consider that if the fur costume shifts its position relative to the body of the mime, and this occurs when the performer moves and bends or twists the body or appendages, a somewhat flexible material that runs into a more rigid material will likely buckle where the softer one meets the stiffer one. So if fur material shifts upward from the waist toward the chest, it will buckle and bend directly below where the chest piece stops being glued to the fur, the rigid point. This was verified in some costume tests conducted as part of the scientific research into the PGF, with performers in costumes made for scientific analysis. When the furcloth shifted up toward the chest, it buckled right below the chest prosthesis that was glued to the fur.

Fig. 8-10

Costume Fabrication Design

TOP ROW - Breast Prosthesis (left) and furcloth (right)

MIDDLE ROW - Furcloth glued to prosthesis (circle)

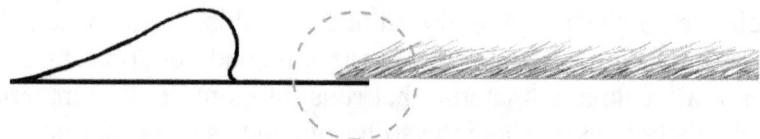

LOWER ROW - When furcloth shifts toward prosthesis, the furcloth will buckle before the rigid zone

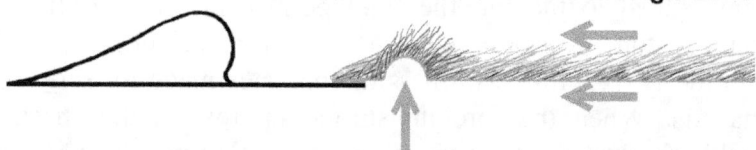

Fig. #8-10 shows a collection of gorilla suit chests as examples, and below diagrams how fur material is glued to a chest prosthetic, and where it will likely buckle if the fur material shifts toward the chest section.

We should see some kind of furcloth buckling on Patty, when she walks and the body fur surface is noticeably shifting from down the leg up to the chest area. But the chest area is flawless throughout the skin shifting. This is certainly curious for costume material motion dynamics. It suggests the entire chest and the entire torso are made of the same material and has the same degree of flexibility and shifting potential. But all traditional costumes do not have that trait. Any molded prosthetic chest surrounded by fur material has an overlap junction that is more dense and stiff than either the fur or prosthetic alone. Needless to say, this is problematic, unless one wants to argue that the entire body is a molded prosthetic rubber body suit and the fur is hair laid up on the rubber surface, or argue, as Janos Prohaska suggested in his "Bigfoot: Man or Beast" interview, that perhaps the fur was applied to the naked body of the performer (like Dick Smith did for the "Primal Man" character in the movie "Altered States"). If so, the PGF figure is a remarkably sophisticated costume effect.

The Armpit Fold

Now looking at the curious fold across the armpit from chest to arm, that feature can be sculpted and molded into a rubber body prosthetic. I made something like it for the foamed latex full-body suits for the movie "Swamp Thing" The attached Illustration shows the body sculpture as a work in progress, and the finished sculpture just before molding and casting. And if we wanted to create an armpit fold in the PGF Figure, we'd need to make something like this for a section of the chest, shoulders and arms like the diagram in that illustration.

But still, there are problems with this method. We create such sculpted prosthetics with a body manikin that has the arms angled out about 45 degrees from the normal arm position straight down. This allows us to get clay into the armpit area angle for good visibility, and the breathing while wearing the mask can be resolved safely and visually correct. No one has ever done it, and I personally feel no one ever will. I'd love nothing more than to figure this one out and make a head mask to match the PGF subject's head, because makeup artists have been trying to create such a mask for about 80 years now, and nobody's done it yet. So if it was done for the PGF, it is a masterful design I would like to learn the trick of. Anyone who figures it out is on the fast track for making the most perfect gorilla suit ever in Hollywood.

But nobody's made that perfect gorilla mask for a human to wear yet, because the shape of a true gorilla head cannot accommodate a human head inside if you want the human eyes to look out the eye sockets of the gorilla head. And the same challenge faces anyone trying to explain the PGF's Patty as a masked performer. The combination of the small head size and the flatness of her head, behind the eyebrow ridge, has never been done with any creature costume we can find.

In the practical world, it can't be done. I cordially invite anyone to prove me wrong, and if someone can, I will be fascinated to learn where my analysis was in error.

The Neck

In 1967, the back of the neck on a fur costume was a pain in the ass. There was no great solution for the problems it cause, except to ask whomever is directing and staging the mime to not show the back of the costume and back of the neck, if it could be avoided. If we had to show the back of the neck, the best plan would be for the mime to stand still, the makeup artist brushed out the fur to be as smooth as possible, and then film without letting the performer move around and especially not turn the head.

Even later, in 1976, with the King Kong suit that was given more talented artists and more money than any prior effort, and

and then get a fairly thick section of mold plaster into the area for a strong mold. So torso and arm sections of a sculpted and molded suit are generally made with the arm in this 45 degree up position.

When the finished costume is worn, however, the arms normally lower to be beside the body, and in most cases, the prosthetic material in the armpit buckles into a crease that runs vertical toward the shoulder. It is diagrammed and examples from creature costumes and molded rubber wetsuits are offered as examples of this crease occurring. Yet we don't see this buckling crease in the PGF Figure. We see the exact opposite, a smooth curved flow from chest to arm.

Fig. #8-11 shows the Swamp Thing body sculpture, top, and second row, diagrams a common gorilla suit chest (center, second row) and the kind of chest needed to resemble the PGF (right, second row) which nobody does for gorilla costumes. Third row shows the sculpture and then the finished suit in use, plus a Star trek suit, and below each, in a black line, we see the folds marked, and in each case of the suit, the fold is vertical when the arm is lowered to a normal posture. The diagram below shows where that crease occurs, and the bottom row shows commercial scuba diving wetsuits that also have this vertical crease. Patty does not have this vertical crease.

One could argue that this armpit fold was somehow stiffened in the casting of the prosthetic so it wouldn't crease, but you can't make the entire piece out of such stiffened material and get the arm swings we see. If just the armpit fold was stiffened, and the rest of the prosthetic was of normal flexibility, then as noted before, when a flexible section shifts toward a stiffer section, the flexible material will buckle at the line when the stiffening section begins. But we don't see any such buckling.

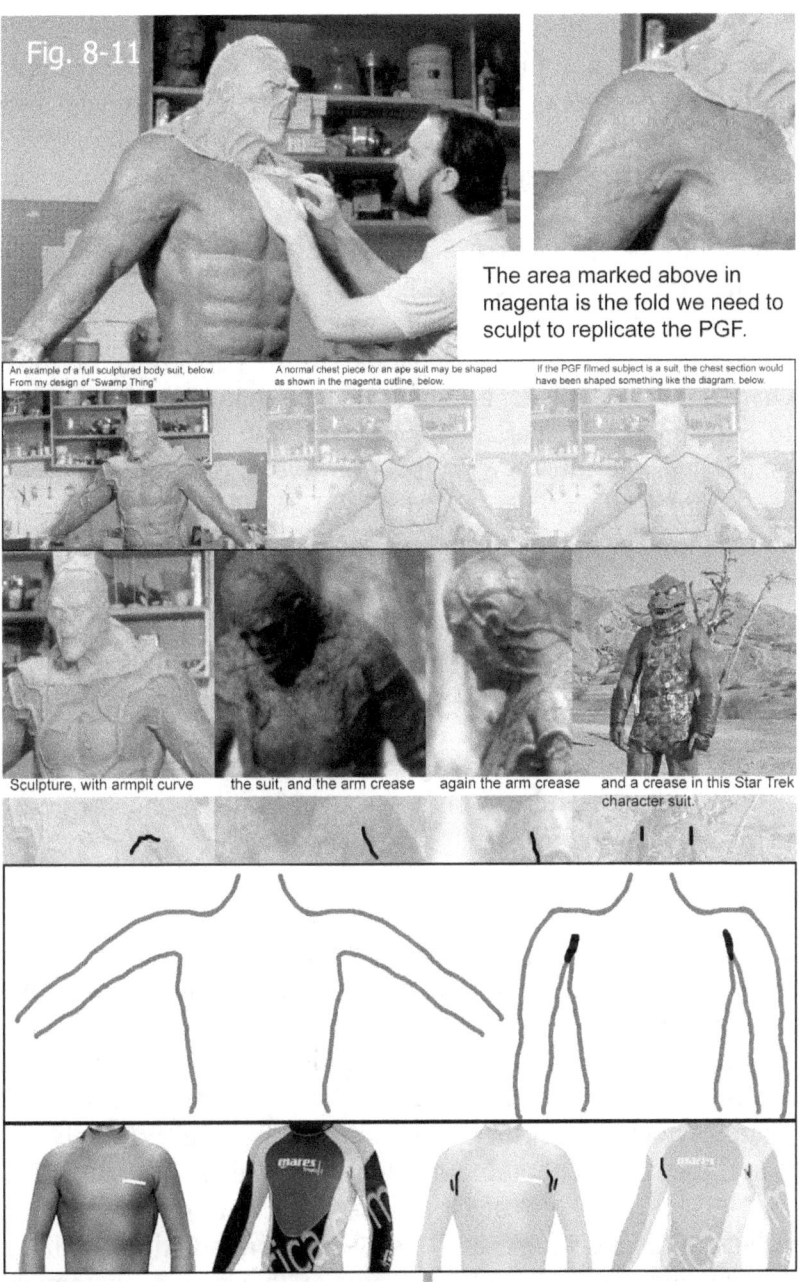

Fig. 8-11

The area marked above in magenta is the fold we need to sculpt to replicate the PGF.

An example of a full sculptured body suit, below. From my design of "Swamp Thing".

A normal chest piece for an ape suit may be shaped as shown in the magenta outline, below.

If the PGF filmed subject is a suit, the chest section would have been shaped something like the diagram, below.

Sculpture, with armpit curve   the suit, and the arm crease   again the arm crease   and a crease in this Star Trek character suit.

There is also a challenge in the hair blending on such a structure. When hair is blended into a section which seems to be more "skin", the two realistic options for some hope of success are the fur lay away from the skin, and you punch hair through the skin in increasingly more and more sparce patterns until there's none, or you set the lay of the hair to shingle over the skin, and maybe punch some to help the transition. But if the PGF figure utilized such a cast prosthetic section to make the armpit, then the thick hair on the upper arm area needs to blend into the sparce hair and skin contour of the armpit where the line of the fur edge runs parallel to the lay of the fur, not shingling over it or sweeping away from it. But such a line where the fur stops running parallel to the hair lay is the most challenging to achieve, because whatever hair material you use to put on the skin part is almost assuredly not the same hair material in the thick fur. When you overlap one hair type on top of another, any disparity in visual characteristics (different color, thickness, texture, or sheen) is softened by the irregularity of the overlap. But if you run two different hair materials each laying side by side but neither one overlapping the other, the two materials will be all the more noticeable as different. It's a hard blend to make look good.

So once again, we are seeing something which simply has no history of successful use on ape costumes up to and including the late 1960s and even into the 1970s.

There is another challenge one must face to plausibly argue this was a deliberate design in a costume. If it was put there, it was put there for a deliberate intent to go against normal costume design and create a realistic armpit skin flap. If so, why was the filming done so 98% of what we see is the subject figure from the back, where we can't see this very substantial deliberate feature which was custom made for this filming? We only see it because of new photographic technology that was unheard of in 1967 as a photo analysis technique. So once again an advocate for a costume is faced with a contradiction. Is this a special and deliberate effort to make this subtle anatomical feature noticeable, when the deliberate staging of the performer's actions made it nearly impossible to see and appreciate this effort? So much effort to make it, and no effort to make sure people viewing the scene saw it? Why make it at all?

Is the proposed hoaxer designing this costume really brilliant at costume innovation, and pathetically ignorant of how to show off what he/she made? It's not easy trying to butter both sides of the toast.

If you want to argue for a costume as explaining the PGF, the bar to make a credible argument just got raised higher.

The Breasts

To properly analyze the breasts on Patty, the right copy was needed, and that eluded us for some time. When I first began studying this in earnest, and relied on the scan images of other researchers, I couldn't find any that allowed for a proper study of the breasts. But Copy 8, made by ANE and including the sharpest 4x zoom in print of the lookback sequence, was ideal, and today remains the high standard for study of things pertaining to the lookback. Aside from the zoom in that holds the detail from the original superbly, the ANE people also made this sequence a 4:1 slow motion copy, meaning they printed each source frame onto four sequential frames of the print. Unknowingly, they gave analysts today a bonus in terms of image data quality.

Unlike digital images today where each dot of color, called a pixel, not only has exact color information but also has a location address in the image, which line across, which column down. So copying a digital image without compression is loss-less. The copy can be as perfect as the source image. Film, however, has grain particles of light sensitive silver halide which are in a random array on the emulsion layer. So the grain pattern on the source film does not coincide with the grain pattern on the copy side. As such, some detail is lost in copying. But when your image is small in the overall frame, and you examine it closer and closer, you are magnifying the grain and at some point, the image just becomes grain pattern and all detail is lost. And these random grain patterns from one frame to the next can actually create a visual "noise" that can sometimes create an illusion of movement, in the highly magnified image examination. Some researchers have seen things about "Patty" which they attributed to anatomical movement but in fact they were seeing grain noise creating an illusion of movement.

But when you have the same source frame printed four times, there is a unique opportunity to use a technique I describe as "grain nullification" (described in more detail in Appendix Five - Image Analysis Techniques) to greatly reduce the effect of grain noise, so any perceived movement is more true to the camera original action that was filmed. Grain nullification in this case was done by taking the four print frames of each source frame, and matching them in two groups of two. In each group, one image was copied and pasted over the other in Photoshop. The two images were aligned perfectly, and then the upper layer was set at 50% transparency, and the layers were melted down. The same was done for frame three and four. Then these two composites were matched up, one copied and overlaid on the other, and the top one set at 50% transparency, and the two melted down to one final image. Doing so minimized the grain pattern and held what truly was common image data from the source. With this lookback set done with the grain nullification step included, any motion we find from one frame to the next is now very reliably true motion of the subject in the film, and not artificial motion from grain noise.

It was this process that allowed me to identify the breast motion in Patty. As she looks back at the camera, her right leg is stepping forward, and when it tries to touch down, the ground is apparently a bit lower than her automatic walk cycle is expecting, because she starts to shift her weight onto the forward right foot, but then it touches down, the shoulder and head drop slightly, and there is a sort of "hard step" of the body, crashing down. This hard step sends a shock wave up the torso, and the breast ripples suddenly, compressing vertically and widening somewhat horizontally. We can rely on this movement to be a true anatomical motion because it matches a physical cause-effect dynamic, the hard step down, the abrupt landing of the foot with weight shifted on to it, and thus a jarring shock when the frame of the body stops abruptly but the fluid mass of the breast ripples because its not rigidly attached to the skeleton and not solid but a fluid-filled organic mass.

Fig. 8-12

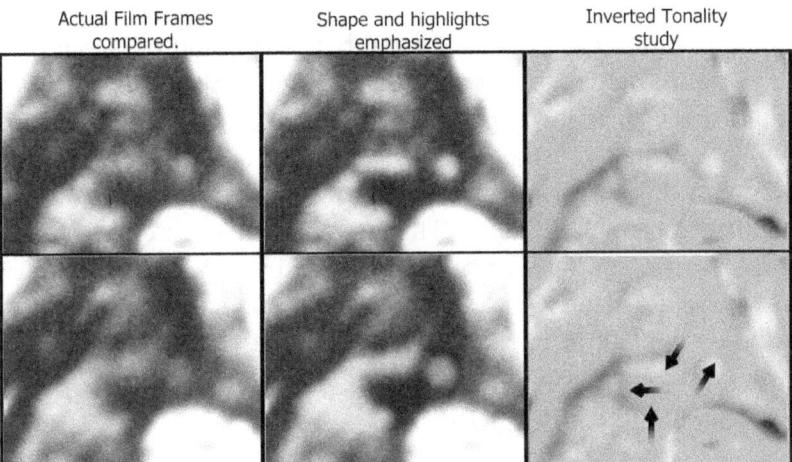

| Actual Film Frames compared. | Shape and highlights emphasized | Inverted Tonality study |

Fig. #8-12 shows the two frames compared (left, top and bottom) where the motion is detected. Suffice to say, it is more clear in a visual animation cycling from one frame to the next, so I regret the limitations of a book published illustration set are not the ideal medium for showing you this. In the middle set, the motion and highlight shift is emphasized, and in the right set, the tonal inversion accents some of the movement, and arrows diagram the motions.

Once this was identified in Copy 8, I formulated a research proposal and was gratified to find that the proposal was accepted and the research was funded (described in detail in Appendix Six - Research Grants). That proposal was to design and build scientific test costumes with breast prosthesis chest pieces, made from the three materials used for such costume work in 1967, and employ female figure models who would permit the project to film them in a manner that we can study the motion of real anatomical breast mass in various walking and stepping down motions. Most importantly, the project allowed the costumes with breast prosthesis and the figure models to be put through virtually identical motion routines, with the same viewing angle, the same photographic setup, the same activity, so we could compare what a costume breast prosthesis does, and what a real organic breast does, and then compare both to what we see in the PGF.

This research was done in September, 2012, and we employed five figure models so we had a good sampling of how firm or fluid a breast mass might be, and we had the three costumes with the three prosthetic breast pieces (made from Slip latex alone, a thin slip latex with a flexible polyurethane foam fill, and natural foamed latex). The figure models were required to submit to a physical breast inspection by a registered female nurse, and the nurse certified that the models did not have any kind of implant or other surgical alteration to their breasts, and we would be assured any motion the tests reveal is truly organic motion with no artificial factor.

Fig. #8-13 shows the three costumes, with the three different types of breast prosthetic materials, and the five female figure models, used for the testing.

The result of this testing showed that the three costume prosthetic breast pieces, made from the materials used in 1967, had absolutely no fluid motion capability. The Slip Latex alone prosthetic was a hollow shell of latex brushed into a negative mold of the breast sculpture. It needed enough thickness to hold the molded shape, and it didn't have any inner mass. So it failed to produce any fluid motion because of the stiffness needed to hold the molded shape. The polyfoam and natural foam latex prosthesis both failed to show any fluid motion because both were foamed materials, meaning they had air bubbles encapsulated into a solid matrix (one urethane, one natural latex) and so the weight of the material was slight and the matrix interlocking was strong enough that gravity acting on the light mass wasn't enough to deform the matrix.

Fig. 8-13

Real breasts, on the other hand are water filled cells in a matrix of connective tissue, and the water weight is more than enough to cause the breast mass to ripple or otherwise show fluid dynamics when acted upon by gravity and the body in motion.

So these costume prosthesis tests were a total fail for any replication of what we see in the PGF figure. People intent on claiming the PGF figure is a hoax with a costume now have a new challenge to deal with, explaining how the costume has fluid dynamics in the breast piece of the costume when the three common materials then used for such prosthetics will totally fail to generate any such fluid motion. In the previous chapter (Seven) in the portion subtitled "The Chest Section", I described how comparisons are made to a reported use of a water pouch in a gorilla suit Charlie Gamora made in 1941. People trying to argue for the possibility of a fluid prosthetic effect in Patty always reference that description, and say, "see, it's possible". A true critical thinker would actually look at that and ask, "Do we have any proof it actually worked?" "Do we have any technical specifications that would allow us to know if the concept would even be adaptable to breasts?"

We don't, in answer to either question. What many people don't realize is that over the years of makeup effects, many artists have tried things that sounded good in theory, but didn't work in reality. Not every invention tried was successful. And sometimes you only realize it won't work when you actually put it all together and try performing with it. So asking for verification, that Charlie's water pouch prosthetic in his suit actually did impart a fluid motion, is reasonable. Thus far, nobody's been able to prove that it worked in a way we can see on film and study the effect of.

The makeup craft today is far more sophisticated than in 1967, and more complex molding techniques are used today that were rare or unknown back then, even to journeyman professionals. So today we can envision ways to produce fluid breast prosthetics, with sophisticated core molding processes, or new resin compounds that are remarkably flexible because of plasticizers added to soften them. But even 10 years after Roger's film, we can't find a single verifiable example of a fluid-filled prosthetic that actually imparted a visually verifiable fluid motion to a costume.

In terms of what we can prove scientifically, we can prove that costume breast prosthetics, made from the three materials used by Hollywood professionals in the late 1960s, do not have any fluid dynamic motion that resembles the fluid rippling we see in the breasts of the PGF subject, Patty.

An interesting counter-argument to this is the claim by skeptics that my experiments are not good science because I didn't test every material. Among those suggested that I should have tested were a bag of coffee grounds, oatmeal (not sure if it was flakes or mixed) and jello. By that logic, one can walk down a supermarket aisle and find hundreds of things that have some capacity to "jiggle" in a somewhat fluid fashion, but it isn't for me to prove every imaginable way to make a jiggly breast won't work. It is first their responsibility to prove someone can even make a breast prosthesis from their proposed technique, attach it to a fur costume, blend the fur into the skin surface of this prosthesis, and then prove it jiggles when the person wearing it walks. I used all the materials by which chest prosthesis pieces have been successfully made from, in 1967. There is no need for me to test things that have never been proven to be capable of making a prosthesis out of. Their claim is extraordinary, and so they must prove it is workable.

The Skin Shifts

The Skin Shift refers to markings and patterns on Patty's body along her right side, as she looks back at the camera and steps forward. We can take Cibachrome images several frames apart and overlay them, and anchor the two image layers at a fixed point of the armpit fold, and we will see a shift in the skin/fur patterns as the right leg extends to step forward. Research studies have verified this is a real phenomenon (described in detail in Appendix Six). The skin of the body stretches almost all the way up the side to the armpit as the leg extends to step forward. Can a costume do this?

Costume tests did not produce the observed effect. Instead, they produced exactly the physical effect we expect when one somewhat flexible material is shifted up against a more rigid material, like where a fur costume material meets a glued overlap of furcloth onto a prosthetic piece.

In the theoretical form, standard old furcloth (from the 1960s) and real fur pelts (like a bear skin) will not exhibit the shifting patterns we see in the PGF. Unfortunately there are very few examples of costumes where we can study them walking exactly as Patty does in sunlight, so we can make valid comparisons. And costumes used for studio work will not have marker notches cut out of the hair as I did for my tests, because such cut notches make the costume useless for any future theatrical use. Any person trying to claim fur costumes can do what we see in the PGF has a challenge in trying to prove that claim. My experiments produced results that would conclude Patty is not a costume.

So I personally don't have confidence that a costume can be designed with 1967 type materials and produce this skin shifting effect.

The Back Contours

We see a lot more of Patty's back than any other part of her body, and most costume backs are rather non-descript. Hers is quite distinctive and unusual.

But before I can discuss the back contours, some foundation information about furcloth is necessary. Furcloth is a woven material more like carpet than common cloth. It has a base weave (the "Base") and it has a "Pile", which are the hair fibers interwoven into the base weave and these fibers rise up away from the base. They are clipped so the fibers end like hair strands. The base may be a fiber-type material, but the pile, the "fur" is usually a synthetic fiber made by resin extrusion. The base tends to be rather dense to hold the pile in place, and that density, plus the non-elastic form of the base fibers, results in a cloth base which has essentially zero elasticity (capacity to stretch) along the directions of the weave (horizontal and vertical) and hardly any potential for stretch on the bias (the diagonal of the weave). If the material does have even the slightest stretch on one diagonal of the bias, it will have compression along the opposite diagonal.

But furcloth also has a hair lay, in that the hair fibers (the pile) don't stick straight out from the base. Instead, they tend to lay down in one direction, and almost without exception, furcloths are woven so the hair lay tends to go down along the vertical weave of the base fibers. The lay becomes important because when we join pieces of furcloth, so one shingles over the next, we must get the hair lay to flow over the tailoring seam that joined the two pieces. Unlike regular cloth and regular tailoring, where you can simply pattern the cloth to the desired shape and assemble the pieces, with no particular regard for the weave directions, with furcloth how you pattern the material does need to consider the lay of the fur, so patterning of furcloth is a bit more complex and skillful than patterning ordinary cloth.

Cloth is woven so it has a flat composition, and it is only totally smooth when it drapes on a truly flat surface. Furcloth, having a stiffer base than ordinary cloth, is even more susceptible to this effect. It tends to buckle with cloth folds over any irregular surface which exhibits compound curves (curves that follow two different dimensional directions at the same time).

Next I do need to take a moment to discuss real fur hides, like bear skins (which have been used for ape suits). Real fur was common for the suits of the 1930s 40s and 50s before synthetic furcloth was well developed, so we see older suits made with real hides. The last known usage of real hides on an ape suit was the 1976 King Kong. A lot of people mistake how hides are used in taxidermy with how hides are used for costume work. The working methods, the tailoring, is totally different.

In taxidermy, the hide is soaked in water and the skin (the hide) becomes remarkably flexible and even somewhat elastic, as long as it's wet, allowing the taxidermist to stretch the hide over a rigid body form and the hide takes compound curves very well. It does drum over concave curves (those that indent inward), so the hide needs to be pinned down to those indented curves. But the hide is actually attached to the rigid body form with hide paste, a glue compound, and any pins or other restraints just hold the hide to the glue until the glue sets up. Then the pins are removed, the hide dries, and becomes exceptionally stiff. Sewing closures with this hide is usually a cross-stitch with the two hide edges butted against each other, fine if they are glued down to a rigid form. The hair is groomed while the hide is wet, and then the hair holds that groomed lay when the piece dries.

People making suits don't do any of this. We don't have a rigid body to pull the hide around, we don't glue the hide to a rigid form, we don't pin it to take concave curves, and we don't use the same sewing stitch, because it easily opens up when a fur shape flexes.

If we work with fur hides, we still must follow cloth tailoring procedures, patterning the hide to the compound curves we need, and joining the edges with glued cloth gussets for a really tight and smooth connection with a perfect hair lay. So, whether using furcloth or real fur hides, the tailoring is actually the same. And real hides, with the skin tanned and dry, doesn't have any significant elasticity like it does when wet, so we classify it as a non-elastic material the way we work with it.

So tailoring furcloth or real fur hides requires some intricate patterning and consideration for the hair lay to be successful and create a finished shape that looks like a real fur-covered creature.

The relevance to the back contours is that they represent compound curves (curves that follow two different dimensional vectors or directions) and neither standard furcloth or tanned fur pelts will drape over a compound curve unless they are tailored specifically to that contour. The random cloth-folding that occurs with loose furcloth when there is motion distorting the support structure (provided by the person wearing such cloth) will not take compound curves the way tailored cloth folds can be shaped. So we can distinguish between random cloth folds and tailored compound curves.

Fig. 8-14 Adipose Tissue Deposits on Back

Patty's back has multiple compound curves that simply won't occur through random cloth folding, so if you want to argue for her being a fur costume, you must concede that these back contours were deliberately designed and tailored to these odd curvatures. These are contours that do occur naturally in real anatomy, but specifically in anatomy that is aged, overweight and perhaps not physically toned. In other words, they look "flabby". So to argue for a costume, you must then argue that a specific, deliberate and time-intensive effort was made to produce a "Bigfoot" costume of a "flabby" creature. Yet no one noticed these features during the four plus years that Roger Patterson survived the filming and tried passionately to convince people the filmed subject was real. Yet they could not exist in a costume unless he deliberately put them on it.

Fig. #8-14 shows in the top two rows Patty's back and on the right, the pictures are marked in white the contour under consideration. The middle section shows real human backs, and the folds marked. The bottom section shows test costumes and white lines where the folds occur. None of the costume folds resembles real back contour, or Patty's back contours.

Fig. #8-15 also shows the back, but focuses on a shape I refer to as an "inverted Tee", a vertical line down the spine in the lower back area, and then a horizontal fold across the low waist section. On real anatomy, we see this, but it doesn't occur in costumes, unless someone specifically tailored it into the costume design. The vertical spinal line is discussed more in the next chapter as well, in reference to the trapezius muscle.

Patterson-Gimlin Film
Evidence and Proof   Fig. 8-15   Inverted "T" Contours

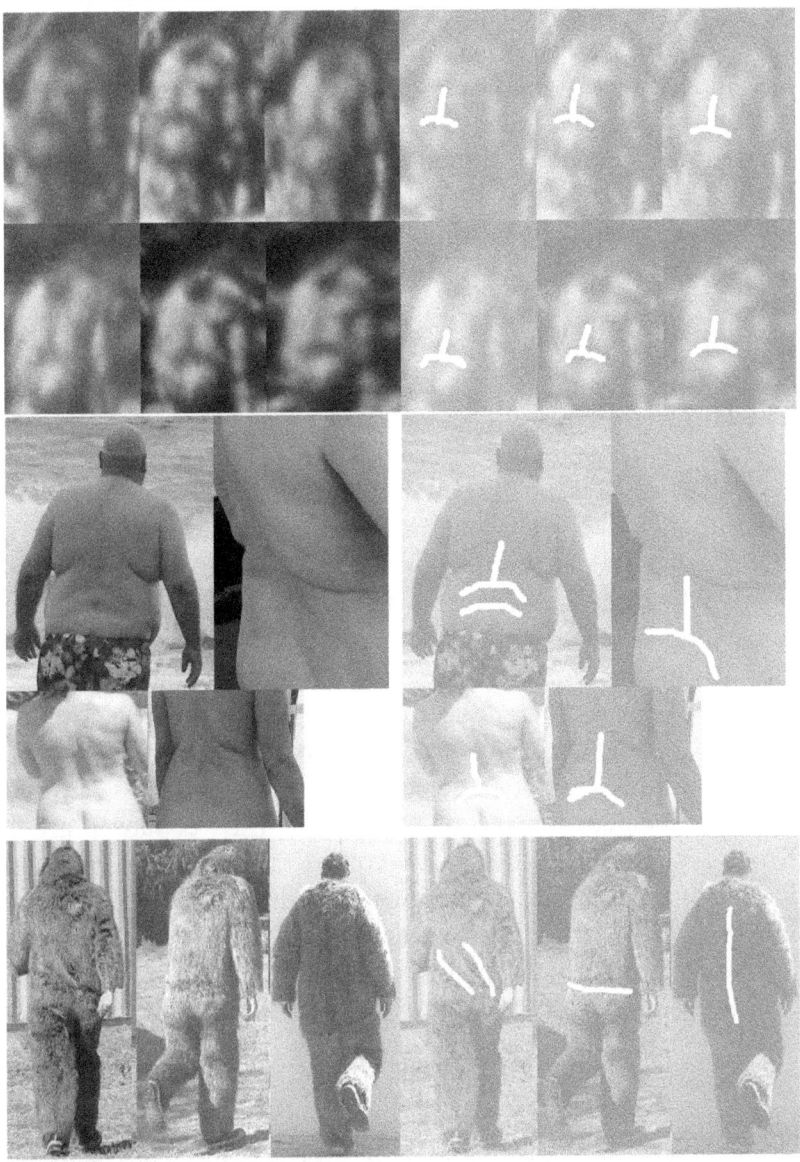

Now in this aspect of the body, there is nothing impossible about the back contours, in terms of whether a costume can be made with them, but one must reasonably ask why such were incorporated into a costume, when there's no history of hopefully realistic monster or creature costumes designed with such features. If you look at the mentality of costume designs, the folds are irrational, unless someone tries to make them a false positive trait, and then, if the false positive isn't seen or figured out by anyone, the hoaxer must draw attention to them to move the analysis to the intended deception. So the problem of the back folds is that they can only exist in a costume by deliberate design and skillful workmanship, but there is no evidence of the deliberation necessary. In that regard, they are problematic to arguments for a costume and a hoax.

The Hip Lines

Patty has an arching line across the hip, back to front, and Makeup Artist Chris Walas in the Bigfoot Forums (in 2004), presented a theory that the arching hip line represents the overlap line between a fur costume leggings section and the torso section which apparently had a wetsuit-like snap crotch closure. In terms of design concept, the technique has some history of use, so there is merit to the proposal from that standpoint. However, a contradiction to this concept is the hair length. Simply put, for such a technique, longer hair would blend the two costume sections better, and if we want to argue for a costume, designed with considerable deliberation and knowledge, we must wonder why the designer didn't choose longer hair to better blend the arching line? Short dense hair is a poor choice for the type of hair that will blend this costume design for a realistic look. Are we going to shift the argument to say Patty is designed to be easily spotted as a fake? If so, why was so much effort expended on executing a hoax so meticulous that it defies every form of analysis that would reveal cheats and staged activities. The short hair and the claimed suit design are at odds with principles of successful design.

Then we consider that this hip line has a curious notch in it, and such cannot occur by accident. On a costume, it must be deliberate, designed and intentionally put there. And that begs the question of why? It has virtually no function from a costume standpoint, and would draw attention to the arching hip line, which the designer should be trying to hide as much as possible. So to argue for a costume, one must explain this notch as a deliberately shaped and tailored feature for what purpose?

Fig. #8-16 diagrams the arch and the notch on Patty, shows real human hip areas with adipose tissue fat deposits and then shows test costumes in motion, which do not replicate any of the diagrammed features.

However, if we examine the film frames and make comparisons of the arching line with two images several frames apart, we can find evidence that the notch changes shape as the leg extends. But if this is indeed the junction of two costume pieces, the torso and the leggings, that junction needs some type of closure seam to hold the torso part securely to the leggings, and such closures tend to be quite stiff. Yet what we see in the film is a shifting form, clearly not stiff. So developing a costume overlap which securely joins the two sections so they can stretch and shift shape across the closure seam, as we see in the film, is an accomplishment the materials of 1967 could not do.

In consideration of all these individual aspects of the body, and the theory and practice of fur costume design in 1967, Patty represents a figure that simply defies just about all the techniques, processes and design concepts of costumes. So the only hope of explaining Patty as a costume is to somehow try to figure out a design so unique and unconventional that the maker was a genius of an order of magnitude the profession has never seen.

Fig. 8-16

Patterson-Gimlin Film
Evidence and Proof

Arching Pelvic Fold

Fig. 9-01

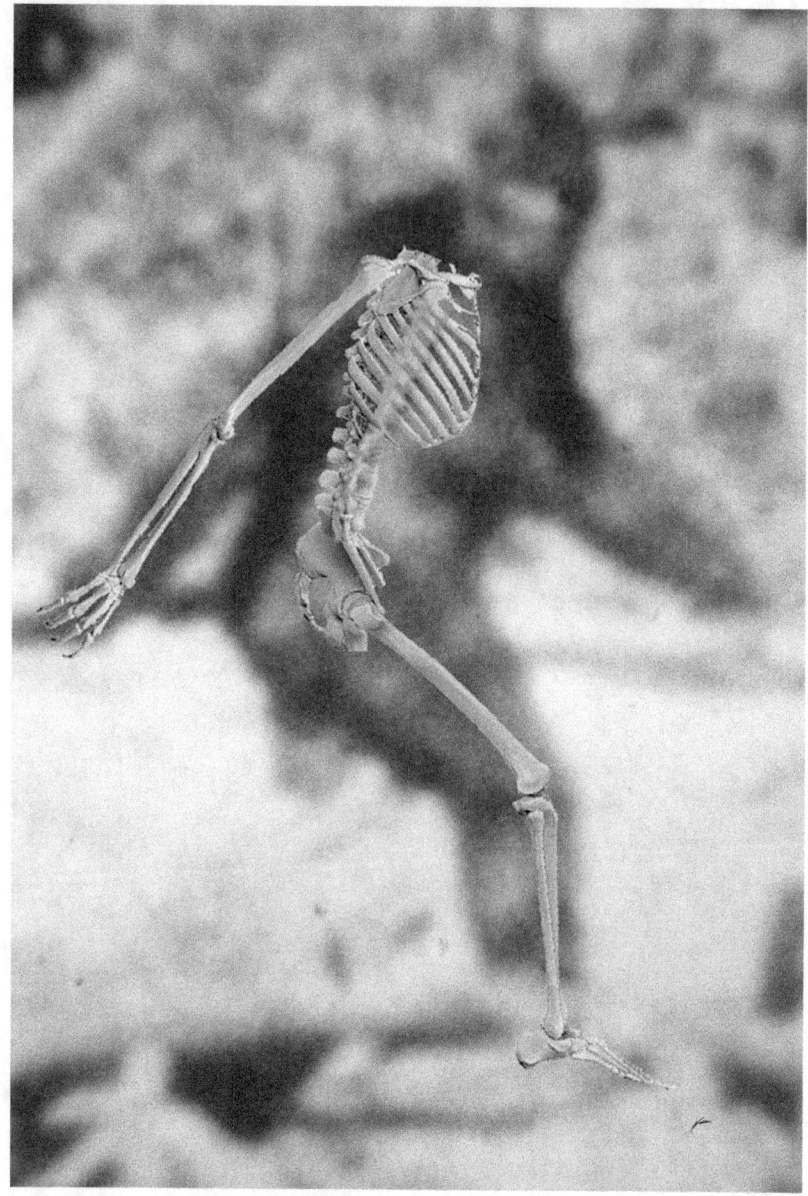

Chapter Nine - Is the Subject Biologically Real as She Appears

The previous chapter looked at Patty and asked, if she's a costume, how would we create this anatomical feature with costume technology and process, and as noted, many of the physical features were either challenging or unprecedented in costume work, meaning she is not a costume as we know they can be made.

An alternate way to try and find evidence of a hoax is to look at Patty's anatomy, and see if we can find things that just are not biologically real. If we find something like that, a feature of her body that doesn't make sense as a real entity, it makes a good case for claiming she's a costume worn by a human, even if the costume is unorthodox in design. Skeptics have been trying this for 47 years, but they have failed over and over because they have closed their eyes to the realities of human and ape anatomy.

In this chapter, I will be making comparisons with the human body, often in the nude, and often a body which is not "centerfold quality" meaning a body that is not young and physically well toned and proportioned. Our culture has been deluded by the media obsession with "perfect bodies" and so we rarely see the realities of human anatomy that are "honest" (sometimes aged, wrinkled, with pockets of adipose tissue such as fat, cellulite, etc.). But any responsible comparison of Patty to human anatomy requires we consider these "honest" bodies, because if she is biologically real, her body likewise would be more "honest" and less perfect.

I simply hope that you, the reader, can look at human anatomy with no expectation that anatomy must be perfect, and instead see that when we speak of human anatomy, that must include the flawed and normal bodies as much as it would include those our culture deems esthetically attractive. We must also abandon the cultural mindset that has distorted our perception of the human body, and vilified adipose tissue (fat deposits). The human culture has been driven to a belief that an attractive and healthy body doesn't have fat, or has very little. Almost all physical training and body conditioning strives to reduce or eliminate body fat. A muscular body is generally defined as much by the absence of fat as it is by the presence of defined muscularity.

But in nature, there is no conflict between muscular development and accumulation of body fat. They serve entirely different purposes and both purposes are valuable for survival. The strong and dynamic musculature allows the subject to move, harvest food, fight (as necessary) and interact with other individuals as necessary. The accumulation of body fat is food in reserve for times when food is scarce, such as harsh and cold winter months. In nature, a powerfully muscled body can have substantial accumulations of fat, and it is not uncommon for animals to "bulk up" in the autumn as they anticipate the food scarcity impending with the winter to come.

The PGF was filmed in late October, mid-autumn, and so the prospect of Patty (as a real biological entity) having substantial musculature and accumulations of body fat (which tends to obscure most of the musculature, actually) is not only reasonable but appropriate. This co-existence of musculature and fat deposits must be recognized because there are traits of the back that show musculature and traits that show fat deposits, and a skeptic ignorant of natural anatomy would easily be mistaken by the muscle or fat "can't have both" cultural obsession and think that evidence of muscles and fat is a contradiction in terms. It is not, in nature. I've seen the argument made in discussion forums and so I wanted to take it head on right away. Strong defined musculature and fat deposits do co-exist in harmony in bodies in nature. When I get to the back, I'll explain this in greater detail.

Over the years, some absurd claims have been made that certain features or traits of Patty's appearance can't be real, but when we examine the realities of both human and ape anatomy, a different story is told.

The Head

As noted in the previous chapter, the head shape is a horrible one to try and accomplish with a modern human wearing a mask. But it has a striking similarity with archaic hominids, especially Paranthropus Boisei, a human ancestor that lived over a million years ago in the Oldivi Gorge region of Kenya, Africa. The skull OH 5 discovered by Louis Leakey is a fine example. It has a cranium that essentially drifts nearly straight back from the brow ridge, as Patty's head does. It also has the suggestion of a modest sagittal crest, and such a feature does cause the fleshed out head to have a more tapering or pointed rear crown of the head than a normally rounded cranium with no crest (like modern humans). And Patty has that apparent pointed rear crown of the head.

Fig. #9-02 shows the OH5 skull and my fleshed out museum recreation of the face of P. Boisei. Top right is the shape of patty's head. Lower left is the Boisei skull overlaid, and below right is how nature would have to modify the Boisei skull to better fit Patty. While speculative, it is anatomically reasonable and does not give any support to the head shape being false or biologically unreal.

So a head shape like Patty's is perfectly consistent with the biological realities of some hominid heads. Whether this would be an ancestral affinity or simply convergent evolution independent of any direct ancestry, one could argue either way. But it is a realistic and fully plausible head shape for a hominid.

Fig. 9-02

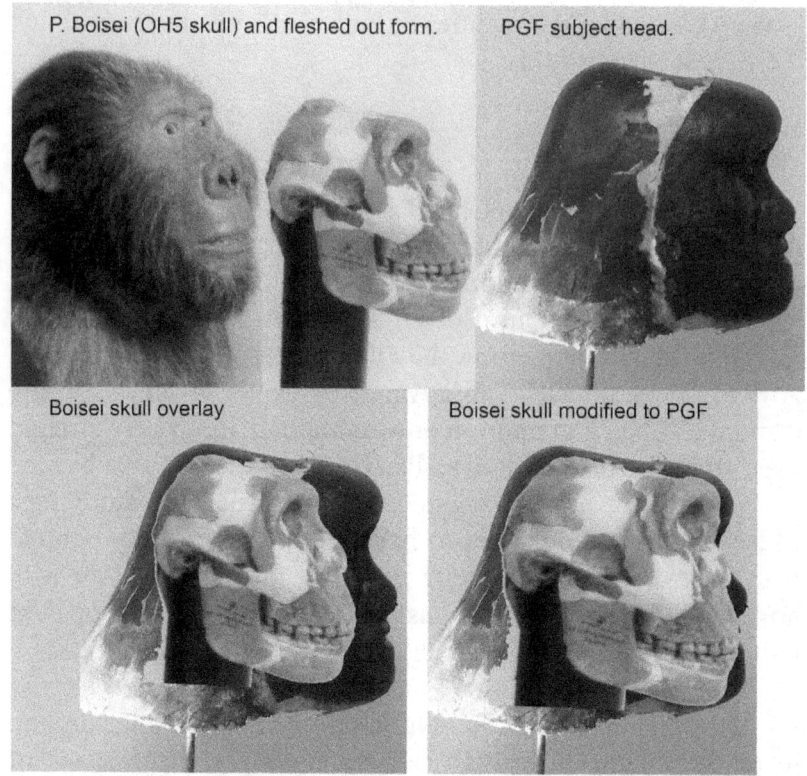

P. Boisei (OH5 skull) and fleshed out form.

PGF subject head.

Boisei skull overlay

Boisei skull modified to PGF

The Neck and Back

Patty seems to have a fairly well-developed trapezius muscle set, particularly the upper trapezius area, and that tends to result in a somewhat minimized neck. We see that in the early frames from her back views. But recently I explored a novel idea and it led to a quite intriguing affirmation of the realistic appearance of the trapezius muscles. I had noted in earlier research a feature of the back I called the "inverted T" (previous Chapter), and I found intriguing examples of this on human bodies, but none on costumes. I used this feature in the scientific paper on the study of adipose tissue deposits. The comparison charts are shown in Chapter Eight, Fig. 8-14 and 8-15).

But as I continued to study human anatomy, I noticed a curious feature on some bodybuilders, from Arnold Schwarzenegger's Encyclopedia of Bodybuilding. What struck me as odd was that in certain poses, the bodybuilders has a furrow up the spinal column from waist to mid torso, but it ceased and the skin drummed over the upper back, with no spinal furrow. So I decided to test this concept when one of my research grants allowed for a second phase of anatomical photographic studies. I had one model, a very athletic woman who was not on the level of muscular development as world-class bodybuilders, but very athletic none-the-less, and she was photographed going through a posing routine of the back with the arms in various postures. When her shoulders were back, the usual dynamic shoulder posture, the furrow runs all the way up the spine. But when the shoulders roll forward, almost like a body slumped over forward, the furrow remains in the lower back, but the trapezius muscles cause the skin to drum over the spine in the upper back, and that erases the furrow in the higher back area up to the neck.

Fig. #9-03 top shows one back with the trapezius muscle overpowering the furrow in the spine. Middle row shows three men lined up, all with varying degrees of the trapezius muscle obscuring the spine. And the bottom row shows my research model, and at left, her spine furrow runs all the way up to near the base of her neck, but as she rolls her shoulders forward, the trapezius muscles eventually drum over the spine and the furrow disappears from the upper back, but stays on the lower back.

Fig. 9-03

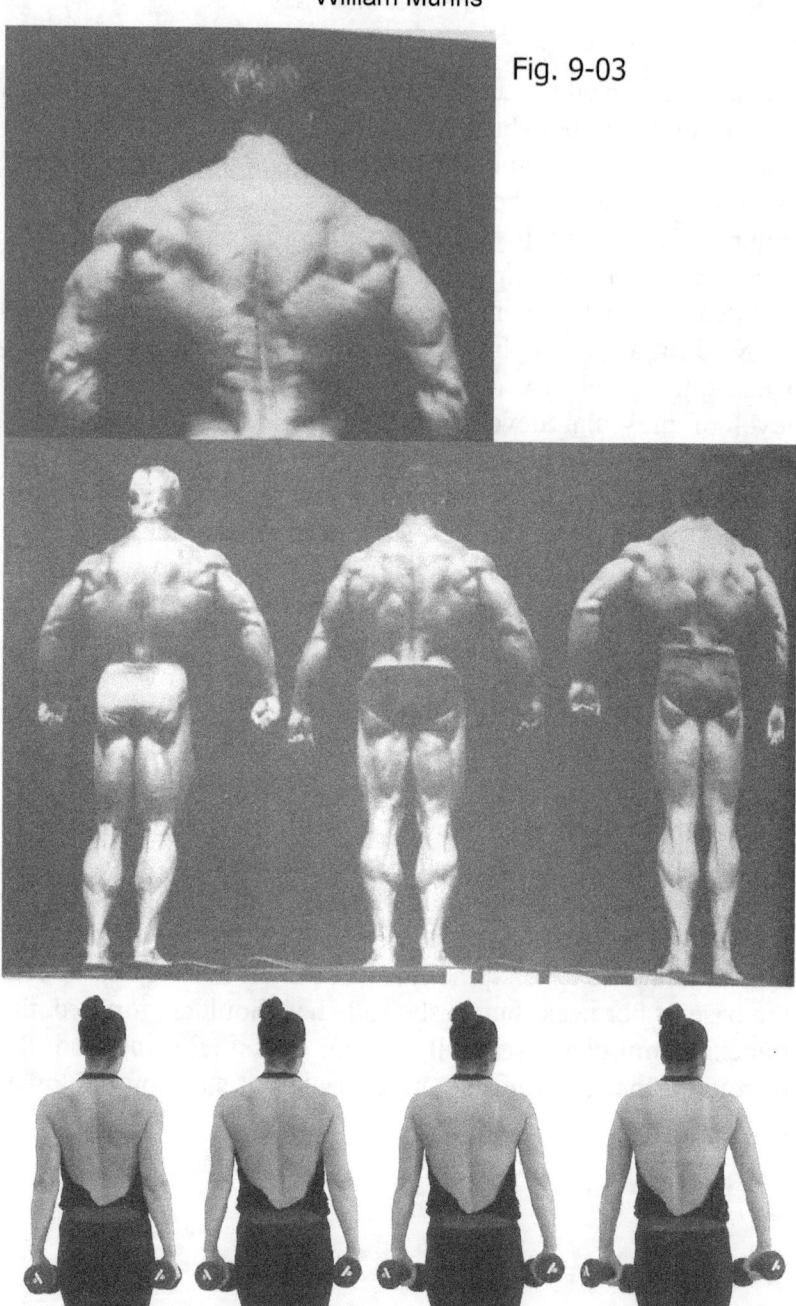

This furrow only on the lower back was what I has curiously studied in Patty, and now through the research photo studies, I had verified that it only occurs on the body when the trapezius muscles are fairly well developed, and the shoulders are rolled forward in a slumped posture. And Patty has quite well developed trapezius muscles and in the early walk away, her shoulders are slumped over forward. It's a remarkably subtle detail, and wholly consistent with real anatomy, but one that no person would ever design into a costume because it is foolish to design a feature into a costume that is conditional on a specific posture, and unnatural in any other posture, because it either locks the costume performer into that one posture, or it is guaranteed to look totally unnatural as soon as the body shifts to some posture other than the one specified. And a slumped over posture is a pretty silly one to design into a Bigfoot creature effect.

But if you wanted to put it in a costume, you'd need to tailor the furcloth specifically to have that furrow only partially up the back, and then you would need to mold or shape the under-padding to also have such a furrow, so the tailored furcloth could sit in that padding furrow. It cannot occur in a costume by accident. It would require considerable deliberate planning and fabrication. But to even put it there, and not worry that the vertical fold would be mistaken by people as a zipper line, you'd have to know exactly what anatomical conditions would produce such a furrow, and then you'd better tell the scientists and anatomists examining the film later why it suggests a real body. And if you knew exactly what anatomical conditions cause it, you'd reject the idea because it limits the posture to a slumped over shoulder, and looks fake if you decide to stage the performer to do something more dynamic.

But getting back to reality now, the feature is definitely real, and occurs on bodies with well developed trapezius muscles and the shoulders rolled forward somewhat. So why would Patty have well-developed trapezius muscles? In the Encyclopedia of Bodybuilding, I looked up what exercises one would do to specifically develop the trapezius, and especially the upper trapezius zone, like Patty shows. It showed exercises that dead lifted weight with the arms straight and the shoulder doing the lifting. Why would Patty do that in nature? The only thing I could figure, on a regular basis would be lifting fallen logs. Why would she lift fallen logs? Well, it turns out that under fallen logs is a very reliable and rich source of edible things, insects, lizards, snakes, maybe small mammal nests with helpless babies, all in all a sort of nature buffet. My research indicated this was an excellent way to locate food for an omnivore, and that bears often tear apart or roll over logs for access to this buffet. Patty might rather lift the log than tear it apart or roll it over, and if she did this on a regular basis, she's giving the upper trapezius muscles a great workout.

Argumentative, to be sure, but remarkably well interconnected from a logic structure. Extremely persuasive for a natural biological entity, and not a costume.

The other fatty structures in the back are the pockets of fat under the arms, which do occur on humans who have substantial fat. Illustration 8-14 in the previous chapter illustrated these. The Scientific Paper on Adipose Tissue also describes them in detail. And because the filming was in October, the bulking up with fatty reserves is logical for a real creature as winter is approaching. So everything rings true for the back.

The Breasts

The Breast Studies are described in detail, especially in Appendix Six, so I'll just summarize here. In essence, Patty's breasts do have a fluid motion that can be documented during the lookback segment, and that fluid motion tends to correspond with real motion dynamics of walking on irregular terrain. The motions correspond to real breast tissue motions (as verified by scientific tests), and costume prosthetic breasts made from any of the materials in common professional use in 1967 fail completely to demonstrate a similar motion dynamic.

Fig. #9-04 shows on top the figure models studied as they stand on a platform that drops abruptly two inches in height, and causes the breasts to ripple. The right tonal inversions show the shift of the breast mass from before the drop to immediately after. The middle row shows the three tested breast prosthetics, and the lower set of three is actually the moment after a two inch drop and hard landing of the test platform, but these prosthesis demonstrate absolutely no fluid dynamic shift in shape. The bottom images show the tonal overlays, and the reason the right panels are near solid grey is because there's essentially no shift in shape despite the drop impact.

So in terms of fluid dynamic motion, they are remarkably authentic as compared to real breasts, more so than any documented effort in special makeup effects. Patty's breasts do sit a bit low on the chest region, compared to the human norm, but this isn't anything unreasonable in the overall mammalian design of breasts (or milk-producing glands, as a more generalized term). Some mammals have them much lower (the obvious extreme being cow udders between the hind legs.) But when you consider that the hominid arms may cradle an infant nursing, the length of the arms bending at the elbow sort of defines that cradle for the infant, and if we follow that reasoning, longer arms would allow for a lower breast to still easily be accessible to a nursing infant held in the arms. Patty does have the requisite longer arms, so there's nothing unrealistic about the breast position in that respect.

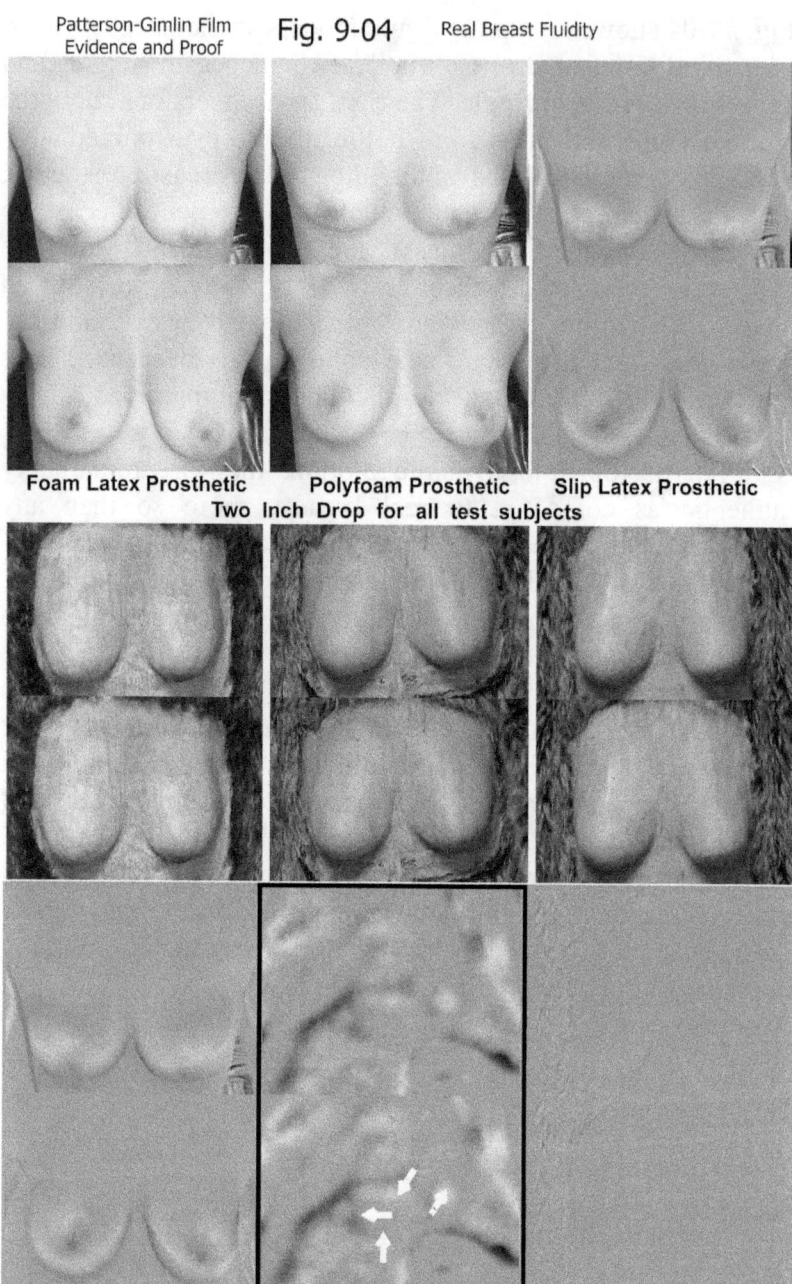

Patterson-Gimlin Film
Evidence and Proof

Fig. 9-04

Real Breast Fluidity

Foam Latex Prosthetic   Polyfoam Prosthetic   Slip Latex Prosthetic
Two Inch Drop for all test subjects

The breasts are usually described as fur-covered, and critics like to say this is unnatural for breasts that expand out from the chest wall. But actually human females do have hair on the breasts but it is the fine and near colorless velus hair that covers most of the human body. The hair follicles are there, and can go active to produce a more robust hair color and growth. But relative to the fur seen on Patty's arms and back, the breast mass and chest wall on the side of the breast actually have more of a skin-smooth tonality, suggesting any hair on the breast and surrounding chest area is quite modest compared to other parts of the body. So there's nothing about the breasts which is anatomically invalid or glaringly false. So there is nothing about the breasts that can be used to argue for a costume and hoax, but there is strong motion dynamic evidence arguing for real tissue and not prosthetic costume fabrication.

The Side Skin Shift

Like the breast motion, the shifting of skin and fur on the torso side from under the arm to the knee is studied in detail in Appendix Six, and summarized here. The shift is caused by the lower leg extending to a straight posture from a bended position as the leg steps forward and is about to land the next step of a walk cycle. This straightening of the lower leg pulls the skin on the underside of the leg and that pull goes all the way up to the mid torso region.

Fig. #9-05 diagrams the skin shift on Patty, and on figure models.

This was documented by photographic studies of female models with marker dots applied with makeup to their bodies from knee to armpit, and they were photographed walking. When we anchor the higher dot markers at mid torso (in photo comparisons of two different phases of the forward step), the leg extension causes the markers down to the knee to shift downward. Patty's skin and fur do exactly the same thing as her leg extends to take a forward step.

The consistency with real anatomy is remarkable.

William Munns

Patterson-Gimlin Film
Evidence and Proof

PGF Subject Skin Shift

Fig. 9-05

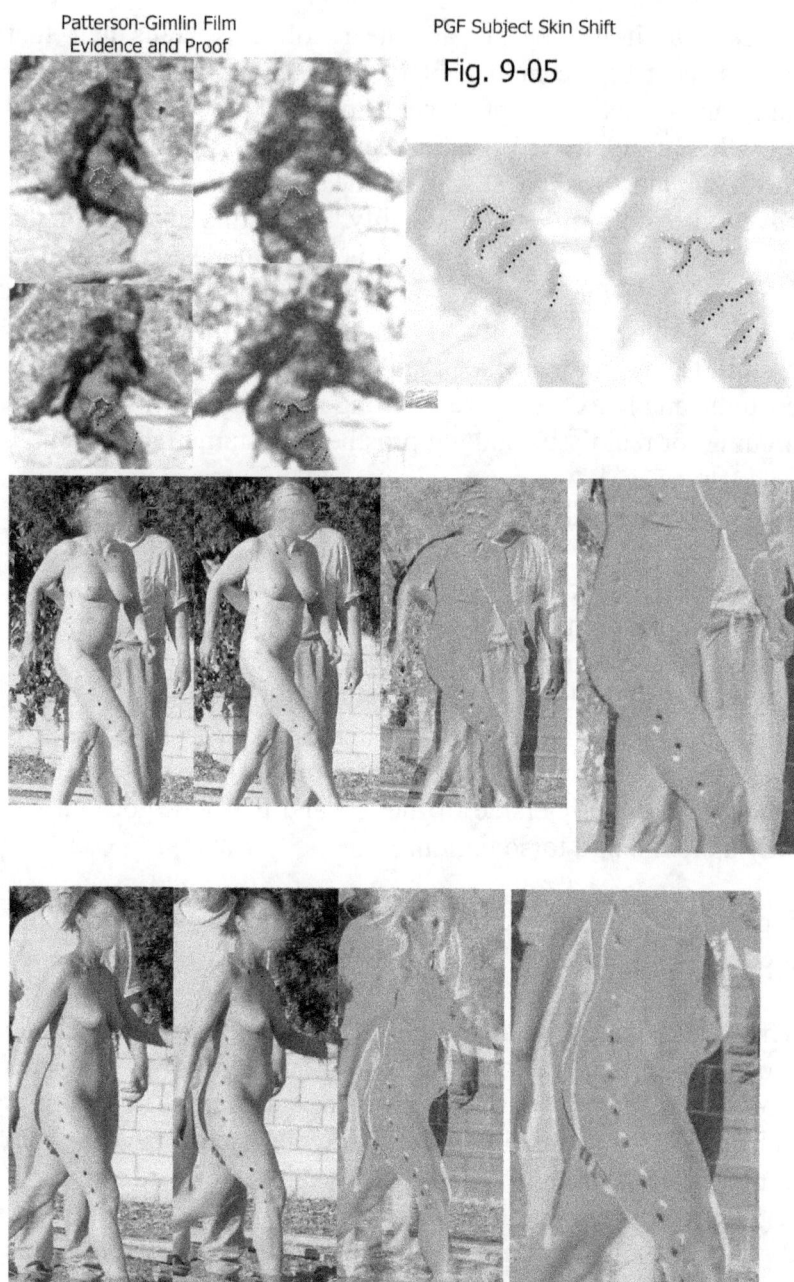

268

The Hip and Pelvic Lines

Critics and skeptics have been using the hip and pelvic lines as one of their "slam dunk" proofs of a costume for years, but examination of real anatomy defeats these claims on no uncertain terms. The same arching hip line attributed to a two-piece costume does occur naturally on human hips. The backward diagonal line, which has absolutely no reasonable costume function, does also match human anatomy. And most curiously, the "notch" of fur on Patty, which defies all reason as a costume feature, oddly has a parallel on the human body, a curious dimple that some people have. So claims that these features must prove a costume are sadly mistaken and simply reflects the ignorance of human anatomy as displayed by advocates for a hoax.

Fig. #9-06 shows Patty's pelvis lines, and then shows examples of real human anatomy with similar shapes, lines, and indentations in the pelvis. The Research Paper described in Chapter Two discusses this in detail.

The Calf Muscle

Patty has a very dynamic calf muscle shape, and costumes rarely have such, because it's a somewhat awkward tailoring task (not impossible but challenging). Her calf muscles are remarkably realistic, and we rarely see any such work on a fur costume. I referenced dozens of costumes and couldn't find a single example. But any reference to real human anatomy will verify a shape that is realistic and appropriate.

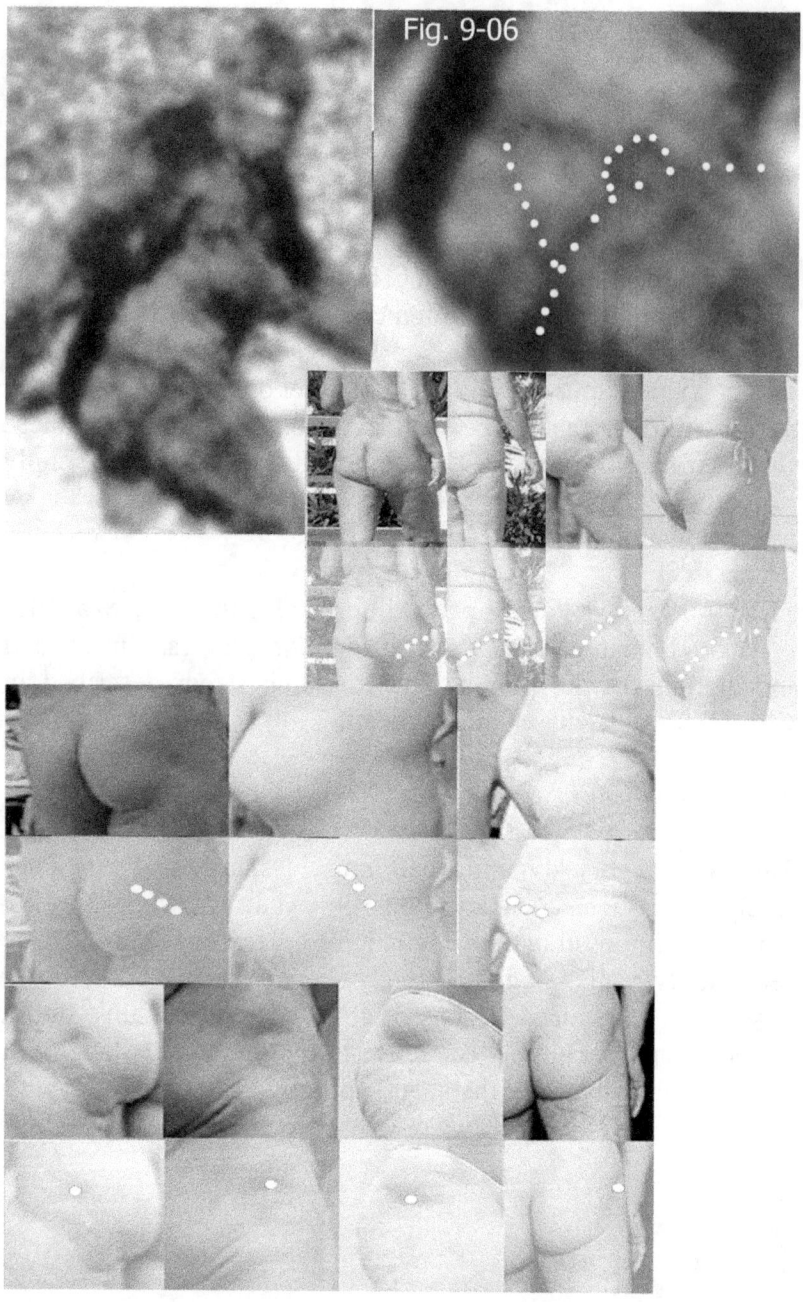

The Thigh Subduction

Skeptics have talked a lot about an effect they perceive where they say Patty's thigh "subducts" (slips under) the buttocks mass, and they claim this is proof of a costume. But in real human anatomy, this effect occurs as well, if the buttocks have accumulated any significant adipose tissue mass. In my research photography work with figure models, who were body painted a grey tone to give the experiment a clinical continuity (the models had quite varied skin tones, so the grey equalized them), during the filming phase the body makeup on one model rubbed off precisely because the thigh was folding under the buttocks mass. This was absolute empirical evidence that the physical action, exactly as claimed by skeptics as "proof" of a costume, does also occur on real biological anatomy.

Fig. #9-07 shows where the figure model's body makeup rubbed off because of her thigh folding under the buttocks mass.

The simple and inescapable fact is there is nothing about Patty that can be argued as evidence for a costume, and can exclude real anatomy as an option. To the contrary, her anatomy is remarkably realistic, and in details which no known costumes have incorporated.

Fig. 9-07

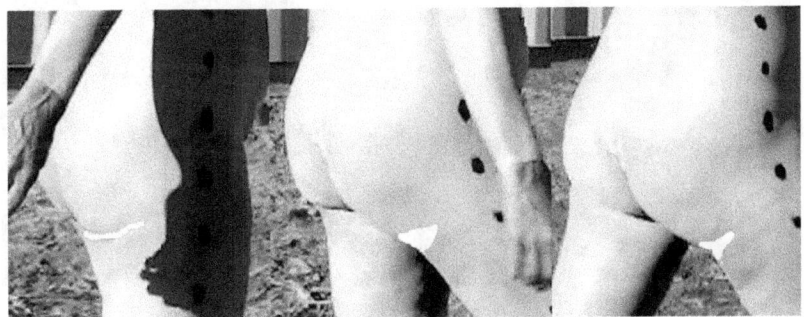

Film Image
Evidence

Fig. 10-01

Chapter Ten - The Swamp

Wading into the cesspool that is the "backstory"

No discussion of the PGF and claims of it being a hoax is complete without at least acknowledging how the skeptical community worships what they call "The Backstory", their intense examination of the life of Roger Patterson before and after the filming happened, and to some extent, examining the life of Bob Gimlin as well. They even delve into the stories of the people peripheral to the story, friends of Roger's, his brother-in-law, Al DeAtley, the other people who went to the site, and other people Roger had contact with doing his book on the Bigfoot mystery.

What's the problem with this material? Better to ask, what's NOT wrong with this material? Almost without exception, it's recollections, hearsay, anecdotes, and gossip. In the world of evidence, this is scraping the bottom of the barrel.

In the book, Abominable Science" by Donald Prothero and Daniel Loxton, one of the few things they got right is in Chapter One, page 14. Author Prothero quotes psychologist Elizabeth Loftus, with this passage: *"Memory is imperfect. This is because we often do not see things accurately in the first place. But even if we take a reasonably accurate picture of some experience, it does not necessarily stay perfectly intact in memory. Another force is at work. The memory traces can actually undergo distortion. With the passage of time, with proper motivation, and with the introduction of special kinds of interfering facts, the memory traces seem sometimes to change or be transformed. These distortions can be quite frightening, for they can cause us to have memories of things that never happened. Even the most intelligent among us is memory-thus malleable."*

Prothero on page 13 quotes science historian Frank Sulloway as saying *"Anecdotes do not make a science. Ten anecdotes are no better than one, and a hundred anecdotes are no better than ten"*. Author Prothero also states in his text, page 14, *"Anecdotal evidence is necessarily regarded as highly suspect in most scientific studies"*. And on Page 16, he also notes, *"What is the proper scientific approach to eyewitness testimony? As we have seen, most scientists give it very little weight unless there is strong physical evidence to support it."*

The irony of these remarks is not that they are untrue, because they are well supported by other scientific sources, but because later in the same book, co-author Loxton uses mainly eyewitness testimony and recollection to try and make his case for the PGF being a hoax. So one skeptical author is using the crappy evidence his writing partner warned us to not rely upon. In my opinion, this is truly exemplary of the hypocrisy of modern "skepticism" as a goal, instead of classical skepticism as a process.

But skeptics still love the backstory material, and often worship it as "God's honest truth", verbatim perfect and splendidly factual. Do we need more convincing analysis as to the unreliability of eyewitness testimony? Google "unreliable eyewitness testimony" on the internet and you'll get more than enough sources for substantiating how unreliable such testimony is, and why it has no place in a scientific analysis.

Example, the magazine "Scientific American" has an article titled "Why Science tells Us not to Rely on Eyewitness Accounts" and the sample lines from that link read *"Eyewitness testimony is fickle and, all too often, shockingly inaccurate."*

Or from www.law.yale.edu/news227.htm, we get *"Eyewitness Testimony doesn't make it true"*

Or scienceblogs.com/cortex/2007/05 we find the remark, *"From the perspective of neuroscience, eyewitness testimony is an extremely unreliable type of evidence."*

I could go on, beating this dead horse that is eyewitness testimony, and the derivatives of same, hearsay, recollections and gossip, but I personally am satisfied that it can be factually stated that such testimony is not reliable for a factual determination about the PGF.

And yet, people who claim to be "critical thinkers" and swarm to the JREF cesspool love this evidence. They wallow in it, and revere it as sacrosanct. Sadly, there is simply no way to avoid this topic in any PGF discussion, because the skeptical "hoax believers" champion it as ironclad "proof" of their imaginary hoax. It's crap, it stinks (as crap invariably does), and they are like maggots who feast on it as if it were prime rib or key lime pie.

So as much as I hate going here, I will face this backstory "evidence" and at least give you a reasonable alternate appraisal of it.

The Timeline Controversy - This is the question of how Roger's camera original film got processed, after filming on Friday around noon, to be viewed as developed film on the following Sunday morning. Well, the film is generally accepted to be Kodachrome type film, and few labs could process it. There was one in Seattle that did process Kodachrome, but their lab manager, Frank Ishihara, is on record explaining they didn't process the Kodachrome on Saturdays. The man who actually arranged for the processing to be done, Roger's brother-in-law, Al DeAtley, did not recall where and how he got it processed, when asked years later by researchers trying to make sense of the timeline controversy. So basically, this "controversy" has no solution anyone can prove. We don't know how the film got processed, and the alternatives seem to be either A. That it was processed on an earlier date by the usual processing schedule, or B. it was processed by some "under the counter" or other "bending the rules" consideration. Now in the real world, both of those are equally viable options, one implying a hoax, the other simply implying a rather strong determination to get it processed fast. But a legitimate Bigfoot filming could likely inspire such an urgency or strong determination. Lesser likely options that have been explored are processing the film by Dynachrome technology, or some type of non-licensed lab that may have done processing films a Kodak lab would not consider "proper" to process. But either "A" or "B" above is the likely explanation.

So where's the actual status of the timeline controversy? We don't know how the film was processed. Did anybody at the time think this would be something crucial to document? Apparently not. Where does that leave us? It leaves us with a question, but not an answer. There is simply no way to extract an answer, especially in any form of rigorous proof to support either proof of hoax or proof of authenticity. People who love suspicions gravitate to this and then take the suspicion to a "red flag" and take the "red flag to up to "proof", but it proves nothing other than we don't know what happened to get the film processed. Good science knows when the data cannot prove something, as much as when it can prove something, and good science would look at this controversy and simply conclude "there is insufficient data to form or support a conclusion." People who actually understand good science would concur.

Claims Roger Patterson was Dishonest - Author Greg Long wrote an entire book "Making of Bigfoot" interviewing people who knew Roger Patterson and it's 466 pages long. In it, it has five pages of "empirical evidence" in the form of photos of documents, such as a contract with a lady named Vilma Radford, who loaned Roger money for the documentary movie he wanted to produce. And it has 461 pages of unreliable crap, eyewitness testimony which I noted above is not accurate enough to be used for evidence in scientific analysis. And what do those five pages of empirical evidence prove? A. That Roger Patterson had an arrest warrant filed because he kept his rented K-100 camera far longer than the rental period, and B. That Roger borrowed money from Ms. Radford to make a movie, and didn't repay it, according to the provisions of the agreement.

So the empirical evidence can reasonably prove that Roger kept a rental too long and that he borrowed money he didn't pay back as agreed. Is that proof Roger pulled off the hoax of the century? Well, sadly you can't even use that to make a case for Roger having the intent to perpetrate a major hoax, and nothing about it gives even an ounce of support to Roger having any capacity to perpetrate a hoax on the level of the PGF hoax claim. And if you don't understand the distinction of needing to prove both "intent" and "capacity" to offer a rigorous proof of a hoax, then don't quit your day job.

The rest of Long's book is the stuff that nearly every authority I searched has judged as being unreliable. And yet skeptics rely upon this book as if it were immaculate truth. It is their skeptical "New Testament". They quote from it like scripture, and revere Long as if he were a prophet.

Roger didn't live a regular life. He didn't hold a regular job. He probably made some bad decisions in his life, and he was probably being truthful when he said, shortly before he died, that "I am probably the worst person this film could have happened to". The flaws of his life have hurt the greater public impression of this film, and he apparently owned up to that unfortunate situation. That does not strike me as the dying regret of a hoaxer.

Al DeAtley - Roger's sister, Iva, married Al DeAtley, and Al became involved once Roger had his remarkable film footage. Al put together a feature documentary and he and Roger took it around the country, showing it in theaters and then letting Roger personally talk to the audience. In that film, Al himself is on screen for most of the first show reel, which runs about half an hour. Mike Rugg of the Bigfoot Discovery Museum in Felton, CA. showed this first reel when it surfaced a few years ago, at a Discovery Day event in 2012. I was a speaker at that event, so I saw the full show reel projected that day. As I understand, it was subsequently given to Patricia Patterson after that showing. So we know Al was involved with Roger's theatrical program and we know Al was the man Roger sent the film to, to be processed in October 1967. A lot of investigation has focused on Al's life, business and personal history, but I personally don't see any reason to go there, because there's no empirical evidence Al was involved in the preparation of the Bluff Creek trip, and he was not a member of the Bluff Creek expedition in October 1967. People who wallow in the Backstory however do indulge in a lot of gossip about Al's life and business.

In my opinion, Al apparently did see the commercial potential of helping Roger do his theatrical showings of his film and so it was just a pragmatic business decision to become partners with Roger in that endeavor. I don't see anything about Al which gives any merit to claims of a hoax, but skeptics see red flags a waving everywhere. I'll leave them to try and make a case for their fanciful theories. But I haven't seen any empirical evidence which a scientist or academic scholar could use to make a proof, or even support a judgment of "probable cause" to suspect a hoax.

The Bob Heironimous Claims - Bob Heironimous was revealed in Long's "Making of Bigfoot" as the man who says he wore the ape suit and did the walk Roger filmed. He claims he is "Patty". Well, if the empirical analysis of Patty shows she simply is not a suit or costume, than anything Bob H. claims becomes a moot point. But because his claims were so popular with skeptics, I took some time to address his claim from a purely scientific method. Bob H. described in his interviews with Long the characteristics of the suit he claims to have worn. So I simply tested his claim, with a fur suit designed as he described it, and with some inner materials as he described them. So we tested a two piece suit split at the waist, with a drawstring waistband, hip waders inside the leggings, football shoulder pads, and we tried a leather old football helmet inside the mask, but it wouldn't fit the head size if the head is scaled to a man 6' 2" tall, as Bob was reported to be.

So the leather football helmet inside the mask was a total fail. The football shoulder pads and no padding filling in the entire back produced a furcloth shift of folds on a diagonal upward to the forward arm in the walking arm swings, and that diagonal fold shifted to whichever arm was swung forward, back and forth. We see nothing remotely like that in the PGF and most of our view is the back of the torso. So that was a fail as well. And the hip waders didn't produce any effect like skeptics love to claim they see in PGF image analysis.

The skeptical response to my experiments was to cry "foul", to say I was not doing the experiment right. I couldn't use Bob's description in Long's book because it might have been figurative, and not literal. It was ironic that the same people who revere whatever Long quoted somebody as saying, it was immaculate truth, but when I took Bob H. at his word and tested what he described, I was wrong for relying on those words literally. And that is the skeptical disconnect from reality, summed up. If they quote from Long's book, it is "god's honest truth", but if I rely on a quote or description from the book, I'm recklessly being too strict in a word-for-word reliance.

Anyways, a lot of the backstory does revolve around the claims of Bob Heironimous. I will simply say that he is not Patty. He may have once worn a suit and indeed, Roger may have once filmed him in such an outfit (and I say this because it's not at all unreasonable for a man making a documentary about the Bigfoot mystery to put someone in a costume to reenact stories of encounters as part of that documentary), but the PGF is not a person in a costume, and so that alone excludes not only Bob

Heironimous, but any other human as well. Bob's own description of a costume he may have worn is further substantiation that he is not "Patty", because a fur suit as he describes will not produce what we see in the PGF.

When empirical evidence is put up against recollection or testimony, and they are in opposition, science must side with empirical evidence. Bob's recollection loses to the empirical evidence.

The "Backstory" has a lot of other lesser topics. There's something about a bent sturrup, a person standing in for Bob Gimlin at the theater screenings, Roger's beard stubble or lack of same in some suspected second reel footage, the Phillip Morris claims (that he sold Roger a costume and it's "Patty"), something about a dog biting the testicles of a suspected Bigfoot creature (or person in costume pretending to be a Bigfoot), Ray "Crash" Corregin's tourist ranch, Correginville, Roger's visits to a Hollywood lawyer, and a bunch of other topics or incidents I can't even keep track of.

So why do skeptics, who champion themselves as "critical thinkers", so love this backstory material if it is so frequently and compellingly classified by all mainstream science as unreliable? Why do they rely on it with feverish devotion? In my opinion, it is desperation, their "Hail Mary" attempt to salvage their belief of a hoax that cannot be supported by any real disciplined science. I sometimes think all this endless backstory discussion is just the argumentative tactic similar to a political filibuster, where the intention is to just keep talking to prevent the issue from coming to a vote and conclusion.

The options to apply good science are tremendous with the PGF mystery, because there is such a wealth of fine empirical evidence. And it is a methodology of good science that will ultimately validate the truth of the PGF mystery. "Bigfoot Exposed" author David Daegling at least did try to work with some scientific discipline, and as previously stated, his conclusion (in 2004) was *"since at this writing thirty-six years have passed, and definitive proof of a hoax has not surfaced."* He is to be commended for not giving the backstory material recognition as scientifically valid, because it is not and never will be, unless some miraculous empirical corroborating evidence suddenly materializes.

On its own merit, the backstory "evidence" (and I use the term very loosely here) panders to the simple-minded and the closed-minded, who are afraid to simply consider that the film may be authentic, and so they embrace a belief that film must be a hoax. In doing so, they likewise embrace any evidence, however poor or unscientific, as long as it reinforces their belief.

Fear

One issue never before explored in the PGF mystery is the intense reluctance of skeptics to give the film and Patty serious consideration of being real. Good science should simply say, what are the options (real vs hoax) and let us consider both options fairly and rationally? But skeptics seem fearful to give this film or Patty the serious consideration that good science requires. We see this as well with many veteran film industry makeup special effects experts, who dismiss the film as a "cheap fake" out of hand, but will not support their opinion with any substantial analysis and certainly not in any form of debate or cross-examination.

This question presents an interesting philosophical issue usually ignored: Who is perceived as being the smarter, the more professionally responsible? Is it the expert who dismisses the PGF and Patty immediately out of hand, with just a casual disparaging remark, or the expert who fairly considers both real and fake options and does not offer any conclusion until a well-reasoned analysis, based on substantial evidence, is formulated?

Certainly in the makeup business, it is a common perception that a true expert of the craft knows exactly how every trick and process is done, and such an expert can spot a fake every time, so by this perception, a true expert should see the PGF and Patty as a fake right away. This PR bravado (paraphrasing the general idea) would claim something like the following, "I'm so good, I can spot a fake every time, and the PGF didn't fool me for a second. It's a cheap fake and I'd fire anybody on my crew who did such a crappy job." Granted, it sounds compelling, but is it fact or career-enhancing promotion? Can experts admit that they are undecided about the authenticity of the PGF and Patty, and still maintain a sense of expert authority, when others in their field dismiss the film and Patty as obvious fakes with confident disdain? Does fear of ridicule cause many experts (skeptic, scientific or makeup effects experts) to take the easy way out and just call the PGF a fake, whether it is or not, to play it safe and look smart, instead of being scientifically responsible and give reasonable consideration to the option that the film and its subject may be real?

When one option (fake) is so easily embraced and makes the expert look smart, and the other option (real) seems so remote and makes experts look undecided or less knowledgeable, does professional fear compel the expert to take the low road and look smart, instead of the high road to proper scientific consideration? How does fear influence the opinion of experts who won't give the film and its subject figure, Patty, a fair and thorough scientific consideration?

The fear factor must be considered and addressed, as a motivation for the intense skeptical denouncement of the PGF as being a hoax, because there is a fundamental contradiction when a subject is so often dismissed casually as a "cheap fake" and yet 47 years of diligent attempts by many knowledgeable people fails utterly and completely to prove the claim of a "cheap fake". How can the claim be so obvious, and the proof of the claim so illusive? The answer may very well be fear, fear of opposing a powerful but flawed belief system that can damage an expert's professional standing. Are these people following the idea of "seek the truth" or "cover your ass"?

The pretentious opinion of absolute confidence the PGF is a hoax is rife in the skeptical community, and any "maybe we should consider the opinion it could be real" opinion will be denounced as stupid or naive, if one foolishly offers that thought in a hard-core skeptical environment where their group-think rules. But this is nothing new, as we sadly see in any polarized niche of our society. Whether it's political, social, cultural, or intellectual, once the group-think polarizes around a point of view, and becomes dogma, the mere suggestion of an opposing idea as having some merit to consider and evaluate, that suggestion is viciously opposed, denounced, ridiculed, and dismissed. Fear of that ridicule can be a powerful form of peer pressure.

That skeptical peer pressure has turned all classical concepts of good evidence upside down, dismissing the empirical evidence and elevating the backstory pile of unreliable evidence to sit upon a lofty pillar of false merit, because the hoax claim cannot be validated by any scientific analysis of the true empirical evidence. The hoax claim can only be falsely given some appearance of merit by using the backstory crap and trying to make this bad evidence look good.

A time-honored folk saying advises us that, "You cannot make a silk purse out of a sow's ear."

Paraphrasing that, you cannot make a compelling scientific proof the PGF is a hoax out of the backstory evidence.

But as long as there are people who think the Apollo moon landings were faked, or who think the World Trade Towers were brought down by a controlled demolition, or think that the "Theory of Evolution" is just a theory, there will be people who think the PGF is a hoax. They will believe, and they will use the backstory as the core evidence for their belief. They will ignore how unreliable the evidence is. They will hate anyone who blasphemes their belief. Nothing I say will affect their belief. I have no such expectation.

Is there any merit to the backstory material? Actually, yes there is. It will never prove the PGF is a hoax, but it is material worthy of consideration if we seek to explore the larger context in which the PGF came to be and was subsequently publicized. The PGF has become a cultural phenomenon, and so the context in which the film came to be is itself a fascinating story. But any evaluation of that context still needs to keep the quality of evidence rules in place, and empirical evidence must prevail over testimony and anecdotal evidence. Eyewitness evidence must be compared to the empirical data, and only when the empirical evidence corroborates the testimony, can the testimony be held as truthful.

We can consider what people said, what they recall, in trying to appreciate the context of the PGF story, but we cannot use those recollections as scientific proof in themselves. We must keep those recollections as a material under consideration, awaiting corroboration.

The backstory, as Roger Patterson so wisely, but sadly appraised it, is the worst thing that could happen to this film, because it is a fertile growth medium for irrational belief, and for 47 years, the irrational belief has had what it needs to grow to an urban legend. That legend proclaims the PGF a hoax, and so a truthful solution to the PGF mystery has a step uphill battle to achieve vindication and dispel the persuasive fantasy of the urban legend.

I wish that uphill path weren't so steep, but it is. So be it.

When Roger Met Patty

Fig. 11-01

Not a Hoax!

Chapter Eleven - Conclusions

Despite the popular skeptical talking point that claims I'm a "believer" and that I don't weigh the real evidence or base my conclusions on that, I take it as a point of pride that I do seek out a rational conclusion entirely based on where the evidence leads, what the data supports. And my conclusion about the PGF was never arrived at lightly, or predetermined by any belief. If anything, it would be quite a respectable accomplishment to prove it is a fake, if in fact it were, because in 47 years, nobody has published any kind of definitive proof of a hoax. So there is ample opportunity to do what nobody before accomplished in 47 years, and if that were truly where the evidence led, I would be proud to be the first to take that conclusion to an absolute determination. I would be proud to expose a hoax and reveal the design, the method, and the analysis process that finally tripped up the hoaxer.

One need only look at another recent famous "Bigfoot" subject known as "Matilda" to see that I was uncompromising in my analysis that the subject figure was a fake, done with a Chewbacca mask with the hair reworked to match a furcloth costume. I don't like fakes any more than any diehard skeptic, and I have no hesitancy in offering such a conclusion if the data supports that determination.

And for seven years now, I have tried to find evidence of a hoax in the PGF film and larger saga. I have been very aggressive in applying new analysis technology with the confidence that any proof of a hoax would certainly be in the analysis processes that no hoaxer in 1967 could have anticipated to exist 47 years into their future.

The inescapable conclusion, however, is that the PGF is simply not a fake, not a hoax, and Patty is something biologically real as she appears, and is not a human in a fur costume and face mask. I have given the concept of a fake every possible consideration I know of to find support for a claim of hoax, or irrefutable evidence of one. And in this chapter, I'll summarize those points of consideration, item by item, like building a home, brick by brick.

1. We know for certain that Roger Patterson, by 1967, was a reasonably experienced cameraman, and he had used several makes of 16mm camera (a K-100 and other models) and he had used fixed prime lenses and zoom lenses. He had used both Kodachrome camera film, and Ektachrome camera film, and had dealings with film labs for both processing and copying films. He shot enough film outdoors in remote locations to know the challenges of filming away from any home base. Roger's own documentary reel of footage verifies all these points. He knew what he was doing with his camera, and any plan to hoax a film would have included consideration for how a location filming can be successfully done.

2. We know for certain that the Bluff Creek location is 100% verified, with many researchers subsequently visiting the site and filming there as well. So we know that Roger had to travel over 400 miles from home, and a hoaxer would, by all reasonable consideration, choose a location closer to home if a closer location would work. There is nothing about the Bluff Creek location that coincides with a deliberate goal of hoaxing. If anything a location like the Mt. St. Helen area, where Roger was exploring earlier that summer, would have been a fine choice closer to home, a place with some documented history of activity, and Roger could have scouted the area for a good filming site that had the right logistics for a successful shoot. In all these respects, any theory that Roger went to Bluff Creek with intent to film a hoax defies all film-making logic, and Roger's own experience should have dissuaded him if hoax was his intent. So the location is illogical for a hoax, but as noted earlier, was reasonable for Roger to sincerely follow up on the trackway found near there in August, two months before.

3. The specific choice of a site with a creek separating him from his subject at first is a major logistical error for a hoaxer. Based on the film image data of his start position and Patty's start position, the creek and the distance would make any communication between Roger and a costumed actor, except big arm gestures, impossible, and because Patty has her back to him throughout all the early activity, a costumed performer couldn't even see arm signals. Having no communication with your performer is genuinely irrational for a hoaxer. And Roger's need to cross the creek in the middle of his filming puts a very real risk of ruining the whole filming if he slipped in the creek and dropped his camera in the water. Instead, a hoaxer would start on the same side of any creek, so he doesn't have to cross flowing water in the middle of his filming. Yet the creek on film is so subtle that there was no intent of showing it as a staged obstacle. Roger's path was only fully understood by image analysis technology that didn't exist when this film was made, so there was no intentional choice of crossing the creek to make his task look more complicated. A hoaxer would not stage his action so his performer has lost all communication with him, and a hoaxer would not choose a setting with obstacles to get past which could potentially ruin the whole filming, and a hoaxer would choose his exact location with care, anticipating what he plans to do. The creekbed setup simply defies any reasonable explanation for a man who is acting with deliberation and planning to accomplish a film with deliberate intent.

4. Roger was experienced with not only the K-100 camera, but with another camera that did not have any ID markings (a few 16mm camera makers of the era elected to not bother with the Camera ID system). And the K-100 camera's unique aperture made it very uncondusive to editing and discrete printing to hide the edits. Modern analysis, unknown then in 1967, shows the specific use of a K-100 camera is one of the compelling pieces of evidence that proves no footage was spliced. If Roger was hoaxing the film, he could have tried some discrete edits, thinking he could hide them, never realizing the remarkable

analysis technology we have today that could have tripped him up. He didn't make any such attempt to edit the film.

5. Roger could have tried a cinema verte technique called "camera cutting", where you meticulously plan each shot and edit the action by carefully planning each segment and then only starting the camera when you are ready to film exactly what action you want. Then you have separate segments combined in camera, and thus, no apparent evidence of editing even on the camera original. But such "camera cutting" requires that each segment be carefully planned and that takes time, and there is no way Roger could have known that 47 years later, we are analyzing tree shadows to show subtle passages of time at that Bluff Creek location (courtesy of John Green, Rene Dahinden, and two cameramen not currently identified today). If Roger were doing each segment with consideration of camera cutting, the filming would take more than about one to two minutes, and we'd find that subtle evidence of time passing with today's analysis. Such evidence of a longer time passage would be irrefutable evidence of a hoax. But this analysis verifies the filming was done as one continuous action, even with the camera starts and stops.

One continuous frenzied attempt is a terrible way to hoax a film, with so many uncontrolled variables, that success would be a long shot. And if that were the case, we'd have to assume that the 76 feet of horse and rider footage was deliberate as well, for some intent, and not only Roger but any realistic filmmaker would plan for a "Take Two" attempt. Driving over 400 miles for a one-take fling at a hoaxed encounter defies all filmmaking logic. But to do a second take, Roger would need to redo the horse and rider stuff on a new 100' film load, for the Take Two reel, and that would waste the remainder of the day, before he could do another encounter take. If Roger wanted to do multiple tries to increase his chances for one successful reel, he would have chosen a magazine type camera, like a Revere, a Keystone, a Kodak Cine Royal, etc, because he could shoot a horse and rider shot, pull out the magazine, put in magazine number two, shoot the horse and rider shot again, and even do a Take Three magazine of the same (to be really careful), then go to the next horse and rider setup, shoot it with magazine one, then magazine two, then magazine three, and so on, until he had only 24 feet left on each magazine. Then if Take One of the staged fake encounter didn't go perfect, he just pulls out magazine one and loads magazine two, and shoots the encounter again. And for a just in case, he could load magazine three and shoot it a third time very conveniently. Each one would have horse and rider footage on the beginning.

With a K-100 camera, however, removing the film from the camera either leaves a light washout (if it is unloaded in low level light as the daylight reel in designed for), or if he unloaded it in a black camera bag, putting it back in for the next segment later would assuredly result in a misalignment of where the last shot ended, because he'd have to get the exact frame position in the film gate, but without seeing what his hands were doing. So, in practical terms, either the shots would overlap in a double exposure (because he set the film already exposed back in the film gate), or there'd be some black unexposed frames between the segments (because he set the end of the exposed film too far down away from the gate). There's no way to mark the film in relation to the gate that he could feel with his hands when he can't see what his hand are doing. There is neither a black gap or a double exposure on Roger's footage. So any idea of using a K-100 camera and doing multiple tries at the hoax, with the deliberate horse and rider footage on the first 76 feet of the roll, is simply a combination of challenges no hoaxer would bother with. Only using a magazine camera would make sense to try this.

Then, if he used a magazine camera and several magazines each with one try at the hoax encounter, once all were processed, he could choose which one of the several tries was the best, most believable, and shown that to researchers and the media as his claimed encounter. If Roger had done so, with a magazine-type camera, we could not have known he did more than one take of his action. And he would have had three versions (or even more) to choose from. If he really wanted to cover all bases, he could send each magazine to a different film lab, so no lab technician could testify having seen several attempts at the encounter. Choosing a K-100 camera, Roger made a choice that essentially locks out that option of multiple takes including horse and rider footage on the front of the roll, a terrible choice for a hoaxer to make.

Using a magazine type camera would also have been a wiser choice for camera cutting, because if any segment went wrong, he'd have several more versions to try and get a segment correct (with three magazines rotated through each segment). But the aperture shape of a magazine camera is so distinctly different from a K-100, there is no way a magazine camera was used and somehow falsified to look like a K-100 camera.

A Hoaxer would not have chosen to film with a K-100 camera under these circumstances. And Roger did use such a camera, to a 100% certainty.

6. A hoaxer would not have bothered with a two frame trigger slip, as we see in Segment Three of the PGF, unless he was well aware of putting it there and thus would have drawn some attention to it when people questioned the authenticity of his film, but Roger never brought that to anyone's attention in the four years he was trying to convince people the film was real. There was no evidence he himself was even aware of it. It only came to light 40 years after the filming, through my analysis. But such a trigger accident, while appropriate for a real filming in an uncertain and spontaneous situation, defies all reason for a staged filming done with deliberation.

7. Roger's filming of Patty in the lookback segment (Segment Five) shows her body from more viewing angles than any other part of the PGF, and during that time, Roger is standing still and holding his camera with remarkable steadiness. All claims that Roger deliberately shook the camera to hide flaws of the

costume fail as logic when we consider the true evidence of the film image data. Roger was trying to get the very clearest images of his subject as he possibly could, and obviously had no fear of that clarity revealing a bad costume with flaws.

8. The behavioral analysis of Roger's movements as compared to Patty's in the location, could never have been analyzed in Roger's time as we can do today, so he would never have considered how he must move and how he must get his performer to move, so the behavioral analysis would look realistic for a spontaneous encounter of a man and something which wants to elude that man. The relaxed posture when Patty thinks the man is far away and on the other side of a creek, the more aggressive power walk when this man is close and in the open and there's no barrier between them, and the more relaxed posture when she's gained a good distance from him and he's no longer chasing her but is standing near the creek debris, these are behavioral clues to a real concern for safety in the situation where a strange man is invading your world. We don't see any dramatized hostility, any wild performance an actor might be inspired to do when the director calls "action", any confrontational behavior. Just simple evasion of a stranger, nothing more. Honest.

9. Patty has a head that makeup effects designers have wanted to accomplish for over 80 years, and nobody has. Was Roger really such a genius at developing a mask that is way beyond the skills of the finest people in Hollywood, with the best crews and the biggest budgets, that still today, nobody can accomplish such a feat? Okay, let's say Roger didn't do it but he hired somebody who did. Why didn't this unknown genius parley that design technique into a very successful Hollywood career? Because it's not a mask with a clever design. It's a real head like some ancestral human relatives.

10. How did Roger the genius make a costume neck better than anybody of the time, and then have the audacity to film most of the subject showing the very neck area we mortal makeup

effects people try to avoid showing with our ape costumes? Somehow, if it was done with fur, he found a way to defy the laws of material physics, and did something we would not see for another 20 years when spandex furcloths were finally introduced to the industry. Simple, actually. Just film a real biological neck, and it'll work just perfectly, in 1967.

11. Patty's breasts have a fluidity that cannot be replicated by any established material for costume prosthetics in the makeup profession in 1967. Did Roger invent a new fluid prosthesis technology which even ten years later, in 1977, true makeup master and genius innovator Dick Smith was unable to accomplish for Katherine Ross' breasts in "The Stepford Wives". If so, why did Roger stage the filming so Patty's back was to camera for all but one mere second of the film, and we might not have even seen that breast motion if it were not for analysis technology today that didn't exist in 1967. So we are led to believe Roger the genius designed fluid-filled prosthesis and then staged his performer so we could only see that fluid effect with analysis technology that didn't exist in Roger's time? There is no logic for a hoaxer to do such, and if a man were such a genius, he'd have a very profitable career waiting for him in Hollywood, instead of trying to make money showing his PGF footage in small theaters and auditoriums he and his brother-in-law rented themselves.

12. Patty's back has multiple lines and folds which tend to perfectly match real human anatomy, and run contrary to all conventional fur costume design, because these lines can't occur on a costume by accident but to do them deliberately take time, skill, and a justification to make the effort. Yet Roger never tried to show or otherwise draw attention to these features to try and sell people on the subject looking real. So the hoax concept fails here because you have anatomical features that could only be accomplished by deliberate fabrication design and skill, and yet when no one saw these features in Roger's lifetime, he never pointed them out to make a claim for an authentic body. The "brilliant hoaxer" sure isn't

trying to convince people his creation is real. But the man who did actually encounter something real was not sufficiently knowledgeable in anatomy to see these things and appreciate how they support biological authenticity. Nobody would see these authenticating traits for over 40 years after the film was taken. Evidence of a hoax? No way. Evidence for authenticity? Absolutely.

13. The funny lines around the pelvis area, and where the leg appears to tuck into the fatty deposits of the buttocks, all are found on real humans. They just aren't found on centerfold models. They are found on more mature humans who are a bit overweight and have accumulated some adipose tissue (fat). In nature, fat isn't an undesirable thing. To the contrary, it's literally a reserve food supply for when something is looking at a cold winter with scarce food supplies, and by remarkable coincidence, the PGF was taken in October, as winter was approaching. Remarkable hoax or natural occurrence?

14. The Skin Stretch - The way Patty's skin (and fur) stretches from her armpit to her knee, as she does the lookback, simply defies all expectation for the fur materials we had in 1967, but it has a remarkable comparative similarity to the studies of real anatomy when markers were applied to the figure models' bodies, and then frame comparisons were made as the forward leg extends to take the next step. Real dermal tissue has a uniformity about it that a fur costume lacks, because the tailoring seams and closure seams are more rigid than whatever flexibility the fur material has at its most loose condition, and so costumes will shift and buckle where the more pliable parts shift up against the more rigid parts, and we simply don't see any such thing on Patty.

The beauty of this determination is how easily both costumes and real anatomy can be tested and we can study both alternatives in a repeatable way. Any person doubting this determination can easily do the experiments themselves and see if they get different results.

15. The Armpit Fold - Like the skin stretch, the armpit fold is a remarkably subtle feature of the body, and was overlooked for over 40 years. Only a person who really has hands-on creature costume making skills can appreciate how odd it would be if made in a suit or costume, and how remarkably correct it is for real anatomy. The combination of how much thought and effort would need to go into the effect, for a costume, and the way Patty is filmed, which makes seeing this so nearly impossible except with sophisticated image analysis that didn't exist in 1967, adds up to a contradiction no explanation of a costume can resolve. But in the context of real anatomy, it is enchanting that we can appreciate some of the subtle aspects of the real body with Patty has.

16. Roger's lack of skill making costumes - Arguing the PGF is a obvious fake with a cheap costume, simply will never rise above wishful thinking by a hoax believer. Any credible claim of a hoax will sooner or later have to face the reality that if Patty is a fur costume and mask, the designer and fabricator (s) is/are brilliant, and Patty is something unparalleled in the history of makeup and creature effects. So to argue such requires one to argue for some genius, and there's not an ounce of proof Roger was even modestly proficient, much less gifted or a genius, at this craft. He didn't know anyone who was, and he didn't have the financial means to hire some such genius. So the concept of a hoax fails in consideration of who had the capacity to make such and how did Roger persuade some genius to do this.

I realize this consideration will be passionately rebutted with sound-byte quotes of famous makeup artists who casually dismiss the PGF and Patty as a "cheap fake", but sooner or later, somebody else in the makeup profession needs to step up and show a rigorous proof of that claim, or the claims will be verified for what they really are, professional public relations bravado, not factual analysis.

The concept that the PGF and its subject figure, Patty, could be real, should not be taken lightly. Every possible consideration should be first explored to try and find evidence of a fur costume and a filming hoax. And in the last seven years, this curious film and Patty have been given every chance to be exposed as a hoax, and my simple and inescapable conclusion is that there is no evidence of a hoax. All evidence points to a real encounter with something biologically real as she appears. It's not 64-40%. Not even 73-30 or 80-20, or 90-10. It's 100% real and 0% hoax.

So now the people who want to keep their hoax belief alive will run to the backstory, and once again roll out the crappy evidence, the unreliable eyewitness recollections and gossip which every scientific appraisal deems unreliable and useless for scientific determinations. I see no reason to rehash their backstory claims of hoax, because I am confident that I have established the overwhelming scientific consensus that such material is worthless for proving anything in a rigorous, scientific and factual way.

There is a logical fallacy that a remarkable thing can only honestly be witnessed by a person of integrity and responsibility. Witnessing a remarkable thing in fact has no preference or prejudice to only occur with such a person as the witness. The integrity or responsibility of a witness to an extraordinary event will surely be a factor if all we have is the person's testimony, his/her claim to the incident. But with the PGF, we have a remarkable body of fine empirical evidence that is absolutely independent of Roger's personal character or any flaws attributed to him. The film evidence has true integrity, and the film evidence is what I rely upon to make my determination. So regardless of what Roger may be accused of, in terms of flawed character, the film is honest, and the evidence is entirely independent of any discussion of Roger's character.

In this text, I have never once quoted Bob Gimlin, the surviving man who witnessed this event, and I want to make it clear that I have every confidence that Bob is truthful in describing what happened that day, what he saw. But I cannot dismiss some eyewitness testimony of others such as in Long's book, calling it unreliable and useless for proving anything, and turn around and use Bob Gimlin's testimony to support the reality of the PGF and Patty. If I dismiss eyewitness testimony as not good enough to base a proof of fact upon, I must exclude all eyewitness testimony from my proof.

Fig. #11-02 was taken in April, 2014, when I had the pleasure of doing a four hour Oral History interview with Bob Gimlin.

I know Bob Gimlin is telling the truth, as he recalls the PGF story, because the film evidence is 100% supportive to the reality of that encounter he experienced. I do not use his testimony to prove the PGF is real. I use the PGF empirical evidence to prove the film is real, and that reality in turn corroborates Bob Gimlin's testimony as truthful in describing the experience he had on October 20, 1967 at Bluff Creek, California. And I find it truly disturbing that fanatics who are obsessed with trying to prove the PGF is a hoax have defamed, insulted and ridiculed Bob Gimlin in their desperate attempts to make their hoax delusion a credible reality. Bob Gimlin is an honest man, who witnessed something very strange, but very real. And I am pleased that I could accomplish this conclusive proof his story is real, because he deserves that vindication. And in the course of this work, over the last seven years, I've become well acquainted with Bob, and I'm honored to call him a friend.

A recently emerging skeptical talking point has grown into persuasive dogma, where the concept of Bigfoot is compared to Santa Claus or the Tooth Fairy in terms of childish fallacy, and a strident claim is often made that Bigfoot as a biological entity is virtually impossible to exist, because it defies some (unspecified) laws of physics or biology. Simple consideration of the larger mystery that we call "Bigfoot" as having any merit to weigh pros and cons, this consideration is now on the verge of blasphemy in the radical internet circles of the skeptical world.

Fig. 11-02

Oral History interview with Bob Gimlin, April, 2014 at Sequim, WA. with the author, Bill Munns

But if we simply turn back the clock to pre-internet times, when publishing any text went through a collective evaluation of science writers, editors, and publishers, we see the subject given quite thorough and well-reasoned consideration, with no allusions to Santa Claus or the Tooth Fairy, no blanket dismissal of the concept as irrational beyond any hope of salvation. Instead, these writers, editors and publishers demonstrate the hallmarks of good science, as evidenced by a reasonable consideration of the physical or empirical evidence, and an open mind to all options of explanation, until some options can be dismissed with proven cause.

The Time-Life publishing empire has produced many remarkable books and libraries over the years, and I literally grew up with my nose buried in such books as "The Epic of Man" and "The Wonders of Life on Earth", as well as the full set of the Time-Life Science Library and the Nature Library. So their credentials for scientific and educational publishing are above reproach. They also did a series called "Mysteries of the Unknown" and one edition in that set was "Mysterious Monsters". From page 98 to 126 they have detailed the investigations into the Bigfoot/Sasquatch phenomenon, the Yeti, the Almas, the Chinese Wildman, etc. They begin the chapter with Roger Patterson and Bob Gimlin's encounter, and recognize how powerfully this film footage stands in the investigation of all types of "wildmen". Their analysis is cautious, balanced, and meticulous, and they don't find anything "impossible" about the creature. Nor do they equate it to Santa Claus or the Tooth Fairy. Just the opposite, they give the idea reasonable consideration and weigh the evidence for and against, with experts expressing a variety of points of view.

You will not find any of today's internet skeptical talking points in it, or any of the dogma that hoax believers keep proclaiming as obvious truth. Are these denizens of the internet really so brilliant that they see truths that the science authors, science editors and publishers of Time-Life missed? No, they are not. They don't see truth at all, but rather the polarized fringe extreme of absolute denial that the concept of Bigfoot could ever have any prospect of being real.

In a Reader's Digest publication, "Mysteries of the Unexplained", Bigfoot-like creatures are described from page 152 through 165, with mention of the PGF on pages 160 and 161. They evaluate the film and description of events with respectful and non-judgmental consideration. They don't feel a reasonable and scholarly consideration of the idea is unwarranted.

In "Unexplained" by Jerome Clark, (1993), he has four pages on the PGF, from page 30-33. The author also acknowledges that the film defies debunking, with thorough consideration of the accusations against the film's authenticity.

In "Atlas of the Mysterious" by Rosemary Ellen Clark, (1995), pages 153 and 154 mention and describe the PGF, and further affirm that it remains the strongest evidence for the reality of Bigfoot.

These evaluations and appraisals of the film and the phenomenon have the measured consideration of sorting through the evidence and seeing where the evidence really goes. And without exception, their appraisals of the evidence do not lead to the hoax that is so obvious to internet skeptics. Neither does my analysis. The real evidence simply does not support a conclusion of hoax. It never has, never will.

So, in summation, my conclusion of the Patterson-Gimlin Film is that it is authentic, a real spontaneous encounter with something that is biologically real, as she appears, and the film and its subject figure, Patty, are not faked or hoaxed in any way. But that conclusion is perplexing because it then follows that the larger mystery becomes more mysterious. Why can't we find physical evidence of this species of hominid we call "Bigfoot" which certifies its existence to the biological sciences? Why can't we even get another filmed or video record of another such creature with similar clarity and analysis potential? Why can't we resolve this mystery?

I'm often asked if I've even seen a "Bigfoot"? And I answer truthfully, that I have only seen one, and it's in the Patterson-Gimlin Film. After 47 years, I am mystified that I have not seen another, somehow, in some form, if they really exist. So I went into this examination of the PGF with hopes of getting an answer, and indeed I did get one. But as the proverbial law of unintended consequences dictates, "Be careful what you wish for?", while I got one specific answer, I also got a much bigger question. Why does a definitive solution to the larger phenomenon elude us?

Is this the end of my story? No, it is not. I will continue to look for the missing films (described in the next chapter). I will continue trying to assemble and archive more information and evidence of the larger story of how the PGF came to be, and what happened during the years to follow. And I have become quite intrigued by the film analysis I had to do, and I think there is a real value in helping preserve the technical understanding of film technology for analysis of any archival film footage.

So I hope to expand my activity to include some type of 16mm film technology Studies, which could help film archivists in the future whenever a forensic examination of any footage is undertaken to resolve any questions about the authenticity of such footage. I have discovered things about 16mm film that no reference text or source even explained or mentioned, and it seems this knowledge is on the verge of being lost. So I hope to contribute to the film archive world with further studies on forensic film analysis.

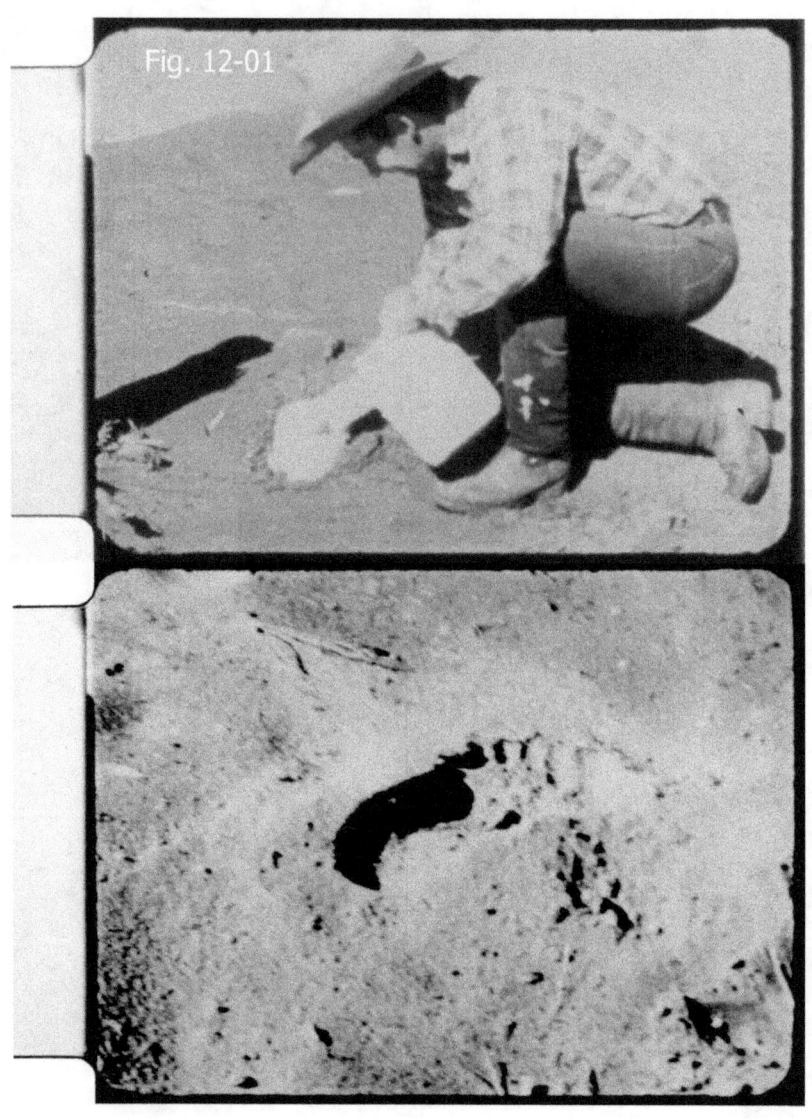
Fig. 12-01

Chapter Twelve - Loose Ends

The PGF has several levels or plateaus of its story. In its pure form there is the actual PGF sequence of 954 frames and the elemental question of whether this footage is authentic or falsified. And that proof also encompasses the question of whether "Patty" is biologically real as she appears, or is a human in a costume pretending to be something more real.

But the total PGF story is a much larger study, once we include the story of the lives of Roger Patterson and Bob Gimlin before, during, and after that eventful moment at Bluff Creek. Whereas the elemental story is more or less a done deal, the film is authentic, the larger story still has a lot of curious unresolved questions. In all likelihood, if these peripheral questions are ever resolved, I don't foresee them changing the core conclusion, but I do think they may change our current appraisal of the larger story, the context of how the film came to be.

These unresolved questions are the "Loose Ends" I refer to herein. And in this chapter, I'll outline the ones which I feel have the most relevance to the larger mystery, the contextual story, and why I hope one day they will be resolved.

Missing Film

Most important to me is the fact that for the wealth of film image data I have accumulated over the past seven years, to compose the largest film archive on this subject in the world today, there are still some film reels that are known to have existed, but are missing now. I hope they can be found and examined, and I will expect each of them may provide empirical evidence to help complete the truth of the larger PGF story.

1. Most important to find is the PGF camera original. Our best information describes it as having been last seen by researcher Rene Dahinden and Bruce Bonney around 1980, when Rene convinced the film vault holding it to release it to him. While it was in Rene's possession, the 12 Cibachrome images were made of Patty, and those Cibachromes we have scans of, so the prints more-or-less verify Rene had possession of the camera original at that time. The camera original was never returned to the film vault and is currently listed as "missing".

If we find it one day, we should expect to see it as only the PGF section, with the horse and rider part cut off. I would be surprised to see the horse and rider footage still intact, but I suppose it's possible. Kodachrome camera original from the 1960s is today still good film, as I have verified buying some Kodachrome home movies for study from the 60s and even earlier. So I would expect this film to still have excellent image data.

It would have the "Kodachrome" film stock latent image on one sprocket side, a date code for 1967, and the Kodak K-100 camera ID notch in the aperture window would be intact. If it were found, an archival scan at 6K would be ideal, with several runs using varied exposures, one balanced for the overall scene brightness, one balanced for the best exposure of Patty, and one balanced for the light gravel of the landscape in hopes of seeing trackway footprints in the sand.

2. The second most important film reel to search for is the Ektachrome master Canawest Labs made for John Green and Rene Dahniden in 1968 from the camera original. That master was done on an optical printer, and has a full first reel copy, then a 2x zoom in copy of the PGF segment, then some slow-motion multiple frame prints, and a freeze frame segment of the lookback. There may be other segments, but the aforementioned are known, but all the known copies are edited into various presentation reels and so we don't have an intact

copy of that master. In Rene Dahinden's book, "Sasquatch" (1974, he described this master as being 320 feet long.

Another related find would be an intact print from this Ektachrome master, all 320 feet of it. But the master itself, in the 2x zoom in segment, would be very near perfect as compared to the camera original for studies of Patty.

Our understanding of the history of this Ektachrome master is that the lab kept the master in their vault and made prints (copies) for Green and Dahinden. This would be common lab practice, to maintain a master in case the customers want to order more prints. And it has been described that the stipulation on the master was that it would be released only with permission of both Green and Dahinden. But Canawest labs went out of business and good business practice would be to contact all owners of masters in their vaults, and make an effort to give the masters to the owners before the business closed. There is no information suggesting this happened. So the whereabouts of this master is uncertain. It's unlikely it would have been destroyed. More likely, any masters which they didn't return to owners, they'd sell to film archives to help with business liquidation costs. So there's a remote chance that the Ektachrome master is still in someone's film archives, maybe uncatalogued and thus not given much attention. If it were found and put up for sale, it would be quite valuable.

All the Green/Dahinden copies in my inventory are from this master.

3. A complete copy of the PGF Theatrical documentary may still exist. Mike Rugg of the Bigfoot Discovery Museum has show reel one (about 30 minutes of footage), and that reel is now described as having been returned to Patricia Patterson. Mike has not disclosed where it came from, and he may not know the origin of it in a manner that can be documented. The reel Mike had was a show copy, set up for theatrical projection. So once Roger and Al made that edited master program, they

had at least one print copy made. It wouldn't be unreasonable to strike off two or even three prints, in case one got lost or damaged. And there would be the edit master as well, so there is hope one of these versions may still surface and be available for study.

An interesting factor is that the second show reel is described as having some footage the BBC assembled for a Bigfoot program, and used Patterson's footage in it. Somehow, Roger and Al made a swap allowing them to use some of the BBC program material in their documentary. I believe Al talks of this when he's on camera describing the film the audience is seeing. It would be fascinating to see this BBC footage, because no copies of their program are currently known.

4. As mentioned above, the BBC did do a program with Roger's footage, and then allowed Roger and Al to use a print of that program material in their own theatrical presentation. This BBC program should be on 16mm film stock, and likely is lost in some archives in the UK. It would be a great find to locate any version of this program, and study the footage. The program title or airdate are unknown.

There is also the description of Patricia Patterson loaning the BBC her master of the second reel, and their not returning it, and it being currently lost. It's unclear if this is the same loan Roger made when he was alive, or a new loan after he passed away. But it is the most common description of what happened to the second reel, and finding that would be a tremendous achievement toward a fuller understanding of the PGF story. Later in this chapter, I elaborate on the Second Reel Confusion.

Continuing with missing footage, we hope to find:

5. The ANE company got Roger's camera original and a lot of his related footage and used it in the "Bigfoot Man or Beast" documentary they produced in 1971. They are known to have kept the camera original of the PGF segment, but may have

kept other footage as well. To make their documentary, they used an optical printer as well, to make zoom ins, slow motion segments, freeze frames, etc. And they apparently used a liquid gate printer because their copies have less scratches than the Green/Dahinden copies, and the liquid gate process does fine scratch removal (on the cell side). Usual copy procedure would be to make the master on the optical printer, then strike a work print for the film editor, and then once the edit is done, do the master editing (normally called negative cutting but this footage was reversal positive stock, so it never was in negative form) and print their documentary program show reels from that edited master. They likely made multiple show prints. I purchased one complete print, three 30 minute reels for the set, and it's in my inventory as Copy 14. So I have one. There should be others. But their edit master would have been in a film vault, and after their bankruptcy, it might have been sold along with other film masters in the vault to film archives. So any material from the ANE company could be valuable for analysis.

I do also have a curious 100 foot segment which I inventoried as Copy 8, and it is another copy from their edit master, but just the PGF parts, and the Janos Prohaska interview cut out, and it curiously has been zoomed in a bit more than the Copy 14 frame aspect. It's four frame slow motion lookback segment is one of the finest for study of the Lookback. Any similar copies would be valuable. We would presume they made more than one such promo trailer copy.

6. The Jim McClarin walk #2 master

Jim did two walks in June 1968. John Green filmed one, and has the camera original, and I have scanned this in full. But for walk number two, filmed by somebody else with another camera, we only have a copy Rene Dahinden gave to Russian Researchers in 1971, so we have no information on the whereabouts of the camera original of this walk #2. Maybe Rene Dahinden has it, since he had a copy to give away? We won't know until we know more about what Rene's son, Eric Dahinden now holds from his father's estate. There have been some dialogues about getting access to those archives but so far, an accommodation has not been reached. But I would expect the Dahinden archives may be a likely place to find the walk #2 camera original. If so, it would be interesting to see if there's any other footage on that reel from that day at Bluff Creek.

So these items are still not located and scanned into the archives. I would not expect any "bombshell" revelations, but they would allow us to assemble a more factual historical narrative of the larger PGF story, and so I will continue to search for these film reels.

Second Reel Controversies

The PGF "Second Reel" is quite controversial, and thus deserves its own section of discussion. First, the basic facts, as best we can determine them. We know conclusively that Roger has a film runout at the end of the PGF encounter, and that he then unloaded that roll of film. That was the First Reel. By both his and Bob Gimlin's accounts, Roger then loaded a second 100 foot reel into his camera and filmed some things as evidence of the encounter.

The actual footage (called Second Reel footage) we have seen is:

1. The trackway footage, showing four complete footprints and a half of a fifth, with plaster poured in footprint #3 of the sequence. That footage in known and scanned, frame by frame, from multiple copies, and the total number of frames verified is 207, which is slightly more than 5 feet of film.

2. Roger casting a footprint with plaster. He's wearing jeans and a pale plaid shirt, a cowboy hat, and he's apparently clean shaven. This sequence is about 5 feet as well.

3. Roger holding up two cast footprints made of plaster, and Roger is standing by a tree. One copy I scanned ran 16 feet of film, and has a camera light wash on the end, meaning it was the last thing filmed before the reel was taken out of the camera for processing. One copy of this footage has the camera original Kodachrome film stock latent image intact, with a 1967 date code that printed through to the copies. It also has a copy film stock date code of 1970, and the footage was described as being given to the Russian researcher in 1971.

4. Unseen but described as being filmed is what is called "The Stomp Test" where Bob Gimlin apparently jumped off a log and stomped down on the ground to see if he could make an impression in the ground as deep as the footprints. There are no known copies of this and no one is known to claim having seen this footage. So there's testimony that the footage was shot, but no evidence for it existing, it's length or content.

That's it, all the described Second reel footage. But what we know of only accounts for about 26 feet, one quarter of the roll. What's on the other 74 feet? We don't know.

But with each of the items above, there are questions.

1. For the trackway, it seems odd there's only 5 feet of it. There should be more. It was very important to document, far more important that Roger holding the footprint casts, and that runs 16 feet. The casts Roger could show anytime. The trackway would disappear at some time, and the film of it and casts would be the only permanent evidence. So there should be more footage, but we can't verify that. Part of my expectation of more footage is that I can't find evidence of a camera start on the first known and scanned frame of the trackway. Without that camera start, we must conclude the first frame we know of is not the true "frame #1". I expect more will be found, hopefully the part with the camera start, and it may explain why we only see this portion. But that's a hope for the future.

Also, we cannot conclusively connect anything in the trackway footage to any landscape element of Bluff Creek in the PGF. All assumption is it was there, but we cannot verify this to a certainty. It would be nice if we could.

Fig. #12-02 shows four sample frames of the trackway, and the lower pictures show a full-scale physical model of the tracks as reverse engineered from the 207 film image frames. The model allows us to see the terrain in three-dimensional form, and better appreciate the way the subject walked to make those tracks. Consideration is now underway to make a donation of this model to a museum, so it will be available for public viewing.

2. The footage of Roger casting a track is problematic. Two such sequences actually exist, the other being on the show reel of the theatrical PGF showing, which Mike Rugg screened in 2012. We don't currently have frame scans of that but I saw it at the screening. So there are at least two different filmed segments of Roger casting a footprint track. It was often suspected Roger made a track himself to demonstrate the casting process, and since the Bluff Creek casting couldn't be attributed to both segments, because Roger is dressed differently in the two, so one is likely the rumored demo casting footage. The problem is, we don't know which one is the demo. Some skeptics love the common-shown Roger casting footage because they claim they can find discrepancies between that and the trackway footage or the Roger holding cast footage, and those discrepancies prove a hoax. But if this footage is the demo, their "proof of hoax" vanishes, so they cling to the idea that the Roger casting images we have, must be Bluff Creek, so they can hold their hoax claim together. But the truth is, we can't prove either one is the Bluff Creek footage today. Nothing is conclusive for either filming material. That "loose end" doesn't bother me, but it infuriates advocates for a hoax.

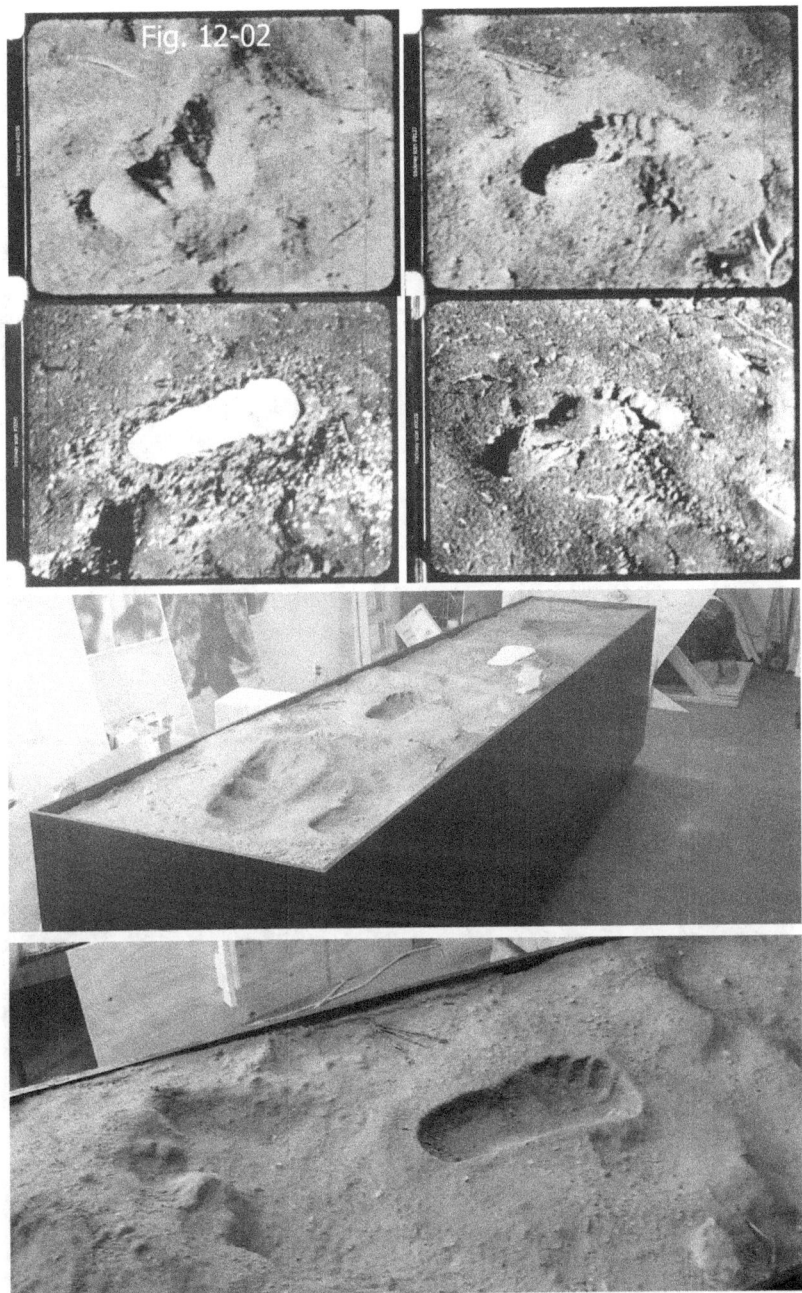

Fig. 12-02

Fig. 12-03

Fig. #12-03 shows the top image as a landscape composite of Roger while he is casting plaster into a footprint track. The composite took bits from many frames as the filming camera shifted a bit up and down, left and right, and each shift caught bits of landscape other pictures didn't have. The goal of this composite was to try and find any landscape element that can be positively identified as Bluff Creek, but so far, we cannot say with certainty that this scene was filmed at Bluff Creek. The lower photo is the one of Roger holding the casts, and the blackness of the shadows is what causes me to consider maybe the photo was taken at night, not day.

3. The Roger holding the two casts, standing by a tree, is problematic because I suspect it was filmed at night, not day, and thus the light source needed to be some type of powerful landscape lighting, like a parking lot light. That would not be at Bluff Creek. I suspect this because the shadows are pitch black, and I have compared over a dozen other films with daylight shadows and I can't find a single example of a daylight shadow that is pitch black. Nighttime shadows are, however. I can't make a conclusion without tests of a scene under both daylight with strong sunlight shadows, and the same at night under a single strong light and no fill light in the shadows, and I haven't been able yet to do that experiment.

So that determination is in limbo at present, but will be resolved in the near future.

Why this poses problems is that Roger has an obvious two-three day beard stubble in these images by the tree, yet he's apparently clean shaven casting the footprints, in the version we commonly see. If these segments were taken only hours apart, his beard could not grow out that much. So there's a contradiction there. But if the casting the track is not at Bluff Creek , or the standing by the tree is not at Bluff Creek in the day, but maybe at home a day or two later, then the contradiction vanishes. Again, skeptics hate to concede the discrepancy between these two segments is because the footage was simply misidentified. That would be an innocent mistake, and not proof of a hoax. Their big "Hoax" claims hinges on a proponent of the PGF authenticity claiming that both segments were filmed same day, same place, and they can use the beard discrepancy to say that proponent claim is a lie so the film itself must also be a lie. So they don't like me stealing their candy, saying one or both scenes may not have been Bluff Creek on Friday, Oct. 20, 1967, and that the footage was simply mis-identified somewhere along the way. The skeptics don't want an innocent mis-identification of the footage. They want a deliberate lie about the footage. But the truth is most likely just an innocent mistake of mis-identification.

So good science would responsibly say that we cannot make a determination or form a conclusion, based on the evidence we have. Leaping to a conclusion would be a flight of fantasy.

4. The stomp test, assuming it actually does exist on film, would be fascinating to see. We have no account or description of why the footage is missing. So all we can conclude is that there is testimony that such was filmed, but we have no empirical corroboration at this time.

So, the Second Reel is a fascinating collection of contradictions. Too much is missing, what we have doesn't add up, and we can't find copies of the missing stuff to clear it up. But the biggest fallacy is that the Second reel will "prove" the PGF is a hoax. About the only way it could do that would be to have footage of getting the mime performer undressed from his costume, and hoaxers wouldn't film that if they had even half a brain.

So the Second Reel does add to the mystery of the larger PGF story. It actually does nothing to prove the real PGF encounter is false, and nothing to prove Patty is fake. The best it could do is prove that for some reason, there are errors in the story commonly recalled by the participants. I won't indulge any "why would they do that?" I personally will wait until somebody can prove it's a legitimate question before I feel any reply (even a hypothetical one) is needed.

The empirical evidence of the first reel is so powerful in validating the event as a spontaneous and truthful occurrence, that the questions attached to the Second Reel are reasonably just a matter of human frailty; confusion, mis-understanding, and assumptions that were incorrect.

Lens Issue

This is a loose end I started back in late 2008, and it still isn't resolved, much to my surprise. It has proven to be a curiously perplexing issue, where my data has some conflict that is elusive to resolve. Now immediately, I want to state that no resolution will prove the PGF is fake and Patty is a mime in a costume. What a resolution to this issue would prove is Patty's height and size, and it would refine our analysis of the Bluff Creek landscape and the paths taken by both Patty and Roger to a much higher degree of precision than we have now.

What is the "Lens Issue"? It's a simple question of what lens Roger had on his K-100 camera that day. So let me begin with how it got started as a topic of investigation. I was thinking about the timeline processing controversy, back in 2008, and wondered if the camera original might have been Ektachrome instead of the usually described Kodachrome, because Ektachrome would be so easy to get processed on a Saturday while Kodachrome wasn't as easy to be done. So I turned to my ASC Manual (the American Society of Cinematographers' Manual, the "cameraman's Bible") second edition, 1966, so I could look up what types of film were used in 1966 or 1967. Looking through the manual for the section on film stocks, I passed the lens section and saw the lens formula, and recalled using it in my early filmmaker and cinematographer days. Immediately I realized that this lens formula can be applied to a question about Patty's height.

Fig. #12-04 shows the basic lens formula and the four ways to solve if you have three numbers and want to solve for the fourth.

Fig. 12-04

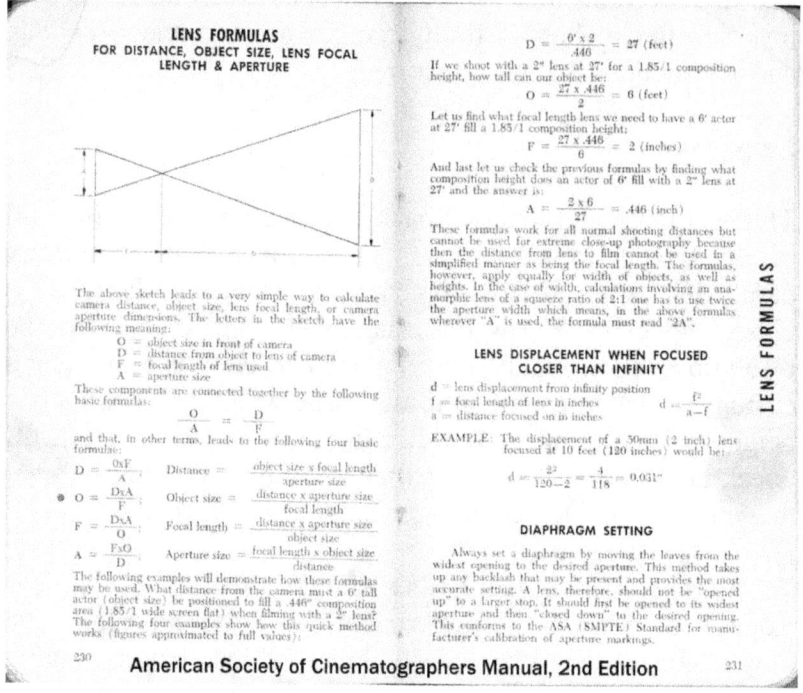

The Lens formula is a simple optical equation, which has four numeric values, the height of a subject being photographed, the height of the subject image as it appears on the film, the focal length of the lens (from film plane to lens optical center) and the distance from the subject to the lens center. If you know three of these numbers, you can calculate the fourth.

So I knew we could calculate Patty's height on the PGF film image, relative to the 0.3000" height of a 16mm film frame image (I had calculated her several times, but the earlier calculations were not with the best frame copies, and so as I obtained better and better frame scans, my calculation did evolve. My best calculation now is 15.8% of the frame height, or 0.0474" high on the image). All current research back in 2008 stated Roger had a 25mm standard lens on his camera, and several researchers who'd been at this for many years

318

before me had estimated that Patty was 102 feet away from Roger at the lookback point. So we had three numbers. We had enough to calculate Patty's height.

So I plugged these numbers into the formula, the distance (102') x the image height (0.0474") divided by the focal length (25mm = 0.9842") equals the subject height, 4.91' (about 4' 10.8"). So based on all accepted research data, Patty was about 4' 11" tall. My keen grasp of the obvious suspected something was wrong about that.

I also tested various distances with the height calculation for each distance, as follows (From my Report Release Document #1C):

If we apply the above F352 subject height standing, 0.0474" and a 25mm (0.984") lens specification (which will yield a slightly taller subject height calculation than the 25.4mm spec would), the following chart shows what subject height we may expect for a given distance from camera to subject in the F352 point of filming.

distance,subject to camera	subject height
90'	4.33'
95'	4.57'
100'	4.81'
105'	5.05'
110'	5.29'
115'	5.39'
120'	5.78'
125'	5.97'

Current investigation of the subject position from camera in relation to a scaled site model puts the subject at less than 105' from camera for F352, **so if a 25mm lens is used on the K-100 camera which took the PGF, then the filmed subject is under 5' tall.**

Now the size of subject image height in frame is the one measurement I could certify as being near perfect, so that number was correct. The other two I was relying on other researcher data, and so those two I had to question. Either the distance was wrong, or the lens focal length was wrong (or both). But reviewing the Bluff Creek site measurements by both John Green and Rene Dahinden, taken separately by different methods, they tended to concur to a large extent, and Patty can be quite reliably positioned by one object in those measurements, so the distance seemed the more reliable of the two in question.

So I turned to the lens. Could the focal length have been different than assumed?

My first consideration was that with a wider angle lens (shorter focal length), Patty will calculate bigger, and so I checked the K-100 inventory of lenses and found Kodak made a 15mm lens and companion viewfinder lens for the camera. But with a 15mm lens, Patty calculates at 8.19'.
(102' x 0.0474 divided by 0.59 = 8.19')

So either way, using the 25mm lens or the 15mm lens, I was getting wild numbers that didn't correspond with the mainstream estimates of Patty's height. I did film tests of a man, a personal friend with a measured height of 6' 3", and I filmed him at a distance of 102' and both the 25mm and 15mm lens on the K-100 camera.

Fig. #12-05 tested both a 25mm and a 15mm lens at a distance of 102 feet, and the man towers over Patty in one, and shrinks well smaller than her in the other.

Then he was compared to Patty, and neither of his filmed images came close to Patty. One was way too big, one way too small. So either way, with either lens, we didn't have a match to the common estimates of Patty's height.

Fig. 12-05

Man 6' 3" tall photographed at 102' with a 25mm lens

Patterson's Creature is reported to be 102' from the camera in Frame 352 shown here

Man 6' 3" tall photographed at 102' with a 15mm lens

But as I continued the analysis, I finally obtained excellent frame scans of John Green's filming of Jim McClarin, and was able to make comparisons to the PGF. I found the camera position for Green slightly back from Roger's position filming the lookback, and a slightly different lens focal length, but the discrepancy between his lens and Roger's was only about 1mm at most. So if Roger's lens was 25mm, John's was about 26mm. If Roger's lens was 15mm, John's was about 16mm.

But I then calculated Jim McClarin's distance, knowing his height (about 6' 5" in shoes) with both lenses, and neither put him on Patty's assumed path. But the 16mm calculation put Jim in front of trees that the image data verifies he walked behind, so the image data absolutely refuted that lens option. That was when the 15mm lens for Roger died as a viable theory.

But the 25mm lens calculations were still problematic, because that lens put McClarin about 25 feet behind the path Patty took, and yet Jim walked the path and saw the footprints days after the filming was reported, so he, of all people, should know where they were and how to walk the same path. Searching for other lens options, one turned out to be a 20mm Anastigmat lens Kodak made and was standard issue on an earlier model Kodak movie camera, the model "E" introduced in the 1930s. It had a "C" mount, which the K-100 camera takes, and it puts Patty right where she should be, and at a height that corresponds to the footprint size, and a lens 1mm longer on Green's camera still Puts Jim McClarin almost dead on following Patty's path.

So in many respects, the 20mm lens seemed the ideal solution. It was made by Kodak, readily found on an earlier Kodak movie camera (Kodak's Model E), fits the K-100 Roger was using, gives a better wide angle view for scenic filming, as Roger intended, and as a bonus, was a fixed-focus lens, and if you filmed outdoors, in bright sunlight, just about everything was in good focus from three or four feet to infinity, without you worrying about setting a focus ring. You just had to set your F-stop correctly for the film speed and light.

Fig. 12-06

Fig. #12-06 shows the original Kodak Model E camera which comes with the 20mm lens, the lens alone, and then the same lens mounted on a Kodak K-100 camera, as Roger used.

The only downside, which is why I didn't consider it earlier, was that Kodak didn't make a viewfinder lens for the K-100 to match a 20mm lens, so what you saw in the viewfinder would not be exactly what the film would actually capture. At first, I thought this would cause a cameraman to pass this lens option over, but in reconsideration, I realized that with the K-100 viewer, it has a very large "safety area" (where you can see beyond what the lens sees, and so avoid something in your shot you don't want). If you had a 25mm viewfinder and a 20mm lens on the camera, you just consider all that you see in the safe area as also being in the film image. If you are not worried about avoiding filming anything like signs, trash cans, crew people standing off to the side, etc., you can make this work just fine.

That put the 20mm lens up as an option Roger may have used. So that's where it's at. Roger may have used the standard 25mm lens or may have used the 20mm lens instead. Once we found out about John Green's real camera, (the Revere Camera) and that a second camera was there with a zoom lens on the Bolex, and both Bolex and Revere take a "C" mount lens, it's possible John didn't use the Revere standard lens but put the Zoom on instead. The zoom had a better viewing scope and could be set to any focal length from 12mm to 120mm, so setting it at 21mm or 26mm wouldn't be at all unusual.

So we have some very plausible explanations for the lens possibilities. But here's why it hasn't been solved yet.

Some calculations of the camera positions and trees measured or surveyed support the 25mm lens. Other tree measures and camera positions support the 20mm lens. Jim's walk path coincides more closely with Patty's path using the

20mm/21mm lens set, and Jim's walk path is wildly off the mark with a 25mm/26mm set.

Somewhere in this wealth of data about the sight, and the camera positions, there's an error. And so the only responsible thing to do is review virtually every piece of evidence involved in the lens analysis and ask, is this data accurate? Every bit of data needs to be rechecked, including a return trip to the Bluff Creek location to re-measure the still standing trees seen in the PGF film. And this review of all the data requires some funding to do correctly, especially the trip to Bluff Creek with a crew and some survey equipment. So the lens determination is on hold, pending this total review of data, and that review is on hold pending some financial resources to accomplish in a comprehensive way.

One of the exciting things that gives me hope is a new alternative to an expensive process. A few years ago, I thought that the aerial photograph I needed of the Bluff Creek site could only be accomplished with a helicopter, and the fees for such were estimated at $8000. But since then, photography by drone has become far more available and effective, and so I can see the aerial portion of the site photography being done with drones, and that makes the goal far easier to attain.

So there we stand. The lens focal length is still undecided, unproven. As a result, Patty's height, by the lens formula calculation, is also undecided.

A lot of people have offered simplified conclusions and think they have solved the matter, but nobody actually has, so please disregard anyone who claims to have the solution. The solution is currently not known, and more work is needed. I will remain committed to finding an answer, because in the old proverbial sense, it was me who opened "This can of worms". So I consider that I have an obligation to find the solution and document it meticulously. I'll give it my best, but it is going to take funding from some source. Maybe my other business endeavors will be sufficiently successful that I can invest my own money. If I can, I will. But for my own desire to get to the truth of this most elusive question, I will persist and eventually find the solution.

Why does it matter? I feel this is the best method for a precise calculation of Patty's height, and that info is important enough to justify this investigation. Solving the lens issue will also provide the most precise Bluff Creek analysis. So the matter has enough importance to be proven to a certainty, one way or another.

Camera Running Speed

This question doesn't really concern me, but it does concern other researchers, and I address it simply because I'm probably the best person positioned to resolve it. The K-100 camera Roger used has a camera running speed dial that starts at 16 frames per second (fps) and ends at 64 fps. Usual professional filming speed was 24 fps, and 16 was common for amateurs projecting their film with no added sound track, like home movies. The question is whether Roger had the camera set at 16 fps or 24 fps. Some researchers think there is value in this determination to calculate things about Patty. But a curious wrinkle in this topic is that one researcher, Bill Miller, found technical data from a Kodak technician that stated the K-100 cameras were tweaked so even when the dial is set at 16 fps, the camera actually runs at 18 fps.

I bring this up because I'm probably the researcher best positioned to resolve this, because I have nine K-100 cameras now, and a digital camera that can take video at 240 frames per second, and that allows me to actually video a camera shutter running at very high video speed so I can actually count how many times the shutter opens in a second on that video. I tested it on one camera, and got 18 fps, but the rest still need testing. But that one test was without any film running through the camera and the final and most accurate testing is to run film through each camera and film a digital clock or stopwatch, ideally with a time readout in 1/100ths of a second, so I can time the shutter by the photographed readout on film. The difference between running film and the motor running free without film is that the film pulldown may extract some energy from the spring-driven motor, and so it may run slightly slower with film than without. Only by testing both can I be certain. What's holding me back is the costs of film stock and processing of the film in a lab.

But that is the ideal way to resolve this question, and I am half way there with owning the cameras, so there's a good sampling of multiple cameras. The rest will be done when the research endeavor has some new funding.

Frame 191 Rotational Motion Blur

This is a curious ,optical phenomenon still under investigation. No conclusion is offered, and I hope no amateurs will cloud the issue with simplistic guesses passed off as facts. Nothing about it will likely have any effect on the determination of the film's authenticity, but it represents a curious photographic anomaly I want to understand, and then I can appraise if it somehow affects the analysis. So this, as curious as it is, remains "unsolved" at this point.

In essence, it is curious because motion blur tends to be global, meaning it is the same across the entire image. If it blurs on a diagonal, for example, everything near and far in the scene has the same length of blurred streaking. Rotational blur is more rare, but this Frame 191 has rotational blur around an apparent centerpoint in the picture, the upper right corner, so it would appear the center of rotation is at that point, which is a very awkward way to rotate the camera. Plus, the right upper corner is actually quite sharp, while the rest of the frame has blur, and this goes against the usual global blur effect we see. So I do plan to continue exploring ways to analyze this curious photographic anomaly.

Fig. 12-07

VFC-2 Frame 191, start of Segment Three, with
rotational motion blur, but sharp upper right corner

Corner
enlarged
at right

Fig. #12-07 shows frame 191 with the rotational motion blur, and the unusual clarity of the image in the upper right corner, magnified in the lower image.

Correction from earlier analysis about Magazine Camera Apertures

Years ago, I suggested John Green's camera frame of Jim McClarin's walk may be explained with a camera modified as a DIY (Do It Yourself) conversion from standard 16mm to "Ultra 16MM", a sort of improvised alternative to suggest a Super 16mm wide frame. However, with the chance to actually use magazine cameras and purchasing many home movies shot with magazine cameras, I can now say that the aperture on John Green's camera is more or less stock for magazine type cameras, because of a curious aspect of the physical placement of the camera's aperture as compared to the magazine aperture. There are technological questions about this design I have not been able to solve, and cannot find any technical literature that explains the design of these magazine cameras, so this issue is still under investigation. But my years earlier opinion that John had a DIY modified camera is incorrect. A stock Revere camera, as he's photographed using, will make the aperture shape of his film.

The Monster From the Id

Chapter Thirteen - The Monster from the Id

Internet Discussion and Defamation

One of my favorite classic science fiction movies is the 1956 "Forbidden Planet". I was eight when I first saw it, and that invisible creature melting through the walls of Krell metal to threaten Walter Pidgeon, Anne Francis, and Leslie Nielsen, scared me so much that I insisted my dad stay in my room that night until I fell asleep, in case the "Monster from the Id" started melting through my bedroom wall to attack me. But I outgrew the fear and the movie came to be one of my true favorites. But the monster would eventually attack me. It just took another 50 years.

In the movie, the story introduced the idea of "The Monster from the Id", a terrifying creature which is the physical manifestation of a person's thoughts, especially their deeper subconscious thoughts, the Id, as it was named in Freudian psychiatric theory of the era. The Id was considered the part of the mind where the deep and most primal or animalistic thoughts or impulses resided, or, as stated from Freud's own descriptions, "an instinct of destruction directed against the external world and other organisms". In the movie, Dr. Morbius (Walter Pidgeon) had unknowingly hooked his mind up to the awesome Krell machine and it was now actualizing the impulses of his own subconscious, which hated that Leslie Nielsen had alienated his daughter from him and disturbed his perfect little world on the planet Altair IV. So now, this "Monster from the Id", his own deep rooted anger, was attacking Nielsen and his daughter who betrayed him by falling in love with the intruder to their world. The Krell machine's power was limitless, and it would keep investing more energy into its goal (to melt through any thickness of Krell metal needed to kill the intruder and the unfaithful daughter) until it achieved its goal. Morbius had no conscious control over this machine, or his own Id. Only by sacrificing himself to this monster could he save his daughter, and finally stop the monster from its relentless action.

In my work on this book, and especially on this chapter, I realized that the internet today has in a way become the Krell machine empowering 'the Monster from the Id". It gives near limitless power to those dark and primal impulses of people's minds. It energizes the deep-rooted hatred in people for anything which threatens their world as they perceive it, or threatens the beliefs they hold as true and sacred.

It's well known that the internet has given anonymous people the power to criticize other people, especially named or known real people ("Public people" as I will call them), with vicious action and relentless opportunity. And the search engines of the internet have no filter to distinguish between truthful statements by responsible people and the vicious accusations or even outright lies by anonymous people intent on defamation and character assassination. Like the Krell machine in Forbidden Planet, the internet energizes both the conscious ideas and the deep rooted primal anger, hatred and malice of the Id. In its search and display capabilities, both the consciously responsible thoughts and subconscious hatred get equal power, consideration, and listing on search results for all to read. By unlocking and empowering the Id, the internet has become the machine of defamation, vicious insults and lies.

I'm certainly not alone in being on the receiving end of such defamation. As I write this, several women celebrities in Hollywood were making news for their decision to stand up to internet defamation, and other news stories had accounts of distraught teenagers cyber-bullied to the point of committing suicide. Oddly, it was my posting of one such news account, from CNN, that resulted in my being banned from the JREF forum, where defamation is raised to a fine art. The posted news account was about a judicial ruling which allowed a woman, a model, to sue a man who was defaming her viciously on the internet, and hiding behind an anonymous screen identity. His anonymity apparently unleashed his own "Monster from the Id", some kind of deep anger and primal need to insult attractive women. He didn't even know this woman personally. He simply hated her because of some dark motive in his own Id we will likely never know the nature of. The internet was his Krell machine. But this court judgment would allow her to force the internet service provider to disclose the true identity of this anonymous internet person so she could sue him and serve him with papers.

Over the last 7 years, I've seen this on various internet forums and discussion venues, where anonymous people, using screen names that can't readily be traced to the true person's identify (unless you can access the forum's registration data) seem to freely unleash vicious criticism and defaming claims about public people, even accusing these public people of crimes like fraud, or psychotic behavior, or posting blatant lies about the public person, simply because the anonymous writer is somehow deeply offended by what the public person says or does. And I found it quite a brutal wakeup call for myself that in the last seven years, I know of more people who truly hate me than for all my life in the 58 years before I began participating in these forums. What did I do that inspired this hatred. I dared to publicly say, "I don't think the PGF is a hoax."

Of course, I'm not alone here. People who think there really is a problem called "global warming" experience such attacks. So do people who think that Darwin's "theory of evolution" is true, and people who actually think the Apollo space program landed men on the moon, or people who think the collision of hijacked jets actually did cause the World Trade Towers to finally fall. Take a stand advocating any of these are real, and you'll likely get someone who believes otherwise and hates you for it.

So, I am apparently hated by some of the people who truly believe the PGF is a hoax, and that idea is so sacred, so immaculate in their minds, that my opposing opinion is almost blasphemous to their belief. And in this chapter, I choose to take a moment to face this "Monster from the Id" which seems quite aggressively intent on defaming and attacking me, as punishment for my horrible transgression of being a "hoax atheist", saying that I do not think the PGF was hoaxed.

Throughout this text, I made occasional reference to my critics, or to skeptics who fanatically believe the film is a hoax. A book author embroiled in a controversy does not traditionally recognize his critics, but I feel any understanding of my research cannot separate my proactive effort from my reactive concerns dealing with these critics. So after much reflection on this issue, the counsel of some people whose opinion I value, and thankfully the freedom that comes with self-publishing my book, I have chosen to acknowledge the criticisms, but not the critics.

Claims of Dishonesty

Now I would have thought that if people wanted to discredit my work, their first line of attack would be claims I am incompetent, stupid, wrong in my analysis. Such claims are made, and discussed next, but I was quite surprised to find that accusations that I am dishonest seem the more prevalent. These claims include the curious suspicion that I don't actually believe what I'm saying, that I'm just "selling" an idea because there are buyers, or that I am deliberately not disclosing some facts and thus acting in what they popularly call a "lie by omission".

So I've been dealing with these criticisms, these accusations, for quite a few years now, and pondering the question, how does one "prove" he or she is honest? Does it matter that my career resume is meticulously honest and after seven years of critics examining every claim and searching the internet for contrary evidence, they have not been able to prove a single career credit or description of accomplishment wrong? And believe me, they have searched. So, my resume is honest.

My awards are documented, my publications (such as my educational series of wildlife re-creations in Breakthrough magazine) and my related achievements are verifiable. None have been invented or misrepresented. My knowledge of makeup special effects, and general filmmaking is substantial and I've written extensively about the facts of the process. I might have made an error or two along the way, about some materials or processes other people used and I read or heard about. Fine, point out the error and I stand corrected. But the general body of professional knowledge is realistic, honest.

These people who accuse me of being dishonest don't seem to realize that if they were in any real evidentiary proceeding, and made the accusation, they would be asked to prove it, and their proof would fail miserably by any evidentiary standard. But of course, the world of internet discussion has no such rules of evidence. It has no process that compels claims to be proven. Accusations can be made, and there is no ruling administrator who requires proof or the claim is invalidated or deleted. The operating mentality seems to be, if they repeat it often enough, somehow the accusation will be transformed into reliable fact.

For the record, I am sincere in what I write. I feel that my analysis is reasonable and supported by the facts. I try to represent technical matters in a factual and correct way. Every claim I make as to my knowledge and work experience is truthful. So in response to the criticisms that I am dishonest, there is no merit to those criticisms, and I am confident that fair-minded people who take any time to get acquainted with my written work, or my life, or me personally, will concur.

Claims of Incompetence

People who try to accuse me of being incompetent almost invariably focus on one movie I worked on, and got fired from, in 1984, thirty years ago. It was a silly little zombie horror comedy called "Return of the Living Dead". I've never misrepresented the fact that I was fired from the movie, the facts have been reported, published, and I acknowledge the situation even on my "Creature Gallery" website. This movie represents about 1% of my career, or less, and even though some of my work is in the movie and works just fine, if you want to call that job a total failure, that would still mean I'm 99% successful in my overall career, and I can live with that.

But some critics seize this one job and act like it is the singular defining experience of my career, and all the success before or after is irrelevant. Critics also like to accuse me of being a "washed up has-been". I don't do makeup effects now, but the way the industry changed and many creature effects jobs just vanished into the world of Computer Graphics Imagery (CGI), I described that in Chapter Four. So many fine makeup artists had to transition into other endeavors. I was not alone. But these people, if they went on to productive careers in the new field, as not thought of as "washed up has-beens". Why should I be so designated?

Just as there are people who live in one town all their lives, and others who move from town to town several times in their life, some people do stay in one career niche all their lives, while others transition from one career activity to another, across the full span of their professional life. There's nothing demeaning or flawed about moving around professionally, and having career changes from time to time. But the critics do seem to fixate on this as a basis of their remarks against me.

I'm quite confident that the combination of my resume and my "Creature Gallery" website images and documentation of my career accomplishments will verify that I continue to look for career challenges and succeed when I focus on a task. I excel on any endeavor I apply myself to, and attain a high level of capability or competency. And even though I am 65 now, I am not retired, but rather look forward to about 20 more years of creative and technical accomplishment in the challenges I still aim to take on. I will revisit the makeup special effects field in some manner, will continue with computer graphics, and expect to continue with filmmaking overall as well. I will continue to invent things and apply for patents on these new technologies.

And contrary to the critics who think this work on the Patterson Gimlin Film is my "last hurrah", my one last desperate attempt to grab 15 minutes of fame, I will simply say that the film is a truly important mystery deserving of several years of my time to solve, and I will be forever proud of my work on it, but my ambitions are so far beyond the work here, that I am confident the final sum of my life will note many accomplishments to come as defining my life far more so than this endeavor will. It will be one of the things I have done. But it will not define me.

Criticisms of Ulterior Motives

Another criticism revolves around critics imagining that I have dark ulterior motives for why I'm doing this work on the PGF. Many of these accusations flat out accuse me of being a fraud, a con man, a charlatan, a manipulative person, etc. My suspected ulterior motives are sometimes described as monetary gain (getting rich), being an "attention whore" (getting notoriety), or being so warped that I like deceiving people.

The problem with these accusations is that you can make the same accusations about any person who has applied themselves to an endeavor, achieved a high level of distinctive accomplishment, and in doing so, has received some measure of public recognition for their accomplishment.

The truth is quite simple and is not dark. I do like to learn, to better understand this world we live in, so I pursue a never-ending goal of self-education. And many things in the world fascinate me. Mysteries and challenges are especially intriguing. And the PGF is a fascinating mystery. So once I decided to take some time and look seriously into it, in 2008, I found the mystery fascinating and decided I would like to invest the time to find a solution. And once I delved further, I realized that I would need to learn, or refresh my knowledge of, many technical subjects in both the creature fabrication process and filmmaking and cameras, lenses, etc. So the goal of solving the mystery ran parallel to my goal of continually learning more myself.

I want to solve the mystery for my own understanding, first above all. So I want the truth. If I find the truth, I would like to share what I found with others, because I've always been confident that sharing knowledge is far better than hoarding it or keeping it secret. So once I did reach a level of knowledge about the PGF, I chose to share it with anyone interested. This book is the culmination of that intent.

I like to be paid for what I do, because living costs money and work you do should compensate you to pay your living costs. There's nothing disreputable about that. I know people who attain some measure of success in an endeavor do get noticed by the media, and I don't consider that in any way disreputable either. But the biggest lie of all is the accusation of some fraud or con, that I am deceiving people. My actions and intent are sincere, and what I write I truly feel is factually correct and morally responsible to share. And the truth of what I'm doing will endure long after the accusations of fraud or con have withered and disintegrated, at least in rational minds. A few dysfunctional people will forever be imagining a PGF hoax, and they will continue to insult me, but like germs and viruses in the world, we'll never get rid of them so we just develop a good immune system to survive them.

There's no hidden ulterior motive to my actions. I want to know the truth of the PGF, and I like to share what I found in my research and analysis.

338

Criticisms of Bad Science

A recent criticism, following my receipt of research grants and my publication of scientific papers with Prof. Jeffery Meldrum, is to say I don't know science. Occasionally they go further and accuse me of indulging in pseudoscience.

To address that, I will start off with a brief mention of my science education. I had four years of science in my three years of High School. I studied a year of biology, physiology, chemistry and physics, and in each course, the second semester was at the honors level. I was on my high school team competing in the American Chemical Society annual High School national competitions, and of the 15 students from my school, I was one of the top four who comprised our school team. I also had three years of math, up to college level calculus, in High School. So I guess you could describe me as a Math/Science Major. Critics like to say that High School science doesn't count, it's not real science education. I'll grant you college and grad school science is a higher level, but to dismiss my science education as "nothing" is a blind denial, and I am confident any High School science teacher would concur.

But for the record, all my formal education in science is at the High School level, and none at the college level. In terms of self-education, I've always loved science and continue to buy, read and learn from science books and magazines, but self education doesn't have the impartially measured levels of proficiency that formal education and college degrees have. It's a foolish thing to think a person without the credential is without the knowledge, but this seems to be the critical perception. I on the other hand do feel self education is tremendously valuable and a person can attain a very high level of knowledge from that process. I also firmly believe that the "lay-scientist", or sometimes called the "citizen-scientist" today, can indeed do science and can solve problems and expand our knowledge of a scientific question.

If a layperson or self-educated person is doing science, it takes longer for the credentialed scientific community to accept said person's work as scientific, so I know I've got a longer haul to get recognition for my work. So be it. But I do love science, I do understand it, and I am doing it with the PGF research and analysis. The future will bring my vindication, of that I am confident.

Claims of Confirmation Bias

Skeptics love to claim that I am a "believer" and so all my analysis is distorted to reinforce the foregone conclusion of the PGF being real. This accusation is one of the major skeptical talking points in superficial criticisms about me. But confirmation bias, as judged in other people, is itself a subjective judgment, and so the accusation toward others is itself as vulnerable to being confirmation bias as any claims. I'm frankly surprised that these critics don't see that. Their accusation can be as much an act of confirmation bias as anything they say I'm doing. They are doing what they accuse me of doing.

If they actually focused on the ideas, the facts, the analysis, and simply strived to put together a concise and coherent proof of a hoax, they'd be able to stand on the merits of their analysis. But they can't. Every attempt thus far has been a failure. So if they can't build their proof of hoax, they tear down my proof that the PGF wasn't hoaxed.

As an argumentative point, if I were a "believer", as they say, I'd be endorsing other evidence as well, to support the belief. But other photo evidence I have studied, I have judged a fake. Footprints and trackways I have looked at, but have not endorsed as real, with the one exception of the PGF filmed trackway, which is real because I've actually build a full scale physical model and so I've seen the terrain as no other person has, and it most assuredly is not a faked trackway Roger filmed.

But for all the various forms of evidence that abound and other researchers judge to be authentic, I don't endorse them, because I personally don't understand how they are authenticated and I won't endorse anything I can't personally authenticate with my knowledge and analysis. Is that the pattern of a "Believer"?

When someone is accused of being a "believer" and the accuser is locked into a mindset of confirmation bias instead of objective analysis, that accusation ultimately fails because the accusation itself can be made with no proof, and a truthful proof of it being false can almost never be made, so it's effortless to make and nearly impossible to refute, in the strict sense. As such, the accusation can be made about anybody, and whether the accusation is true or false has no bearing on the enthusiasm of its being made. Thus the accusation itself has no inherent integrity. Ultimately, all truth of any accusation someone's a "believer" is found in an analysis of the person's evidence, their analysis method, and their basis for forming a conclusion. If the analysis can be verified by other people, the "believer" accusation is diminished, and ultimately defeated.

Time and continued scientific analysis do eventually vindicate the reasonable and objective person. The vindication may be some time in coming, but it usually does arrive eventually. So in this criticism, this accusation, I will simply let time be the deliverer of my vindication, that my analysis is a result of objective analysis of the facts, and not some delusional product of bias and belief.

The Investigation of My Life

As I have noted, people who are now obsessed with discrediting me have been digging into my life for seven years. They have found remarkably curious facts and details I would never have imagined even existed. My credit in the crew of the ABC television production of a special program (1985) based on the Teddy Ruxpin talking doll is in my resume, but somebody found a screenprint of the actual end credits of that program, and sure enough, my name was there (much to their disappointment, because I'm sure the only reason they were looking was to hopefully say my resume credit was a lie).

Another intrepid researcher dug into the film history of an obscure movie I did called "What Waits Below" (1983), found the name of the writer, found some man she was connected to, found this man had something to do with Bigfoot or a program about same, and thus tried to say I was a "believer" because I worked on a movie written by a woman who was with a man who made a Bigfoot program. Pretty convoluted as proofs go, but it shows the length people have gone to trying to discredit me.

So I expect that with this book, the intensity of digging into my life will increase. About the only "stone unturned" is the 45 year old saga of my involvement in the populace movement to try and stop the Viet Nam War when it was becoming tragically obvious it was a war we would never win.

After much reflection, I have decided to simply and factually disclose this chapter of my life not yet examined and debated, because it will come to light one day and I'd rather proactively disclose it now, than wait for others to dig it up and distort it, forcing me to reactively try to respond and set the truth straight. The only relevance it has to my analysis of the PGF is that it is a reflection of my character, and my moral priorities, and skeptical people have invested an immense amount of time and talk attacking my character. I will let you, the reader, be the judge of whether it has any relevance to my work finding the truth of the PGF.

The Viet Nam War was a sad time for America, perhaps the first war fought for no actual need to defend our country from any real threat, and no real gain for our US citizens. I have the upmost respect for all the men and women who served in the armed forces during that conflict, and sacrificed so much in the call to duty.

But duty takes many forms. The duty to obey is one and they answered that call. But the duty to disobey wrongful orders is also an important position (even established in Military Protocol), to refuse to be part of an endeavor or activity which is morally wrong. By 1969 and 1970, the Viet Nam War was becoming painfully obvious as an unjustified and unwinnable war, a war capable of causing great casualty but no benefit or gain in American defense or security. It was inevitable the war must end, and the casualties must stop, but the government was simply incapable of taking the courageous step of ending the war at that time. Many citizens of our country did see the truth, that the only way to stop the casualties was to simply stop the war, and so populace movements toward that goal were increasingly strong and outspoken.

At that time, 1969, I was deemed eligible for military service. But I believed the moral imperative was to help stop the war and stop the casualties, and so I filed for a Selective Service status as a Conscientious Objector. The Selective Service Board refused to grant me that status. So when I was given my induction notice, I refused to comply. I was not influenced by any concern for my own harm, since I would have been classified as a "sole surviving son" in my father's family, and men so classified were not assigned to duty in combat zones (the movie "Saving Private Ryan" is based on this policy).

With my refusal to report for induction, I was arrested and brought to trial. During the proceedings, I was repeatedly assured that if I simply consented to join the army, this prosecution would all "go away". Time and again, I was presented with a choice, to accept the position of joining a war effort I truly believed was wrong, or stand by my principles and be part of the solution to end the war. Time and again, I chose not to take the deal, rather to stand by my principles and do what I believed was morally correct. I wanted to be part of the effort to end the war. I was convicted of violations of the Selective Service Act, and sentenced to prison, for a year and a day (minimum sentence).

I served my full sentence (about 8 1/2 months, with time off for good behavior), at a Federal Prison facility in Arizona, being released in August, 1973, to resume my life. The war had ended, and as I and so many other people foresaw, there was no victory. There was only an end to the casualties.

In early 1976, President Gerald Ford issued a conditional pardon for people convicted of Selective Service Violations, to be reviewed on a case-by-case basis. I applied and was told I had not yet fully "repaid my debt to society" (because of my short sentence) and was ordered to perform some civilian community service. So I spent weekends as a volunteer at Los Angeles County General Hospital, for about 8 months.

Then President Jimmy Carter was elected, and took office on January 20, 1977. He immediately issued a blanket and unconditional Presidential Pardon for all people prosecuted under the Selective Service Policies during the War. I was included in that Presidential Order, and I was subsequently issued a full and unconditional Presidential Pardon. It effectively nullifies the conviction. If you've never seen a Presidential Pardon, the accompanying photo shows you mine. Also shown is the letter from the Attorney General verifying my inclusion in the pardon act.

The Viet Nam War era was a troubling time for America. For so many young men of military age, there was the classical sense of duty so many answered the call to, and there was the dilemma of a war that wasn't worthy of their honorable sacrifice, a war that inflicted much casualty with no prospect of victory and no real defense of any true American substance. Sadly, governments are not infallible, and while we must respectfully abide the actions of our government in general, we must occasionally put what it truly and morally right at the pinnacle even above the dictates of government. But if we do, we must accept the consequences of our decision, even if it means going to prison in defense of those principles.

JIMMY CARTER

PRESIDENT OF THE UNITED STATES OF AMERICA

CERTIFICATE OF PARDON
ISSUED TO

WILLIAM HOWARD MUNNS

I hereby certify that the above named person, who was convicted, irrespective of the date of conviction, of any offense against the United States of America committed between August 4, 1964 and March 28, 1973 in violation of the Military Selective Service Act or any rule or regulation promulgated thereunder, was pardoned by the Proclamation of Pardon of January twenty-first, 1977.

In wittness whereof I have signed my name and caused the seal of the Department of Justice to be affixed below.

Done at the City of Washington, District of Columbia

This Twenty-fifth *day of* January, 1978

Pardon Attorney

United States Department of Justice
Office of the Pardon Attorney
Washington, D.C. 20530

EFFECT OF PARDON

A Presidential pardon is a sign of forgiveness. It does not erase or expunge the record of conviction nor does it, in itself, restore civil rights which may have been forfeited as a consequence of state law. It is usually helpful, however, in obtaining licenses, bonding or employment and in having any civil rights lost by reason of the conviction restored by state authorities.

A pardon by the President relieves the recipient of all legal disabilities in the nature of punishment attached to his conviction by reason of Federal law. For example, a pardon remits an unpaid fine. However, most civil disabilities attendant upon a felony conviction arise as a matter of state, rather than Federal law. For example, the right to vote, to hold public office under the state, to sit on a jury and other rights or privileges that are forfeited as a consequence of state law may only be restored by state action. Although a pardon removes any Federal firearms disabilities resulting from the conviction to which the pardon relates, there may be other firearms limitations arising under state or local laws.

On any application or other document which requires the information, a pardon recipient should disclose the fact of his arrest and conviction. However, the information that a pardon has been granted should be included and the certificate of pardon may be shown. The Federal Bureau of Investigation will be notified of each grant of pardon and will enter on its identification record pertaining to each grantee a notation that such grantee has been pardoned.

ADDRESS REPLY TO
UNITED STATES ATTORNEY
AND REFER TO
INITIALS AND NUMBER

WDK:RAS:alb

United States Department of Justice

UNITED STATES ATTORNEY

CENTRAL DISTRICT OF CALIFORNIA
U. S. COURT HOUSE
312 No. SPRING STREET
LOS ANGELES, CALIFORNIA 90012
April 13, 1977

Mr. William Howard Munns
3039 W. 7th Street, #3
Los Angeles, California 90005

Dear Mr. Munns:

With reference to your letter received in this
office on April 13, 1977, please be advised that our
records indicate that you do come within President
Carter's Proclamation.

By a copy of this letter, I am requesting the
Pardon Attorney to send you an application for
individual certificate of pardon.

Very truly yours,

WILLIAM D. KELLER
United States Attorney

RICHARD A. STILZ
Assistant United States Attorney
Chief, Criminal Complaints Section

cc: Lawrence Traylor
Pardon Attorney
Washington, D.C.

I believed the war was wrong and that the casualties would
only end when people stood up to the government and refused
to be part of something wrong. I made a moral decision to
refuse to be part of a problem, and accept a prison sentence
rather than compromise my principles or my dedication to a
just world. If I am faced with such a decision again today,

347

having to choose between an action I believe is morally wrong, and being punished or imprisoned for choosing what is morally right, I will have no difficulty making that decision again. I will stand for the principles and take the path I believe is morally and conscientiously right.

Has this any relevance to my Analysis of the Patterson Gimlin Film? Given some of my most severe critics accuse me of being unconscionable, this may be a fly in their ointment. Personally, I do think it does offer an example of my determination to do what I feel is right and morally proper. I do apply that same determination to the analysis of the film, and the aspiration to find the truth of it. I will not do something I believe is wrong, regardless of the benefit such an action will provide, and I will not compromise on my choice of a rightful path or action, even if threatened with punishment that could ruin my life. I learned as a child to "let your conscience be your guide". It still is.

I will be monitoring the internet machine and will be intrigued to see how this information impacts on the consideration of my PGF analysis, and debate about me. It will be a fascinating study in human nature, and an intriguing experiment to see how the Monster From the Id responds. But whatever the consequences, I have no regrets for my choices in life, and am willing to accept the consequences for my actions. And I have no regrets for my analysis and conclusions about the PGF. I honestly and proudly know what I have written is true.

To Quote or Not to Quote, That is the Question

One curious thing I see occurring again and again, is that my critics will not quote me, or source the claims they make about me. They will "explain me" (in their own words), they will describe what they perceive as my ideas, my efforts, my motives, or my thoughts, but they will not disclose what I actually said, wrote or did that can be verified. To me, this is truly the most profound failing of theirs, and the ultimate vindication of my being both truthful and sincere. They cannot use my own writings, my own speech, and show what their criticism is based upon, or point out the specifics of where my remarks are in error, because if they could, they would quote me and source the quote.

In my opinion, this is the most damning characteristic of the JREF members, collectively, as deluded people lost in a mindset of criticism masked as skepticism, and malice as their goal. They cannot quote me correctly word for word and then dissect my statement and show you the error of my ways with a specific rebuttal. They will not reference my published documents by title, by actual content, or location on the internet so you can go to the source and see for yourself. Instead, they must describe me with their vocabulary, they must "interpret me" as they see me, and then attack what they interpret, demean me for their choice of words and ridicule not the truthful me, but the perception of me, in their warped minds. They try to insure that you will only see me through their eyes, not through your own. And they consistently try to hold me responsible for their errors, their ignorance, their prejudices.

Somehow, my simple statement "I do not think the PGF is a hoax" rattles their world enough to turn their brains to cottage cheese and the only thoughts that stay focused are the most primal ones, the fear of the threat to their hoax myth, and the desire to counter-attack the source of the fear.

So I will simply say that those who criticize me, if they will not quote me and source the quote, so you can go to the source and see what I actually said or wrote, in context, then these critics are attacking a figment of their imagination they have projected at me. I can't stop them from doing so, but I will trust that rational people who do understand the importance of actual quotes will see through their deluded or manipulative "interpretations" of me.

Taking on the Monster from the Id

I continue to be intrigued by this phenomenon, the internet as a modern day Krell machine, unleashing and empowering the "Monster from the Id" of lonely and disturbed people, and I expect as long as I am considered a "public person" (one who's work and ideas are of public interest) this monster will continue to attack me. I am tolerant of it simply because I currently have no capacity to fight it. The old and well-worn saying, "Grant me the strength to change all that I can, the patience to endure all that I cannot change, and the wisdom to know the difference." comes to mind.

I do wish to change this situation, but it will take a civil lawsuit against anonymous people who defame a public person on the internet, and a judgment which sets new definitions on what behavior defines defamation. But most of the prospective defendants, those defaming someone, are likely not financially well-off (deep pockets, as the legal profession describes them) so any lawyer qualified in internet defamation suits would likely only take such a case on retainer and cost, not on contingency. So while I'd like to take on my critics and force them into a court of law and see how well they can actually prove their claims against me, it will likely cost me somewhere between $50K and $100K to do so, and that is money I do not have.

So if I have the "strength" (the financial resources), I will change all I can. Now, I have the patience to know I cannot change this yet. And I do understand the difference.

But one day, I do hope to take on the "Monster from the Id", in court, because I am sufficiently confident of the sincerity of my effort to expect I will prevail, and by prevailing, I can set a judgment or judicial president that will benefit other people who also are subjected to attacks from the "Monster from the Id", anonymous internet defamation.

When Roger Met Patty

Chapter Fourteen - Why Claims of Hoax will always Fail

Claims that the PGF is a hoax have dogged this film for 47 years, and there seems to be an entire subculture that believes the hoax idea with such fervent devotion that they are passionately offended by anyone who says otherwise, and these devout believers of a hoax have developed their own alternate reality where their "proof" makes sense to them. It will never make sense to any impartial scholarly analysis, but fringe beliefs never do. I must acknowledge their passionate dedication to this fantasy because it is so prevalent in the PGF saga and it does muddy the water, so to speak, and make getting to the truth more challenging for many people.

I've been dealing with these delusional people for seven years, and have repeatedly asked to see the proof of hoax, and they have never delivered such. What am I asking for that is so unreasonable that they cannot provide it? Well, for starters, I'd like to know who wrote the proof? A real and verifiable name, not an anonymous internet screen name in a chat room. I'd like to have some reference to the person's credentials or qualifications, some descriptive information about what may qualify the person to do an analysis. Third, the title of the book, document or published paper or article would be required. Last, I want to actually read this document, this proof, and see the evidence used, the method of analysis, and the reasoning applied to the evidence to arrive at the conclusion.

That should not be an unreasonable request to comply with, if proof of a hoax exists. Seek and you will not find. It doesn't exist.

What does exist are the following:

David Daegling's book, Bigfoot Exposed, and even he concedes, on page 106, about the PGF, *"since at this writing thirty-six years have passed, and definitive proof of a hoax has not surfaced."*

Abominable Science, by Daniel Loxton and Donald Prothero, finishes the PGF analysis with *"If Roe's report is a hoax, we would be compelled to conclude that the Paterson-Gimlin film is also a hoax"*. Given William Roe's event is simply a man's testimony about an encounter, and cannot ever be proven true or false, that admission by Author Loxton is essentially an admission that the PGF cannot be proven a hoax either, by his criteria. And the "if one, than the other as well" concept is actually just wishful thinking, not a concrete cause and effect connection. Proving Roe's testimony false cannot change all the empirical evidence of the PGF.

Greg Long's book, "The Making of Bigfoot" is 466 pages of personal claims, recollections, memories, and various hearsay and gossip, and I've already cited multiple sources (including skeptic Donald Prothero, author of the above mentioned Abominable Science) as to why such memories, recollections and similar testimony are poor evidence science cannot rely upon for proof of anything. And in Prothero's book, the author quotes science historian Frank Sulloway as saying *"Anecdotes do not make a science. Ten anecdotes are no better than one, and a hundred anecdotes are no better than ten"*. That just about sums up Long's book, and no matter how many anecdotes he presents, the volume does not improve the quality of the poor evidence. It should also be noted that none of Long's testimonial accounts have been cross examined, or otherwise vetted by independent review, making them even more unreliable.

So all of these publications at least meet the criteria I requested as to author identity, title, some description of the author's qualification, and I can read these "proofs". Two admit they fail to prove a hoax, and the third author claims to prove it but with evidence that all good science rejects as unacceptable for factual proofs. All in all, the advocates for the PGF being a hoax are failing miserably to find a champion who will prove their belief in a concise and definitive way.

I have often seen such advocates for hoax claim the "proof" is in internet forum or chat room discussion threads, and they are apparently oblivious to how truly irrational that claim is. The phrase "It must be true, because I read it on the internet" has become a cultural joke, yet these die-hard hoax believers act as if this statement were impeccably truthful and profound. There cannot be a more absurd claim than to say the "proof" is in a discussion thread of a chat room, when no one of them is willing or capable of taking said chat room postings, edit out the useless idle socializing, and verify the factual claims, and put the "proof" into a concise and structured scholarly document we can study.

So, what would actually prove the PGF is a hoax? Here's a list of things that would scientifically and factually prove it:

1. Prove the PGF was edited before it was copied. This would powerfully impugn the film as evidence of authenticity. No one has done so. The only comprehensive analysis of this issue has been done by me and I can certify there is absolutely no proof of editing the original before copies were made.

2. Prove that the elapsed time from segment one to segment six is more than a few minutes. This would be ironclad proof of a hoax, because a spontaneous truthful encounter must occur in such a brief time, and a longer duration would only occur if filming were deliberate and staged. There is no such proof of longer elapsed time, and the film has ample evidence to examine this issue, using analysis technology that Roger Patterson could never have anticipated to exist 47 years in the future.

355

3. Prove that Roger Patterson had some actual ape suit costume making experience and show that his skill level is on par with what the PGF subject looks like. Even Long's exhaustive examination of Patterson's life failed to turn up any such proof, any evidence, even the poor testimonial evidence, much less some good evidence. And Phillip Morris' claim about providing the "costume" has never been verified or endorsed even by the makeup artists who denounce the PGF.

4. Prove there is any kind of "cheat" routinely done in staged filmmaking. New analysis technology should be able to find such if it exists, but it does not.

What does not, and never will, prove a hoax:

1. Trashing the life of Roger Patterson. Regardless of how convoluted you twist the events of his life to try and paint a picture of him as a dishonest man, proving a hoax is like proving a crime, and one needs to prove motive, means, and opportunity to do so. Opportunity we will concede, and motive (or intent) is arguable if you like to wallow in character assassination, but means (or capacity) to perpetrate a hoax one simply cannot prove with Roger Patterson, and the often claimed "but he was an artist and inventor, so of course he was capable" fails miserably when one actually studies how ape suits are made and how a person develops the skills to make one that rises above a cheap fake, and the PGF subject effortlessly rises far above any such simplistic claim. If you cannot prove means, or capacity, you cannot prove the hoax.

2. Looking at statements made by Roger Patterson and Bob Gimlin and trying to find discrepancies in what they describe, and trying to expand that into proof of deception, and then proof of hoax. Such discrepancies are more likely just human frailty, and until you can show cause why that option can be confidently excluded, these "discrepancy" claims will fail as proof of hoax.

3. Sifting through the whole PGF saga looking for "suspicious" events or occurrences, like the film processing timeline issue, and trying to build a mountain of "suspicions" by recklessly disregarding the more benign possible explanations for each suspicious event or occurrence. Failing to consider and show cause to exclude the benign possible explanation is the surest way for such a proof to fail on review by impartial minds aware of how real evidence is used to actually prove things.

4. Starting with the absurd claim that Bigfoot does not and cannot exist, so the PGF must be a hoax because a real creature is impossible by the laws of physics or biology. The foundation (Bigfoot does not and cannot exist) is impossible to prove, actually, and the concept of Bigfoot (if you define it as simply another hominid co-existing with us today) is not impossible by any laws of physics or biology. So using the claimed "impossibility" of Bigfoot existing to try and prove the PGF must be fake is a massive leap of irrational reasoning to the pretense of seeming logical, and will fail miserably as an enduring proof. But advocates for a hoax do love to fail, and they embrace this bit of intellectual flimflam with sincere devotion.

But the final and most compelling reason why attempted proofs of hoax will certainly fail is because the PGF is not a hoax. The truth of that bizarre but honest encounter will always trample over convoluted and contrived attempts to prove a hoax was staged. The truth will win, at least in rational minds.

Endangered Species?

Feral Human?

CO-Existing Hominid?

Ownership of Habitat?

Chapter Fifteen - Implications

This brief chapter is simply my thoughts on the future impact of this determination, that the film and Patty are authentic. I am speculating and would welcome hearing or reading alternate ideas about this.

First and foremost is the question of why the entity we call "Bigfoot" continues to be so elusive. Aside from Patty, I actually do know about people who have seen these things, and I find them credible, but it is for them to decide if they want to put their experience into the public mind and endure the skeptical antagonism, or keep their experiences private until such time as the entity is fully verified beyond any doubt. If I had a fortune, I would not hesitate to finance more field research to hopefully verify the entity's existence. But at present, that's just wishful thinking, as I do not have a fortune, and the way I priced this book, the royalties certainly won't bring me one. But my life has seen several phases where the work was very lucrative financially, and some work like that may come again for me, so if it does, I will gladly invest in field research, because I feel it is a worthy endeavor.

So, knowing Patty is real, knowing some very credible witness experiences that suggest the entity is alive and well, even if Patty herself may have passed away, I think that I will see the day when the anthropological sciences must come to terms with this entity. And I will be fascinated to see what is determined when they do.

One of the most intense debates I would expect to see is the question of whether this entity is "human", because that raises the question about whether anyone has a legal right to collect a "live specimen", (or kill one) because we obviously can't if its human. No doubt someone would illegally, but that would need to be kept quite secretive if someone did.

Then there would be the question of whether this entity is a feral human population or a race or tribe needing legal protection. There is the question of whether we have any right to try and make contact, attempt communication, and introduce them to our modern technological world.

Then there would be the question of their habitat, their land. Suffice to say, any consideration of their habitat will lead to some intense controversy about what is theirs, what is ours, and now, since nothing is theirs, that means we will lose some substantial land, and somebody won't like that. So I wonder how that controversy will play out.

Another potential controversy that could make a royal mess of the land issue is the question of whether they might be designated an endangered species, because that could also impact on the land ownership and land use idea in a powerful and perhaps inconvenient way.

In a way, it's funny that a final proof they exist is fraught with problems that in all likelihood the scientific community, the business community and even the government bureaucracies do not want to deal with, and there are of course fanciful conspiracy theories that tell of secret government agencies intent on hiding Bigfoot bodies so the species is never proven and these problems can be avoided. I don't buy into these wild theories because I don't have that much confidence that the government is that good at hiding things. But I do think that any such final proof of existence will come with a lot of baggage, as much as it will come with the enchantment of finally understanding the mystery.

So I do wish for that future day when I may see one with my own eyes, instead of just through Roger's film.

When Roger Met Patty

Fig. A1.01

Appendix One - Film Frames and Footage Copies

A detailed explanation of the film and photo analysis material

As I have stated from the introduction on, the film itself is by far the best evidence and so I give that evidence priority over the circumstantial trivia and the so-called "backstory". The film data is truly the more impartial, the more factual, and the more reliable.

But there is a misconception held by many people that "the film" is one whole entity and any image from it, any copy of it, is the same as another. This is emphatically wrong and can lead to errors or misunderstandings, as well as flawed analysis.

So I wish to divide the photographic evidence into three categories, for clarity:

1. The Film - This will refer to the fundamental film data which is constant regardless of what copy or what frame in being studied.

2. The Actual Analysis Film Components - This will refer to specific frames, copies, or image material derived from the film in a certain specification.

3. The Ancillary Film Data - This refers to film of related use in the analysis, and includes other footage on Roger's reel, other films Roger took, and other footage of the Bluff Creek site taken by other researchers over the years.

One - The Film

The film itself is generally known by these specifications. It was 16mm color film, generally described as of Kodachrome film stock (a fact attested to by others over the years and apparently verified by markings on a film can from John Green's archives, when he had the original and Canawest Labs made his copies. That film can has the markings as follows, on the can:

"Canawest Sasquatch Orig. Kodachrome 6533A". This would give support to the description of the original being Kodachrome.). The Kodachrome film type is a vital fact in discussions of film processing and suspicions the film was shot on a day other than described (October 20, 1967), because of how Kodachrome is processed and the questions of what lab did the processing.

The film was a 100' roll on a daylight reel, as evidenced by the light washout found on the last few frames (a common occurrence unloading a daylight reel) and all known descriptions of the reel say that the footage before the Bluff Creek encounter was simply multiple shots of a man on horseback pulling a pack horse and pack through woodland scenery. As we have the footage, this description is reasonably reliable. The horse/rider material takes up about 76 feet of the 100' reel.

We also have a camera identification notch on the film verifying the camera used was a Kodak K-100 model camera, and that type takes a 100' daylight roll of film.

The Bluff Creek Segment, which we will refer to as the PGF, is currently known to have 954 frames (by the Verified Frame Count System, Version 2 that I have developed. See Appendix 2) and at 40 frames per foot, that calculates to 23.85 feet of footage.

There is an error in the original frame numbering, and we actually do not have documentation of who did that first frame number system, but that numbering system has been used for most of the PGF history, so chances are most other books or written references to frame numbers will use this incorrect system. The old frame numbers are off by 2 frames and likely the error was cause by using a copy that started at frame #3 and mistakenly thinking it was frame #1. So the famous "Frame 352" (which is the only film frame widely accepted as being in the public domain) is actually frame #354 by the Verified Frame Count V2 system.

So the current status, as this text is written, is VFC-2 system and 954 frames truly identified.

Appendix Two describes the Inventory System in more detail.

The film has six filming segments that can be identified, and two frames which are suspected to be single frame exposures but not certified as such yet. The known film segments were determined by a combination of subject change from one frame to the next that cannot be accounted for by camera motion alone, and an over-exposure of the segment's first frame, which is evidence of a camera start. So, basically, Roger Patterson started his camera 6 times to film this famous footage.

Focus vs. Motion Blur

The film is often described as "out of focus" but it is not. It has a lot of motion blur, and some people do not know the difference. When film is out of focus, the image gets fuzzy, diffused, and often is more intense as a factor of distance, so the closer things could be fuzzy and the distant things sharper, or the closer things sharp and the distant things fuzzy, but we do not see this. To the contrary, in the frames that are sharp, we see the nearest objects in foreground and the furthest landscape objects both equally sharp, indicating the film was in perfect focus. Also, we see changes from blurred to sharp occurring in a matter of a few frames, and a camera operator cannot change the lens focus setting that fast.

Motion blur on the other hand, does exist in the film, and that is caused by the shaking of the camera by the man holding it as he filmed and ran to chase his filmed subject. So the frames can get blurred and then sharp and blurred again, often in fractions of a second, as shaking vibrations and the camera operator's attempt to steady the camera were in a constant battle.

What was largely overlooked for most of the last 47 years is the fact that with a hand-held camera and a camera operator in motion, chasing the subject, the film footage actually tells us quite a lot about what the camera operator is doing, while we see what the filmed subject is doing in the film. This relationship between what the subject does, and what the camera operator does, has a connection that is quite fascinating and revealing. It was never properly analyzed before, because the analysts did not have the camera operator experience I had to draw from (described in Chapter Three).

Film Grain and Claims the PGF is "Grainy"

Film grain is an issue in the analysis, and the film itself is often described as "grainy". This is incorrect and oversimplified. All film has "grain" which is the common industry term for the silver halide particles which are light sensitive and thus change depending how much light strikes them, and in color films, depending on which color layer they are embedded within. It is the size of these grains of silver halide, as compared to the dimension of the film, which determine the characteristic we call film grain.

A second factor in general is that "fine grain" films tend to require more light for a correct exposure, (they are called "slow films") and less fine grain films which require less light for a good exposure are usually called "fast films". A common example is three of Kodak's classic black& white films, Pan-X, Plus-X and Tri-X. Pan-X was the slowest, and the finest grain, but needed the most intense light for good exposure. It had an ASA of 32. Plus-X was a medium grain film, a good utility film, and needed less light than Pan-X, having an ASA of 125. Tri-X was the "fast" film, very grainy and had an ASA of 400. So you could choose your film based on the degree of fine grain or less fine, and a common rule was that the smaller the ASA number, the more fine grain the film was.

Based on transparencies taken off the camera original, we can confidently say the PGF original is as fine grain as any color 16mm film can achieve. The best image benchmarks for analysis show the PGF is right at the top of Kodak's image resolution threshold, as good as Kodachrome film can resolve. Indeed, its quality of fine grain would make any professional cinematographer proud.

Processing also affects grain, and every film has a recommended exposure and a standard processing routine. When done this way, it achieves its finest grain structure (and most image detail). But if film is deliberately underexposed and then over-developed by the lab (a process commonly called "pushing the film" and a phrase like "push one stop" means to over-develop enough to compensate for a one F-Stop under-exposure of light), the result will be grainier than the same film at optimum exposure and correct processing time. Another phrase to describe this "pushing" is "hot soup".

Why is the PGF so often called "grainy" when it is not? The actual likely cause for this belittling of the image quality is the fact that most people don't look at true full frame images. If you saw a true full frame image of a sharp frame (no motion blur), you would see it is very high quality, and not "grainy" at all. But most people simply look at the highly magnified images of the subject "creature" and that subject was generally not more than 1/6th of the frame height at its closest. So the magnification to make an image of just the film subject is a 6x magnification larger than what the film was intended to be viewed at, so you are seeing the film's grain magnified 6 times. Under those conditions, yes, you would see the grain with any film, not just this one.

So the "grainy" description is incorrect in the true film sense, but still widely used to disparage the film and suggest there's no precise image data to analyze.

The film's grain is, however, a factor in analysis of the film but in a way usually not considered or properly described.

The degree that the film grain affects analysis is in direct relationship to the image area being studied. If you study the full frame image, the grain has the least impact, indeed essentially no impact to be concerned with.

If you study a zoomed in 2x magnified portion of the film frame, the film grain is now twice as large in comparison to the image, and may impact on the analysis slightly.

If you zoom in to a 4x magnification of the film frame, the grain may start being noticeable and may definitely affect what you are analyzing.

If you zoom in to an 8x magnification, you will see grain very obviously and it will be so obvious that it may actually introduce false image data that could lead to errors in analysis.

In essence, the smaller the image area of the true original frame size that you try and study, the more film grain will affect the analysis and possibly distort the findings or introduce an error. So it is very important in any analysis to evaluate what is being studied and how large that item or study material is in the true full frame size.

As an example, if we look at markings on a tree as seen in the film, and we try to measure the tree vertically using distinct markings as our measure points, those points may be 2400 pixels apart (using a true rectified frame scan where the image is 3000 pixels high and scaled to 1 pixel equals 0.0001" of actual image size on the 16mm film). Film grain would have very little impact on this measure. We would consider it relatively reliable.

If we measure the same tree across the width of its trunk, and get 100 pixels (for example), the grain is 24 times more likely to cause an error in that measure, and we would consider that measure far less reliable. This would be what is often referred to as a "error analysis" or "margin of error calculation". The first example would have a very low margin of error. The second measure would have a much higher margin of error.

This comes to factor into the analysis of the anatomical features of the filmed subject. If you try to measure the subject head to toe, you get a reasonably reliable measure. If you try to measure just the head, the measure is far less reliable, because the grain is relatively bigger and throws off the accuracy. If you try to measure the features of the face, like the size of the nose, the film grain is so large in relation to that measure, the measurement is nearly useless. Same subject, same frame image, but the smaller the feature being measured, the more the film grain becomes a factor of influence, and the less reliable the results are.

So in this matter, the film's grain does impact analysis.

How do we factor the film's grain into our analysis?

The general rules (briefly summarized) are:

1. The larger the image area in relation to grain size, the less the grain is a factor.

2. If a shape or feature has a continuity from one frame to the next, it is more reliable image data because grain is a random pattern and it changes from one frame to the next in random ways, lacking continuity of pattern. So small features with shifting shapes are likely caused by film grain. Small features that have continuity one frame to the next are less likely to be influenced by grain, and are more reliable.

3. If a motion is studied and the motion has the following criteria, it is more reliable: A. Appropriate anatomical motion. B. Companion anatomical motions occurring in concert. C. Some anatomical cause/effect which influences the motion. D. Appropriate timing of motion to be anatomically reasonable.

If grain is causing an illusion of motion, it should exhibit the following traits: A. Motion which seems odd and not anatomical. B. Companion motions occurring randomly, not in concert. C. Motions occurring with no cause/effect influence (the best example of this is background objects which are fixed and inanimate, but the grain causes apparent motion, frame to frame even though we know the object has no motion). D. Apparent motion that has unnatural timing based on frame count.

So film grain is a factor we must consider, but we must consider it correctly and in context, and not use film grain as a reckless excuse to just dismiss evidence or belittle the film.

Copy Process and Impact on Analysis

As the true camera original of the PGF is not currently available for study, all analysis is done with copies of some form. So understanding copy process is important to any analysis. The most basic concern of any copy is it's generational level. When film is copied one to one (the image is the same size on the copy as on the original) then the grain patterns of the source film and the copy film stock may misalign, since grain patterns are random This causes image detail loss. The copy simply is not as sharp as the original and some fine detail is lost. If the copy is copied again, even more detail is lost. With each generation of copying, the image simply gets poorer and poorer. So knowing what approximate generation any given copy is (based on some benchmark of image sharpness) is important for foundation of the analysis.

But if an original image is copied to a larger film format, the effect is different and the grain is less damaging to the image detail. The best example of this is the transparencies, made by Kodak from the camera original that Roger Patterson had then. These transparencies were 4x5" and so the full film frame was magnified from true 16mm frame size (approximately 0.3"H x 0.4"W) to 3.75"H x 5"W, a 12.5x magnification. Thus the copy transparency had grain about 12.5 times smaller than the source film image, so the image detail was not diminished in any noticeable way. These transparencies generally represent our finest evidence of the true original film quality and are the benchmark standard for image clarity that other copies are graded against. Unfortunately, only a few such transparencies were made. They have been scanned for analysis twice in recent years, once by a prior researcher and once by myself.

The second type of enlarging copy was done by John Green through Canawest Labs when John and Rene Dahinden acquired Canadian Rights for the film. They had their lab make a copy of the original that included a full frame pass of the entire first reel, and then a 2x magnified pass, then a 4x magnified portion showing the lookback part, then a slow-motion pass of the lookback (by printing each original frame 4 times), and a freeze frame pass, where frame 354 (called 352 by the old numbering system) was printed over and over so on projection, it just seems frozen as a still image. There may have been additional print versions, but these have not been verified yet.

The full frame pass would suffer the normal image detail loss of a one-to-one copy. The 2x magnified pass would have far less detail loss because the copy film's grain size was now 1/2 as large in proportion to image detail, so that pass is clearer and more detailed on the creature subject, but it is cropped so we don't see the full scene. The 4x pass of the lookback is potentially an even higher level of detail, near the true original in grain sharpness because of the magnification making the grain relatively 1/4th of the original, and thus capturing more original detail well. But this segment is only for the lookback sequence, and has even more surrounding landscape cropped out. Also, a variable is how accurately the Optical Printer was set up for this. I have seen two versions of the 4x enlargement done by different people, and one is far sharper than the other, because of printer setup precision.

This copy group done by John Green and Rene Dahinden is potentially very useful for analysis, but within this copy group, we again must consider the generational level of any given copy film strip. Current research indicates the copy process they used was as follows:

Camera original source to copy interpositive master on Ektachrome film stock. The copy I have scanned included edgecode for Ektachrome stock, and typical Ektachrome edgecode numbering. This master would have each type of print version (full frame, zoom in, slow motion, etc) and would be considered a First Generation Version.

Interpositive to multiple projection positive copies (exact number not determined.) These would generally be fine copies. They have a second edgecode with a "Print Film" latent image on the opposite side as the source Ektachrome latent image that printed through onto these copies. These Projection Positive Copies would be considered Second Generation Versions.

Copies of Copies - If the original positive copies were edited into presentation or program reels and a new copy of the presentation were thus printed in a direct reversal positive print, possibly with a sound track added (as one example), these copies of the PGF parts would be lesser quality than the Original Projection positive copies above. Such Copies of Copies would be considered Third Generation Versions, and the full frame segment would be mediocre at best, but the zoom in segments would still hold good research quality.

The genealogy of these copies is sadly not well documented, and I am currently engaged in trying to establish a reliable copy quality benchmark system for them.

For analysis of the creature subject, during the lookback segment, clean and sharp 4x zoom printed segments would be by far the best quality for the creature analysis. But it had never been scanned at high resolution before I began doing copy scans of this film. It was scanned at TV resolution for a DVD program of "Legend Meets Science" but the reduction of the scan to TV size lowered the potential image quality, plus the TV scan must add frames to get to TV standard of 30FPS, and these added blend frames diminish quality. Never before has any researcher scanned the 4x magnified segment at high resolution (4K) and at true frame scan, with no frame blending for TV conversion. So these scans are the finest research and analysis scans we can hope for, unless by some stroke of luck, the original turns up and can be scanned itself.

I am the first researcher to do these 4K scans of true frames of the 4x zoom segment.

All of this copy material is obviously quite technical, and the reader may not be able to follow all the details of the matter. Generally a solid education in film photography and processing is needed to fully grasp the significance of this material. But the important thing to keep in mind is that any film researcher or analyst who tries to show images from the film and prove anything, or argue anything in discussion forums, that person should know the copy history of the image being used as foundation for the discussion or proof, because if the analyst discussing the film does not know this, the analyst is simply unqualified to offer any conclusion and show proof of anything.

This issue is precisely what undermines the remarks of so many Hollywood Makeup Artists who have from time to time offered opinions about the film's creature. We do not know what they actually looked at, what they actually studied, to form their opinions, and so we actually don't know if they looked at the best image material or if they looked at flawed and inaccurate copies. Thus the very foundation for their opinions is lacking and so their opinions will fail to hold merit in any rigorous proof. Sadly, I have actually offered numerous makeup artists access to my image database (the finest one currently in the research community) so they at least have the appropriate and best image data to examine, and base their opinions on, and yet none who were offered has taken me up on that offer.

Two - The Actual Analysis Film Components

So now, I will list the general categories of Copied Image Data useful in research and the considerations for use of each. A more detailed history and specifications of each PGF Copy is shown in Appendix Two.

The 4x5" Transparancies - As there are only 3 that I can verify, they have very little analysis usage except for setting a benchmark standard for image quality. They reveal to us how much image detail was actually in the camera original film for frames that didn't have any motion blur.

The 12 Cibachrome prints - These have very high detail of the creature subject, but also are known to have some image anomalies which make them challenging as research tools for analysis. They have the potential to show evidence and help present research issues clearly, but they alone cannot be used for proof of anything. Plus the fact that they are scattered frames not in exact sequence means they are not useful for anatomical motion studies by themselves. If used for motion studied, we must factor in the frame gaps and adjust the time of the apparent motion shown by switching from one to the next.

A true full frame First Generation Copy - even though a first generation is generally fine, if it was a true full frame copy, the creature subject does lose considerable image detail because of the film grain on proportion to the creature image size. These types of copies are better for general analysis of the Bluff Creek landscape, and ideal for photogrammetry analysis of the site.

Second and third generation full frame copies - These have little research value other than as cross references for image anomalies, where we examine multiple copies to see if the anomaly is in all copies or just some. Poor for studies of the creature subject.

Second and third generation 2x and 4x zoom copies - These are the finest for the creature study. The LMS DVD shows these, but the quality was diminished by both the downscale to TV resolution in the scan, and the frame blending necessary for the 30 FPS TV format. They still hold good general observation quality when watching the DVD and thus can illustrate some creature studies in general terms. But they should not be used for any formal proof. Better used would be true 4K true film frame scans. To date, I am the only researcher who has done such scans.

ANE Zoom In Copies - The American National Enterprises (ANE) company had Roger's original and printed some zoom in copies for their documentary, "Bigfoot Man or beast" (1971). Their 4x zoom in of the lookback is the best version for studying the lookback phase.

The Noll frame scans - researcher Rick Noll did microscopic photography of a 2x zoom copy and produces over 500 3K frame scans of the creature subject. Distribution of the scans is informal and generally by reference or personal assistance of one researcher to another. The generational copy level of the source film hasn't been certified yet (if they were second gen or third gen). We simply know they were from John Green's archives and John has both second and third gen copies amongst his holdings. The Noll Frame collection is widely used by researchers and for many years was the best scan set for analysis, so we see many analysis efforts based on that scan set of images. The M.K. Davis image stabilization work was done with this scan set.

Unverified Copies - occasionally researchers use copies that are not verified as to source, generation and quality. Use of these copies is problematic and undermines any analysis or argued proof that is based on them. An example of this is some frames published by researcher M.K. Davis, using copies which were curiously cropped and heavily distorted in spatial accuracy, as if there was a mis-alignment of a copy device lens as compared to the source film plane.

Images from the internet with no copy history and no foundation of image processing - There are a multitude of images on the internet, some horribly distorted, and useless for analysis. They clutter up the debate and mislead people more so than they help or prove anything. But they enjoy popularity with what can only be candidly described as "the lunatic fringe" because they often show strange things which foster claims of strange explanations of the PGF story. Sometimes these are frame blends from TV scan conversions and they are false images. Sometimes they are image variations on one specific film copy which are mistaken as being on all copies, and used to support odd conspiracy theories.

Such internet images with no copy history should be avoided by any responsible analyst.

Three - The Ancillary Film Data

Aside from the actual PGF (which is a 23.85' segment from a 100' roll shot by Roger Patterson), other footage does factor into a proper analysis and full understanding of this mystery. Most of this footage was known for many years, but its proper relevance to the PGF is usually misunderstood. So the goal here is to describe this footage and try to put it into a proper perspective as to content and relevance.

Reel One Other Footage - As noted previously, Roger's camera held a 100' daylight reel of film. The PGF Bluff Creek encounter that forms the heart of the mystery is only 23.85 feet of that 100' roll. The remaining 76.15' is generally referred to as "First Reel Footage" and is generally described as just shots of either Roger Patterson or Bob Gimlin on a horse, pulling a small white pack horse through the woods. Such footage exists and has been scanned for study, so there should be little dispute that this is the earlier first reel footage. But skeptics love to dispute everything, and so they raise suspicions that the first reel footage has something on it that may expose a hoax and thus was edited out. They clamor for an intact camera original 100' roll to "prove" what all was in the first reel, but such isn't available for inspection. And such skeptics fail to understand in any filming, the film's owner may reasonably look at the footage and judge some material as useful and some material as "outtakes" and simply cut the outtakes off the roll and show and copy the useful material. So if the camera original does turn up and can be examined, and if it has the earlier material cut off (as we expect it will have), skeptics will have more suspicions of why. But there is nothing inherently suspicious about such separating the useful footage from footage perceived as "outtakes" because Roger could never have imagined that his film would be debated with such intensity and microscopic inspection today as we do examine it.

But this issue continues to be one of the many "red flags" raised by skeptics to try and fuel suspicious agendas. One of the more curious things about this footage is that people who study the film impartially do not have much interest in this rider and his packhorse footage, because it really doesn't prove anything significant about the PGF segment. We examine it primarily because skeptics use the footage to make suspicious claims and so we study it to refute those suspicious and paranoid claims.

I recently examined one reel of film from John Green's archives, and on it was what appears to be an intact full first reel. It runs about 100', has the horse and rider stuff on the front, a full frame PGF copy on the last 23.85 feet, has camera starts for every segment except the very first (which was likely torn, as the edge is jagged, and a splicer cuts a clean edge. Also missing is the Canawest headleader the copy should have if fully intact), and this footage has the correct history, made by Canawest labs in late 1967 when John Green and Rene Dahinden purchased Canadian rights to the film from Roger and borrowed the camera original reel to have their lab make both a straight through copy as well as some enlarged and slow motion versions. I have done an archive scan of every frame of this intact first reel copy, and so, as far as I'm concerned, the whole issue of suspicions the first reel was somehow spliced, tampered with or edited, has no merit at all. But cynics indulging paranoid fantasies will still try and claim the PGF was edited. Nothing rational will shake their delusions.

Reel Two Footage - This is described in detail in Chapter Twelve - Loose Ends

And like the first reel footage, this footage is of little significance to people trying to impartially understand the PGF, but is widely used by skeptics to try and find more red flags to wave. So researchers study this also just to refute skeptical claims and delusions of hoax.

Its only real value is the trackway footage itself, and the potential to develop a photogrammetry analysis of the terrain and footprints in 3D. Technology to accomplish this has been under consideration for some time, and the solution is a work in progress.

The Green/McClarin Re-Creation Filming

Jim McClarin was a researcher who went to the film site shortly after the event was reported, and Jim described that he could still see traces of the original trackway to get a sense of where the creature walked. He and John Green would return to the Bluff Creek site in the next summer (of 1968) to film a recreation of the walk and lookback. John would film, and Jim would be the subject on camera doing the walk. As he recalled about where the trackway was, he tried to walk a similar path. And John, having some stills of the PGF footage tried to find a camera position which very closely approximated where Roger had stood filming the PGF lookback part.

This footage was scanned for TV once for a documentary called "Legend Meets Science". And in January 2009, I scanned about 35 frames of this walk at 4K high resolution from John Green's camera original footage. Other than that, it is John's testament that he has not had this footage copied in any other form.

All things considered, John did a highly admirable estimate of camera position (being off by only about 4-5' further back than Roger was), and his site measurements of some of the trees, from that position, helps us to develop an accurate site map. And Jim did a fairly good attempt at walking a similar path. But neither was perfect, to today's exactitude, and so comparisons between the McClarin walk footage and the PGF footage have caused more bad analysis than good, over the years.

But for many reasons, this filming by John and Jim was truly one of the most beneficial film supplements to the PGF ever done, and if properly analyzed, have great potential to help us clear up the PGF Mystery. The challenge with this footage, is to first establish the proper foundation for comparisons, and that was not done by prior analysts.

Jim's second walk was filmed by a different camera (as we can determine by the different aperture shape) but from the same spot, so it was likely mounted on the same tripod John Green set his camera on to film walk #1. This was likely the Bolex we have photos of at the site. The cameraman is not currently identified, and the camera original is not currently sourced. All we have is a copy of this footage held by a Russian Researcher and this footage is described as being given to the researcher by Rene Dahinden in 1971.

Walk #2 footage has a camera stop about a third of the way into the walk, and apparently Jim was asked to pause and restart his walk. The physical copy has the editing of the two segments with something cut out, and when Jim re-started his walk, he pushed off on the opposite foot, so the second section of this walk has his legs in opposite posture as to which is forward as he passes landscape markers. As such, the full walk can't be reconstructed as one panoramic photo montage start to end, as Walk #1 can for study.

The value of Walk #2 as compared to Walk #1 is in the study of tree shadows and passage of time at the site, and comparisons to the PGF tree shadows not showing any passage of time.

Dahinden 1972

This footage quite literally took us all by surprise, because no one seemed to know it existed. It was well established that in 1972, a research trip undertaken by researchers Rene Dahinden and Peter Byrne photographed a man known as Michael Hodgson at the site, holding an 8' white measure bar vertically, to scale the site. Several photos existed, showing Michael at various places along the walk path holding this measure bar, and for many years these black and white photos were used to try and scale the PGF creature, often leading to height estimates of over 7' tall.

What nobody knew was during that same trip, Rene Dahinden himself was filmed by another person there, and Rene is in the scene holding the white measure bar at 7 or 8 varied positions along the estimated walk path. Rene apparently never felt any need to share this processed footage with anybody, and it sort of "vanished" in that no one was aware it existed, and those on site that day apparently forgot about it. It wasn't until a research source offered to let me scan some old footage that the owner thought was something valuable that Roger had given him. The owner himself did not know exactly what was in the footage, just that it was in a really old film can. So it was sent to me for examination and scanning.

Once I had sample scans, I circulated the images of the man (seen in the film) to various researchers, and got a positive ID on the man as being Rene himself. With further analysis, I was able to determine that the camera was exactly where Rene had estimated the camera to be on his one site measuring work (which disagrees with Green's camera location estimate by 20', Green being the further back). But examination of this footage of Rene's also allowed me to determine his camera had a zoom lens on it, which is why he could pick a spot much closer and then still frame his camera lens for a similar horizontal angle of view, and think he was in the right spot.

Interestingly, there was a famous still color photo of the site, and one of Rene's sons as a young boy at the site, taken that time in1972. In that photo, we see some metal rods with red/orange cloth flags on them, which Rene used as markers for his site survey and measurements. One of these markers is near the boy, and doesn't correspond to any on-camera object in the PGF. But having Rene's footage of himself at the site, we can now determine that this odd marker was where he put his 16mm filming camera to film himself with the measure bar, and where he estimated Roger was for his site measurements. So it was remarkable that this footage would surface after being literally forgotten, and that it would clear up one important issue of the site measurements and site analysis.

It would likely never have been known, scanned, and factored into the analysis effort if I had not undertaken to scan and study as much ancillary footage as possible, and built up a trust with the various researchers so the film could be loaned to me for scanning and analysis.

Roger's Other Documentary Footage

This reel of about 550' containing five 100' rolls and about 50' of a sixth roll represent examples of footage Roger Patterson was taking in the months (and maybe years) before his Bluff Creek encounter and the famous filming. These rolls show a lot of mundane landscape filming, a few glimpses of his "Bigfoot Expedition 67" VW Bus, some shows of an old man named Fred Beck, who figured into a much older Bigfoot encounter described in Roger's book on the Bigfoot phenomenon, and curiously, some footage of a man with his daughter sprinkling Kellogg's Corn Flakes on his back yard lawn to attract deer to the property, and one appears and is filmed by Roger.

Nothing seems to have any direct relevance to the PGF but in unexpected ways, it does gain relevance in some of the PGF controversies. One example is the dispute about what type of lens Roger had on his Kodak K-100 camera while filming the

PGF. Conventional descriptions for nearly 40 years said the lens was the standard 25mm lens that comes as default primary lens with the camera. I raised the question that maybe the lens was not a 25mm, because such a lens and the site analysis done to that time did not add up to a complimentary conclusion. But when I raised the question about the lens, possibly not being a 25mm, I was met with a legion of skeptics who insisted that camera rental documents and the police report for the long overdue rental, were "proof" that was the lens. Well, in truth, such documentation at best only proves what lens was on the camera at the time it was rented, and that was 5 months before the PGF filming. One can switch a lens in about one minute, so it did not prove what lens was on the camera that day in October, 1967.

But skeptics pressed the issue by claiming there was no documentation Roger used any camera or lens other than this one rental, and so it must be the one. But Roger's documentary footage that I scanned and analyzed shows Roger used at least two different types of cameras, and a zoom lens on each, at some time, so now we have proof he has used cameras and lenses for which no rental documentation exists. With that data, one can no longer say we must assume the rental documented is the limit of Roger's equipment specifications. So that skeptical claim fell apart once this footage was analyzed.

Roger's Fictionalized Bigfoot Film

Still photos from this filming effort are occasionally shown and discussed, but their relevance is very thin and argumentative. As of this writing, I am not aware of any actual footage being held by any source for examination, just still frames from the footage circulated on the internet. I cannot appraise any potential value to the PGF debate based on the material I have seen thus far.

Blue Creek Mountain

In August of 1967, there were reports of footprints found in the woodlands not far from the PGF's Bluff Creek location, and investigators John Green and Rene Dahinden visited the site, examined footprints, and took still photos and some 16mm movie footage as they studied the evidence. This footprint activity is described as one of the possible influences on why Roger Patterson chose the Bluff Creek area for his search two months later.

But this incident and the 16mm footage also has become a component of an absurd claim commonly called "The Massacre Theory", and said theory tries to say that this footage and the PGF footage are part of the same event, that this footage shows bloody evidence and possibly a body, etc. As noted in Chapter One, The Massacre Theory tries to prove that many of the principles in the PGF incident and investigation were actually hired by lumber companies to massacre a group of Bigfoot creatures and bury their bodies so the lumber rights for the area would not be restricted by any claim of the woodlands being habitat to these creatures. Frankly, this theory is so absurd on many levels that I find it hard to follow the claims and the fantasy explanation. I try to keep my focus on showing why film evidence used in the claim is flawed analysis, such as claims of a "shooter" (a gunman) and gun or rifle muzzle flashes (from discharging a weapon) are not in the PGF as claimed by proponents of this theory.

If the opportunity presents itself, I would hope to examine, scan and inventory the actual footage taken on that day in August 1967, but the opportunity has not yet opened up to do so. Until I can examine the footage, I tend to keep my focus on the PGF footage and any false claims using PGF footage as a basis of the claim.

Still Photo Evidence

Most Still Photo evidence is useful for analysis of the Bluff Creek Site itself, and I have used it extensively to try and map out the actual site in one unified site map, because any individual film frame shows only a portion of the full site.

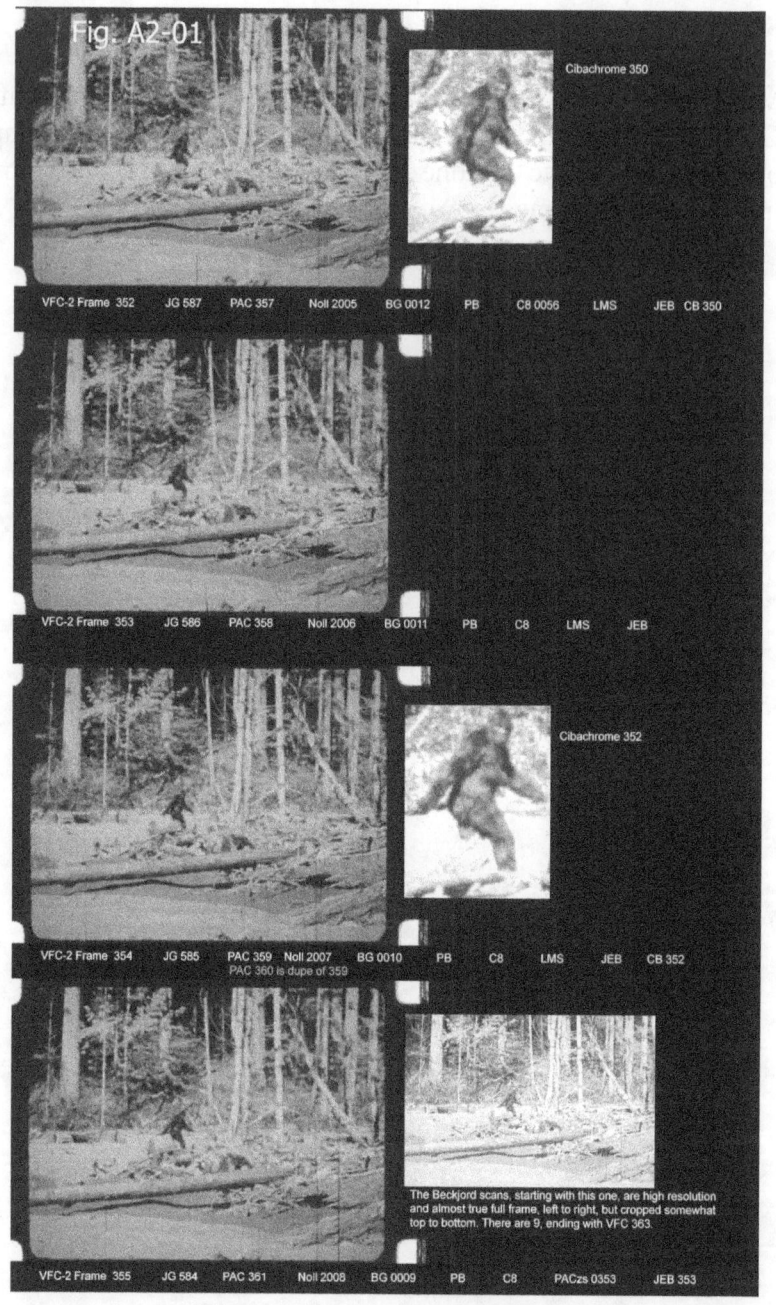

Fig. A2-01

Cibachrome 350

VFC-2 Frame 352   JG 587   PAC 357   Noll 2005   BG 0012   PB   C8 0056   LMS   JEB CB 350

VFC-2 Frame 353   JG 586   PAC 358   Noll 2006   BG 0011   PB   C8   LMS   JEB

Cibachrome 352

VFC-2 Frame 354   JG 585   PAC 359   Noll 2007   BG 0010   PB   C8   LMS   JEB   CB 352
                              PAC 360 is dupe of 359

The Beckjord scans, starting with this one, are high resolution
and almost true full frame, left to right, but cropped somewhat
top to bottom. There are 9, ending with VFC 363.

VFC-2 Frame 355   JG 584   PAC 361   Noll 2008   BG 0009   PB   C8   PACzs 0353   JEB 353

386

Appendix Two - The Verified Frame Count System and PGF Copy Inventory

Two things I feel I may rightfully be proud of accomplishing in my PGF research are the facts that I was the first researcher to actually inventory every single known frame of the PGF film footage, and number them correctly, and I was the first researcher to appreciate the difference between various PGF copies and inventory these copies for comparative study and reference.

The Frame Inventory is important because it gives a true positive identification to every single frame, and that allows us to weed out TV conversion false frames some researchers have mistakenly used for analysis that was wrong. This Frame Inventory can identify image artifacts in some copies that weren't on the original and thus led to false conclusions when the artifact was mistakenly assumed to be real image data. It also allows me to easily reference one copy scan to another, for frame comparisons, and it allows me to easily search the inventory for any specific frames I need for some specific topic of analysis.

The PGF Copy Inventory has proven of tremendous value in comparing copies to distinguish real original image data from copy artifacts and copy edit errors (like one copy that has the first frames flipped left to right, and that has confused a lot of amateur researchers into thinking the film captures two creatures, one walking right to left, and the other walking left to right later. It's the same figure, with just the film flipped, left to right. The analysis of multiple copies verified this). The Copy Inventory has also allowed me to identify the best copies for any specific type of analysis, and also to weed out copies that have some distortion in the copy process and thus are poor for analysis. The Copy Inventory also has been vital in developing the true copy genealogy of the PGF films, which copies are related, which ones are made by which printing process, etc.

So regardless of any criticisms of my PGF analysis, these accomplishments will always be a major contribution of mine to solving this intriguing mystery.

The Frame Inventory and Verified Frame Count

What I have done is sequence and index every frame of the PGF known, all 954 of them, with a large full frame image example, and then a true frame number, and the image scan number of the copies I have scanned. I reference four scan copies now, but my Copy Inventory has expanded considerably and I will have to eventually revise the Inventory to cross reference the newer copy scans as well. It's tedious work, but necessary.

The Inventory is set up as Photoshop image files, 5000 pixels high and 1500 pixels wide (except for the first set, where it's 5500 pixels high to include one frame of the tentpole trees , the sequence immediately before the true PGF starts). In these files, each has ten PGF frames set at 500 pixels high, so Inventory Panel One is frames 1-10, Panel Two is frames 11-20, etc. So there are 96 such files for the 954 frames.

Fig. A2-02 shows the first three inventory panels, with the first having eleven frames to include the last landscape before the PGF starts. So these are the first 30 frames of the PGF illustrated, in three panels of ten each.

On the left is the frame sample at 500 pixels high and at right, some have an identical frame in another format, such as a Cibachrome, or a Noll zoom it, or the light flare that inspired the silly "massacre theory". I put these related reference images alongside the master frame reference image for some study topics I tend to frequently refer to.

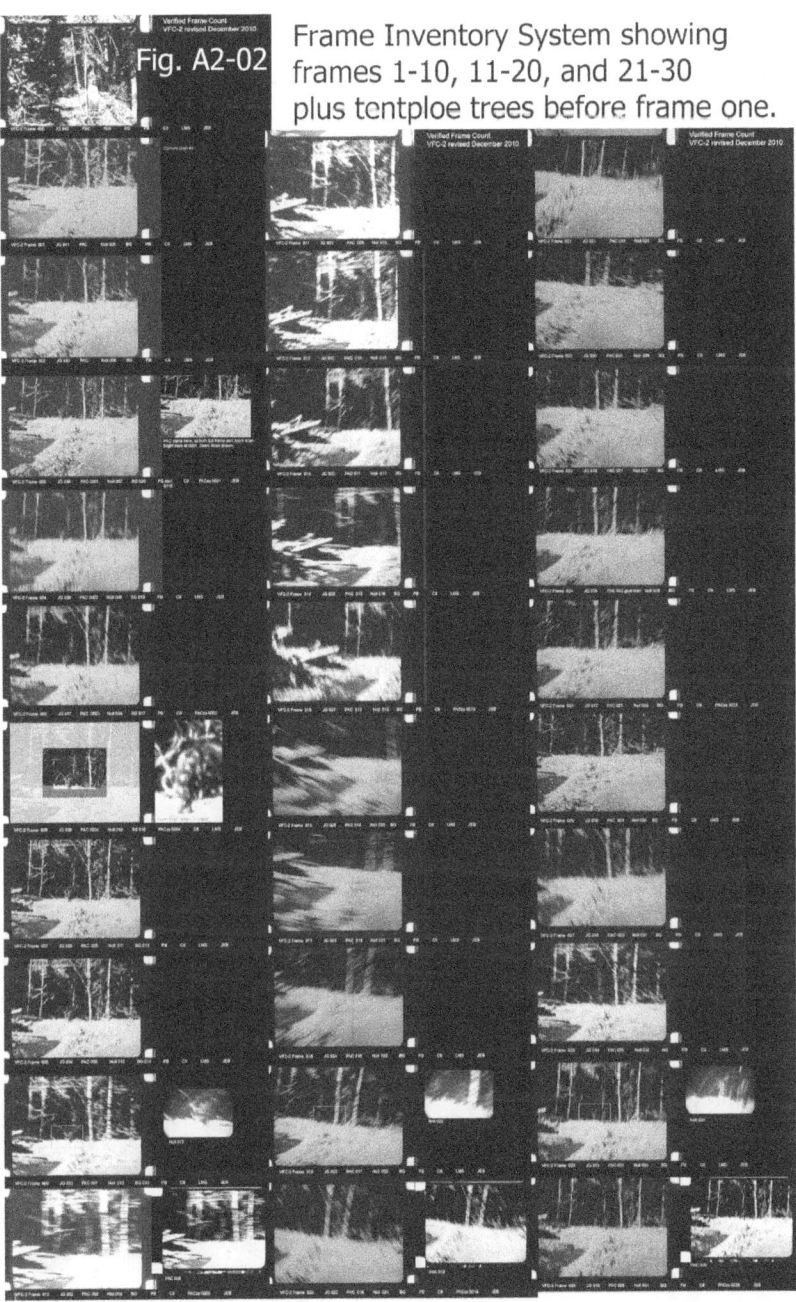

Fig. A2-02 Frame Inventory System showing frames 1-10, 11-20, and 21-30 plus tentploe trees before frame one.

In the early scan sets, I was hand advancing the frames and the work was tedious, and I missed or duplicated a frame, which I later corrected when I assembled the inventory. So in the section where I note the original image scan file number for each scan set, I have a red indicator when a frame has been duplicated or skipped in one scan set. Once the projector was motorized, it was easier for me to scan a PGF copy with an exact frame count in the image numbering, and many of the later copies scanned were done this way. Their scan number matches the Verified Frame Number exactly.

Fig. A2-03 is enlarged to show four consecutive frames and the numbering of each referenced to the early PGF scan copies.

The Inventory is due for a major overhaul, to include the newer copies scanned, and reference their file numbers, and as there are talks about finally setting up a permanent archives for other researchers to access this data, I will need to revise the Inventory to make the referencing of scan sets easier for future research from that archive. It will be publicly announced when it is set up and open for researchers.

The System is noted as VFC-2 which stands for "Verified Frame Count - 2" and that's because my VFC -1 system only counted 953 frames, and had an error in a skipped frame from one copy. So the VFC system corrected the frame error and also acknowledged frame 954 the first time I found it. I have subsequently found several examples of it. There is a possibility for another revision to VFC-3, if I find any more frames in the future.

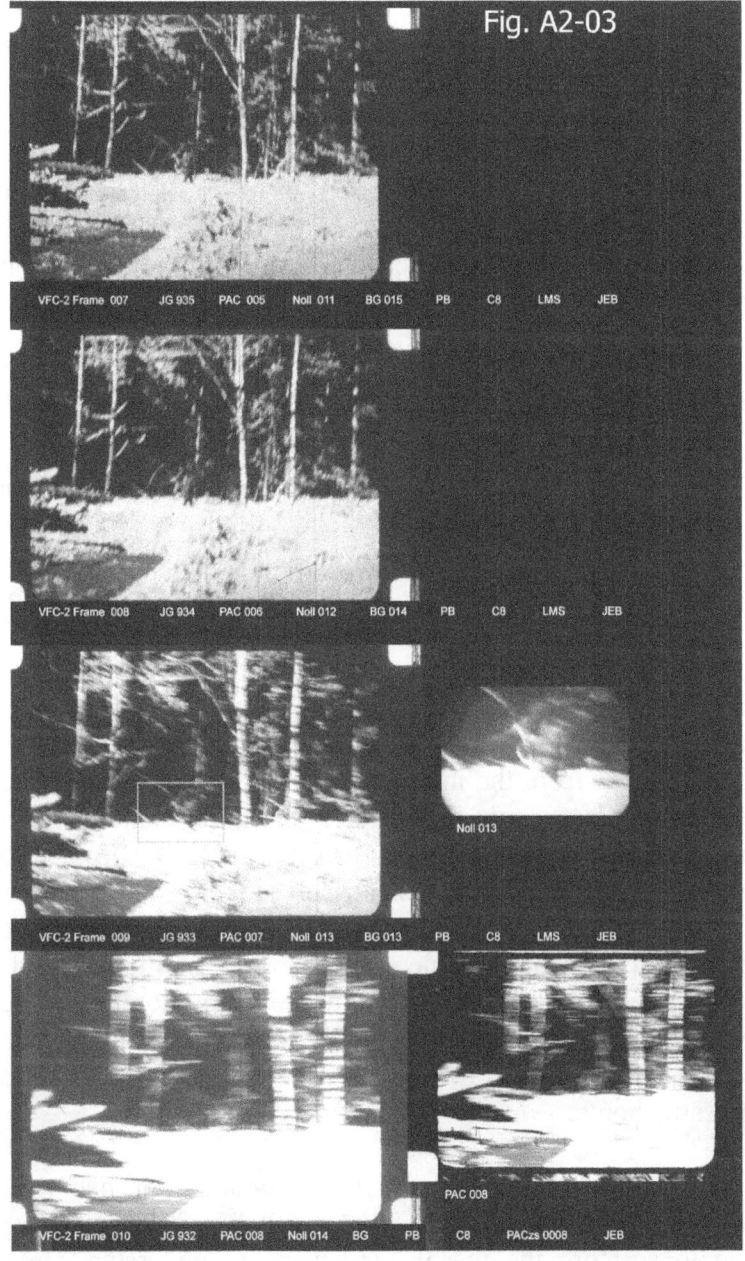

Fig. A2-03

The frame numbering system is two frames different from old
frame numbers, such as the Cibachrome numbers, because

apparently, whoever did the original numbering used a copy starting at frame #3 instead of frame #1 to count. So the classic "Frame 352" (the best lookback) is actually frame 354 in the VFC-2 numbering system, and all the Cibachromes are all two numbers off. Cibachrome 61, for example, is actually VFC-2 frame number 63. So the conversion of any old frame numbers to the true number is to add 2 to the old number to get the correct one. We do know some copies (like the PAC group) start at frame #3 instead of Frame #1, so likely the person who first counted frames did so on a PAC copy.

The PGF Copy Inventory

When I started, I never imagined I'd be assembling a copy inventory. It started accidentally, and I think you will find the story intriguing, as an example of how fate sometimes takes us on a path we don't appreciate when we started down it.

I wanted to get a scanned full frame copy of the PGF to try experiments in a stereo-photogrammetry study of the Bluff Creek setting, and all the scans I could get from other researchers or from the internet were cropped. So once I had a deal to produce some physical sculptures and computer animations for a Monster Quest show, I used my profits from that to arrange a trip up to Harrison Hot Springs (in British Columbia, Canada) to visit John Green and scan one of his copies. I needed the portable scanner to take on the trip, and so I could scan the copy as John sat there in my hotel room, and he could keep a watchful eye on his copy. I put it all together, went to Canada, and on February 1, 2009 I did the full scan, and came home.

Much to my surprise, the full frame version John had wasn't true full frame, but about 96% of a full frame. That's when I tried to sort out the reason, and came to understand John's copies were made on the Optical Printer. That wasn't ideal for my photogrammetry goal, and so I still needed a true full frame copy to scan. Patricia Paterson's archive copy she keeps in the bank box was the best option to pursue. So I began a dialogue with Patricia to come to Yakima in May, 2009 (while I was there for an event honoring Bob Gimlin) and scan her copy. But the negotiation hit a snag and Patricia decided to forgo the scan. So in May, when I went there, I brought my scanning equipment in case there was a change of plans, and I did get to scan parts of Bob Gimlin's personal copy, a Green/Dahinden copy family version, while I was there.

Then a company making a new documentary for the National Geographic Channel, a program called "American Paranormal-Bigfoot" began discussions with me about producing a fine image stabilization of the PGF for their show, and they negotiated with Mrs. Patterson to finally make a deal whereby they licensed the TV rights to show the PGF on their program, and I would be allowed to come to her home and scan her archive copy, with the show filming me doing the scan, and I'd image stabilize it for the show. My only compensation for this work was that I could use the scans for my research as well. And the show producers paid my travel expenses to go to Yakima and stay three days. So that happened in late June, 2009, and I finally got a true full frame scan of the PGF, because the PAC copies are true contact printed full frame copies of the camera original. Thus began my copy inventory, with one each complete scan of a PAC group copy and one complete Green/Dahinden group copy, plus samples of another Green/Dahinden copy (Bob's).

So from these I began to organize the frame inventory, and began my copy genealogy analysis. Then I had the opportunity to scan sample frames from a copy Peter Byrne had in a program reel he had made years before, and that copy had frame 954 which I saw for the first time, and it had excellent evidence of the light washout at the end of the reel, because the copy had faded to a blue shift and the orange washout thus was more noticeable.

My unique success doing this all derived from my choice to make a scanning system portable so I could take the scanner to the copy, instead of asking anyone to send the copy to me. Once my copy inventory expanded and people saw how unique it was, it was easier to ask people to send me their copy, I'd scan and appraise it, and send it back with a full set of digital scan files for the copy's owner to have. And thus, the PGF copy inventory was born.

As I continued to appreciate the unique value of this, I expanded the inventory to known images of filmstrips, and TV versions in various documentaries, because each had some distinguishing characteristics that made each one unique. The Copy Inventory is currently up to 21 copies, and there are two more copies I know of but haven't inventoried yet, so I haven't assigned them a Copy Number yet.

One is the PGF copy in the Library of Congress, Copyright Office, and I hope to get to Washington D.C. to see it and add it to the inventory. The other copy is one Grover Krantz had and it went to the Smithsonian when he died, and now is in the collections of the Smithsonian's Human Studies Film Archives. I hope to view and catalog that as well, and add it to the inventory. And, of course, I still am very excited to hear of any other copies found and that I can examine, appraise, and catalog in the inventory.

So that's the story of how this unique PGF Copy Inventory came to exist. Now, let me describe each copy:

As it appears now (as of July 2014), the copies tend to fall into three groups or families:

The PAC family are full frame contact prints, apparently derived from the very first contact print copies made in either later 1967 or 1968. Copy #3, #5, #9, and #11 are in this grouping.

The Green/Dahinden family, copies made by John Green and Rene Dahinden, where the copy lab service used an optical printer (so even the "full frame" versions are slightly cropped) and zoom in prints (like the LMS version) freeze frame prints, and slow motion prints were made (if you don't know, slow motion prints simply repeat a frame two, three four, or as many times as desired, and produce a slow motion effect when projected at normal 24 fps). Copies #1, #2, #4, #6, #7, and #10 fall into this group.

The ANE Group, with currently only Copy #8 and Copy #14 in it, were apparently made by the ANE producers from the camera original Roger furnished them. However, Copy 8 itself is still a 2nd, or 3rd generation copy, because of the editing and program assembly, and the copying of the program release copy. But Copy 8 is unusual in that it seems remarkably scratch free, and most other copies are scratched quite a bit, so I suspect the ANE people made their master copy from the original using a liquid gate printing process, because the liquid gate process works for most scratch removal (for cell scratches, but not emulsion scratches). Copy 8 was also cropped in a way different than any known Green group copies were cropped, so this also justifies setting the ANE copy 8 into its own family.

There is a separate PDF document (on the Munns Report website) describing Copy 8 specifically, in far greater detail, if you are interested in delving into this further.

A lot of the copy genealogy is still a bit confusing and convoluted, because for each new copy I inspect, I see things which clear up one issue, and raise another question in the process. I thought the genealogy issue would be one of the simpler and more easily resolved matters of the PGF, but it has proven to be one of the most challenging and surprising.

Copies are now numbered for a simple reference, as well as to respect the confidentially of several copyholders, who specifically asked that I not disclose their names as copyholders.

Copy #1 Family: Green Frames Start #001 End #0939
Description - Slightly cropped "Full Frame" version. The scan set includes a few frames of the "tent pole" segment before the Bluff Creek segment starts. Green copy archive reel in newsprint

Copy #2 Family: Green Frames Start #003 End #952
Description - Slightly cropped "Full Frame" version Gimlin copy

Copy #3 Family: PAC Frames Start #003 End #953
Description - True Full Frame contact print, clipped Camera ID Notch apparent, and end has sprocket burn through indications and some edge washout

Copy #4 Family: Green Frames Start #003 End #954
Description - Slightly cropped "full frame", copied onto a presentation reel with other footage, but has the finest evidence of the end edge washout from unloading the camera on the filming day. Only copy with last frame, #954 found so far. PB Copy

Copy #5 Family: PAC Frames Start #355 End #364
Description - Very high resolution scans from a true full frame contact print. Beckojrd scans

Copy #6 Family: Green Frames Start #001 End #950
Description - This is a TV scan, used for the LMS program, and as such only the start and end frames can be determined with any certainty. TV frames in between are mostly frame blends for the 30fps TV frame conversion. It also contains a brief portion of the segment before the Bluff creek footage, a segment often referred to as the "tent pole" scene

Copy #7 Family: Green Frames Start #003 End #896

Description - This is a TV scan, used in the A&E documentary "Bigfoot", part of their Ancient Mysteries series. Like the above, as a TV scan, only start and end frames can be determined with any reliability

Note: A&E is the "Arts & Entertainment" cable channel, and not to be confused with ANE (which is "American National Enterprises", a film production company which made Copy #8 below)

Copy #8 Family: ANE   Frames are not numbered as above because the 100' reel that makes up Copy #8 has multiple segments of the PGF on it, so many frames are repeated, in slightly different croppings and formats. Copy #8 is described in greater detail in a separate PDF document on the Munns Report Website

Copy #9 Family: PAC Frames Start #003 End #953
Description - A true full frame contact print, with some damage, at least two film tears that were repaired with splice tape. It is, however, a full width print-through contact print, and as such, has some edge markings identical to Copy #11, which helps us in determining more about the PAC family. Bruce Macabbe

Copy #10 Family: Green   Frames Start #001 End #502
These are generally referred to as "The Noll Frames" and they were scanned by researcher Rick Noll. They stop at the end of the look back and walk into the trees, and do not contain the end walk away segment. They were highly cropped close on the filmed subject, and made with a camera attached to a microscope

Copy #11 Family: PACTime Life strip Start #351 End #362
This is a single photograph of a strip of the film, published in the Time/Life book "Mysterious Creatures" and it has scratched edge markings suggesting a frame count was done on this copy of the film. It has traces of the clipped Camera ID

edge notch. It also has a 1967 date code, a square and then a triangle, before the latent image "safety film". The origin of this photo, and the copy that was photographed, are still uncertain, as I write this. The photo of this copy is shown below:

Copy #12 Green Archive reel, with complete Reel One

Copy # 13 Green Archive Reel, second full frame copy

Copy #14 ANE Full program copy

Copy #15 Green "Enhanced" Reel

Copy #16 Green Archive reel, zoom in lookback segment

Copy #17 Discovery Program Flipped Frames

Copy #18 PAC Group, Beckjord estate Copy frome F0003 to F0953, with Kodachrome latent image print-through (with a 1967 "square & triangle" date code), 1977 (square only) copy filmstock date code, edge scratches like Time Life strip and Copy 9. Second or Third gen, on reel alone, no other footage. Possible print-through tear at F347, but no break at F355 like C9 and C19

Copy #19 PAC Group, Beckjord estate copy, from F003 to F953, and also has first reel horse and rider segments, and has Beckjord 1982 ape suit recreation. Copy is curious because it seems to have been made with an inter-negative, because printer device has a roller that holds down the bipacked film at a sprocket hole, and masks light from exposing portions of that edge, but the holdback area is white, while it's black on all other PAC copies and thus could not be seem as different from the black edges in general. So this copy is the first to define the printer device and show why some areas on other copies are darkened. Does not have the edge scratches of the Time Life, C9 or C18 copies. Does have breaks in the film, one at F355, like C9 has, but the breaks are apparently in the source of the copy, not the copy itself, so one gen back. Two breaks in total F0032, F0355

Copy #20 - JG Family, 2x zoom in, held by Russian homoligist, Igor Borstev. Has excellent detail, low contrast and good color, superior to Green's "Enhanced Copy" #15. Appears to be complete from Tentpole trees, and PGF Frame 1 to frame 954

Copy #21 - Complete first reel converted to SD video and shown on YouTube by Alex Hearn. He describes having found the video on the internet but does not recall the source or link or site. Likely the Green/Dahinden family, similar to the full first reel I scanned in Copy 12.

Fig. A3-01

Appendix 3 - Editing Analysis and Questions About Same

Editing Studies

A common skeptical talking point is the accusation the PGF has been edited for some deceptive purpose. Ask for proof though, and you'll get one of two explanations. Either someone will show a frame from some TV conversion where that particular TV program edited some of the footage, or you'll get some circumstantial and convoluted story about claims of footage length or running time, and how those don't match the known segment length, so something must have been edited to hide the incriminating stuff. Neither explanation has any merit.

The fact of a TV producer getting a copy of footage, and editing it into their program, has absolutely no relation to the question of the camera original being edited before copies were made. And any copy a TV producer gets is the copy we all have seen, so what they get has nothing in it needing to be hidden. This concept is popular with people who are woefully ignorant of TV and documentary production realities.

The second explanation usually starts with someone's quoted comment about the PGF, how many feet it was, or how long it ran when projected. The concerned person takes that quote as absolute truth, and then finds some discrepancy with the known 954 frames, 23.85', and uses that discrepancy as the basis for the editing suspicion. Such persons won't allow for the simple prospect that the person making the original description just got their run time or film length wrong.

When these two suspicions are defeated, the usual skeptical fallback is to run to the "invisible splice" idea, that there's no way we'd see splices on the copies we have, so there's no way we can prove the original wasn't spliced. And if we can't prove it wasn't, they can claim it was.

The problem for them is, we CAN (emphasis mine) prove the PGF was not spliced when it was copied. But diehard skeptics continue to rant their passionate belief that I am wrong and splices can't be found on the PGF copies, or any film copies. Even though I have covered this in a "Munns Report" Pdf document, and in one of the RHI Scientific papers as well, and the text of both is 100% correct on review, I have continued to listen to these misguided criticisms and continued to investigate the subject, and so I am pleased to bring even more new evidence to substantiate my assurance that yes, we can prove conclusively the PGF was not edited when it was copied.

But for clarity and having the essential information in one discussion, I will review the material that sets the foundation for this analysis and determination.

First is the definition of the word "Splice". As defined in Webster's New Explorer Large Print Dictionary, a Splice is "join (2 things) end to end". Splicing is not the same as editing, which is re-arranging the sequence or altering the length of film segments. You can splice without editing, but to edit, you join the edited segments back together using a splice.

Splicing can be done in the lab to join undeveloped film to other film or leader to run the undeveloped film through the processing machinery, and once processed, the film is cut off the reel system and usually lab head and tail leader are spliced onto it for delivery to the customer.

Splicing can be done to assemble several short rolls of film (like the 50' rolls that come from magazine type camera loads) onto one larger projection reel, so they can be projected continuously.

Splicing can be done to assemble edited segments.

Splicing can be done to repair film that breaks or tears.

There are several types of splices.

The Butt Splice takes two pieces of film, both cut on the frame separation line, and butted together in the splicing block, and a clear adhesive tape is put over the two pieces to join them. One may use a tape that is as wide as two image frames, so the tape edges tend to fall also on frame separation lines (and are less obvious) or one can use Kodak Press Tape Splices which extend beyond the frame lines and into the image area. This is also called a "Tape Splice".

Fig. A3-02 shows a standard butt splice or Tape Splice, using Kodak Splicing Tapes.

The Glue Splice requires some overlap of one film cellulose base on top of another, so while one segment is cut on the frame line, the other must be cut into the image area, so they overlap, and the glue is applied to that overlap.

But recently, I also found examples of a double overlap, where the film on both segments was cut into the frame area, and neither was cut on the frame line, and this doubled the area of the overlap, but made cut lines in both the frames being spliced.

The Tab Splice has a curiously curved tab shape, and I have only found evidence of this in lab splicing.

Finally there is Diagonal Splicing, which runs contrary to professional practice, but some people have done it none-the-less, and I have found examples of it in the movie footage I acquired for analysis. The diagonal cut is originally designed not for film image editing, but rather for magnetic coated film used for audio recording. But if a person had a diagonal cutter only, they could (and some people obviously did) use it for splicing picture footage.

Fig. A3-02

Tape Splicing Description

Assume the two segments at right are to be spliced together.

1. Each piece is cut precisely on the line between the frames.

2. Below, the two pieces are butted together.

The splicing tape looks like this, but fully clear. It has sprocket holes just like the film.

3. At right, you can see how the splicing tape is placed on the two film segments to join them. It may be put on one side only, but putting a piece of tape on each side, to "sandwich" the film between two tape pieces, is generally the most secure splice.

4. The result of this makes a strong splice but it leaves a line across the film image on each end, through the visible portion of the frame (marked with red dots to show you which lines I refer to.

This above is the double fine line of the two tape pieces, one on each side of the film, from the end of Bob Gimlin's personal copy of the PG Film.

Fig. A3-03 illustrates all these other splice types.

Glue Splice and Overlap   Glue Splice and Overlap

Tab Splice   Diagonal Splice

Repair Splicing usually attempts to put back the film without cutting any frames away, and tears can sometimes be in the picture and can run through more than one frame. So one version of repair splice tape is also made on a roll (with sprocket holes in the tape) so you can cut any length of tape to cover as many frames as are needed to make the repair.

All splicing has evidence of the process, but film editors do sometimes like to put that evidence in the frame line between the picture areas, for a cleaner splice look. The Glue splice, the tab splice and the diagonal splice all put the splice evidence clearly in the image areas, and even the Kodak Splicing Presstapes do as well. But using roll tape run across the film 16mm width and exactly two frames wide (0.6"), it's possible to put all the splice evidence in frame lines between the image frames. This technique has sometimes been described as an "Invisible Splice", and PGF skeptics have seized this idea as if it was an answer to their hoax dreams.

The error of such thinking seems to be that if the splices are "invisible", then we can say the PGF has been edited and spliced and when we have no proof of splicing, we can say, "that's because the film was edited with invisible splices, so of course we can't see them, but they are there." It's like saying that an invisible man is in the room with you, and you can't see him because of course, he's invisible, but he's definitely here, in the room.

Fig. A3-04 shows the technique for the "invisible splice" and why it actually isn't invisible, even on contact printed copies.

The problem with this delusion, as applied to 16mm film copies, is that we can see the splices, even the so-called "invisible ones", if we know what to look for. We can easily see all splices on the original actually edited, but we can also see the splices on copies made from an edited original. This claim seems to infuriate skeptics, but it is none-the-less true. The explanation gets a bit technical, so please bear with me.

Fig. A3-04   The Invisible Splice

5. The concept of an "invisible splice" cuts the splicing tape exactly across the frame line, so there is no tape line in frame to see. It is hidden in the dark line seperating frames.

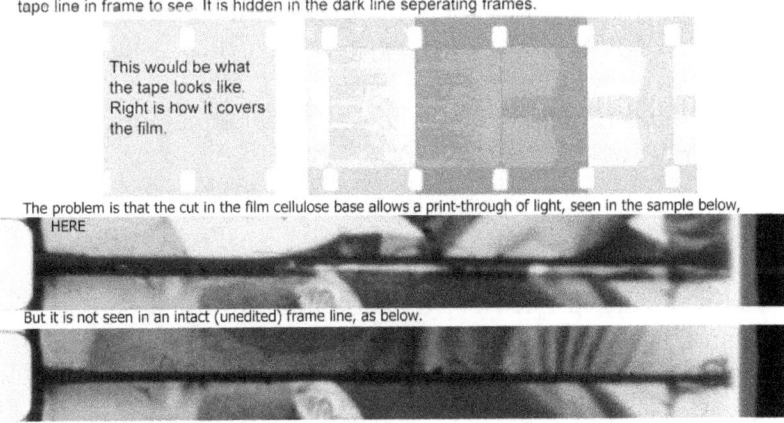

This would be what
the tape looks like.
Right is how it covers
the film.

The problem is that the cut in the film cellulose base allows a print-through of light, seen in the sample below,
HERE

But it is not seen in an intact (unedited) frame line, as below.

First, we have two generalized ways to copy film, a contact printer copy and an optical printer copy. The contact printer simply bipacks the source film (to be copied) with raw film stock (the copy) so they are firmly pressed together, and then run through a printer light gate which shines a light through the source film onto the copy stock, in a continuous rolling movement. As such, the contact printer exposes both the frame images and the line between the frames from the source film to the copy. The copy is one-to-one in image size and content, frame for frame, and frame line for frame line.

The optical printer however, is a projector coupled to a camera, and each can be run at a different rate, so the source frame can be printed multiple times for a slow motion effect, or so many times that the frame appears to freeze, or the projector lens can zoom in so the copy image is an enlarged portion of the original frame. The optical printer also does fade ins and fade outs, and cross fades (also called dissolves) from one segment to another so the transition isn't abrupt.

The reason this matters is that the contact printer keeps the image aperture shape (and some Camera ID markings) intact on the copy, while an Optical Printer has a new aperture shape on the camera side, slightly smaller than most camera apertures, so it masks off a bit of image area and doesn't print the frame line between frames from the source film. And with an "invisible splice", the evidence of the splice is in the frame line between image areas. So a contact printed copy does print the evidence of a splice into the copy, while the optical printer does not.

So if you want to look for evidence of splicing in a copy, when the editing was done on an earlier generation source film, you need a contact printed copy to examine. It's that simple, and it's an irrefutable fact. Why does the copy retain evidence of a splice? Because the edit and splice requires a physical cut of the cellulose film base to be made, and the two cut ends butted together, and so at that point, the cell base has this physical line where the cellulose base ceases to have a solid and unified integrity. Light shines through this physical cut in the cell base, and that light leak onto the copy is the evidence of a physical cut and splice.

I've scanned several professionally produced and edited 16mm documentaries and the film I examine in each case is a contact printed copy from an edited master. And the splices show on the copy, because of the light leaking through the physical cut in the cellulose base of the edited master a generation or two back from the copy I examine. I currently have over 100 examples of this taken from multiple copied documentary prints to prove this point, and some are illustrated here.

Can we tell if a film was edited and first copied on an optical printer, and then subsequent copies contact printed, as a way of hiding the splice evidence? Yes, we can tell if such was done.

Each make of camera has a somewhat distinctive aperture window (which lets the light through to expose the film) and some are more distinctive than others. The K-100 camera Roger Patterson used has one of the more distinctive aperture shapes, both for a camera ID notch and for a curious little curved narrowing of the frame line on each side top and bottom, on many models of the K-100. It's like a bowtie pinching effect. In examination of a lot of footage with the K-100 Camera ID notch, the aperture shape does change somewhat on different cameras, and the bowtie shape is absent from some. However, when it is present, it further substantiates that the camera was a K-100, and that the copy of the film was contact printed because only contact printing will keep the original aperture shape, and optical printers don't have that bowtie effect.

Fig. A3-05 K-100 aperture study.

An optical printer has a projector coupled to a camera, and that copying camera has its own aperture, slightly smaller than filming cameras do. We know this because the Green/Dahinden copy family was made on an optical printer and we can measure the loss of image frame area on such as compared to the PAC family copies or the Kodak Transparencies Roger made. The optical printer copy will hide all camera aperture frame indications. Once those are hidden, no subsequent copy can restore them. So, if a print of the PGF has the original K-100 camera aperture, it CANNOT have been made on an optical printer, at any point in its copy history. The PAC family of copies have the K-100 camera aperture intact. We can say with 100% certainty that they were made by contact printing only, regardless of generation, not on an optical printer at any point of copying. The enlarged transparencies also verify the K-100 camera aperture, as do new filming tests with a K-100 camera I did for my research.

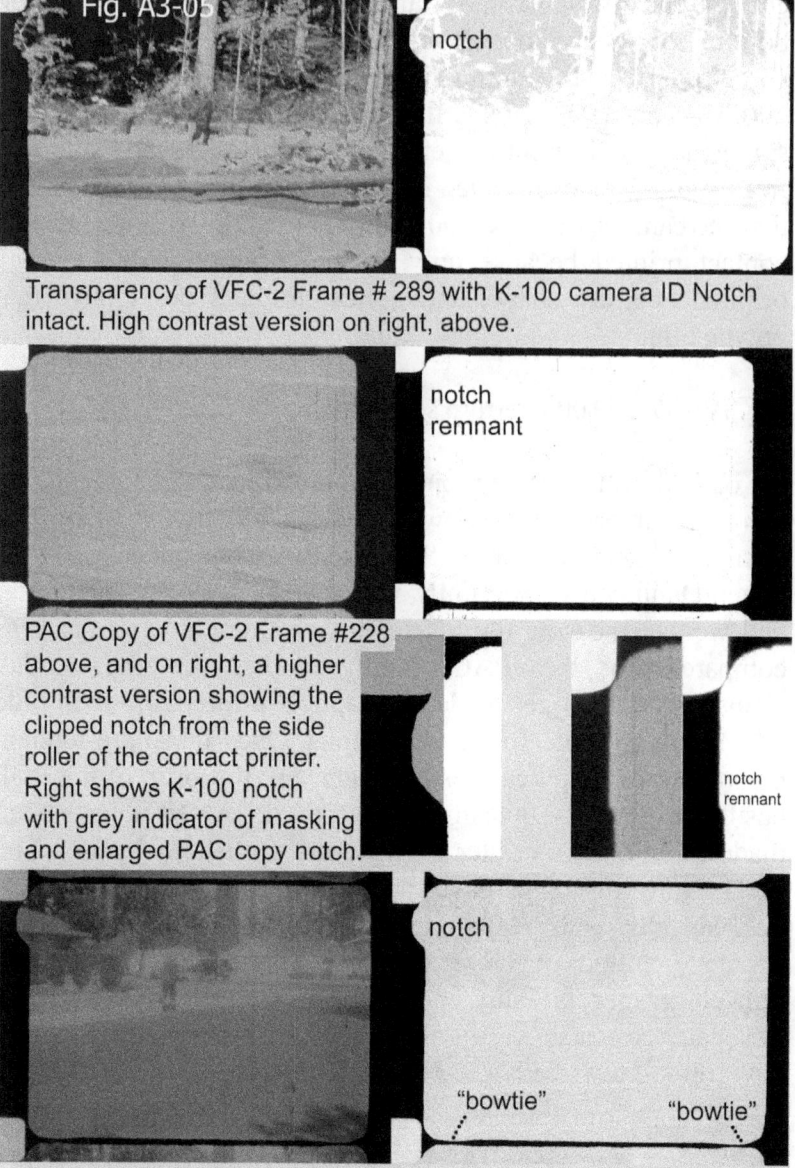

Fig. A3-05

notch

Transparency of VFC-2 Frame # 289 with K-100 camera ID Notch intact. High contrast version on right, above.

notch remnant

PAC Copy of VFC-2 Frame #228 above, and on right, a higher contrast version showing the clipped notch from the side roller of the contact printer. Right shows K-100 notch with grey indicator of masking and enlarged PAC copy notch.

notch remnant

notch

"bowtie"

"bowtie"

K-100 camera test in May 2009 with camera owned by Daniel Perez showing intact camera ID notch and "bowtie" pinching of frame line on each side. PAC copy above shows some of same bowtie effect.

EXAMPLES OF SPLICE CASE STUDIES

When we look for evidence of a film cut and splice on a copy from the edited source film, the first thing we look for is any disruption of the frame image continuity (the way each frame image flows to the next, in terms of either remaining the same or having a smooth and continuous motion or shift of the image frame after frame. Disruptions are where edits are usually found. But for forensic analysis where there's suspicion of editing, every single frame and frame line needs to be examined individually as well.

What we look for is first, and most important, the light leaking through the physical cut in the film's cellulose base, because that light is white and the frame line is black. So it's quite obvious. If we don't see the light leak, it is most likely because the splice was an overlap splice, in which case we see the overlap line in one image frame. Either way, the splice can be seen.

Other indicators of a splice are a change of the camera aperture shape (since footage may have been taken with different cameras at different times), or any mis-alignment of the aperture edge into the black of the film sprocket area (caused by one piece of film sitting slightly off in relation to the other piece in the splicing block when the tape is applied).

So the light leak white line (or the overlap lines) is the primary indicator, but the different aperture shape or the frame edge mis-alignment are also secondary indicators we would use if the light leak white line were subtle.

To study this, first is a pair of film examples where the scene changes but there's no edit, no splice, because these are examples of filming where the camera stopped, and then later the filming was done of a new and different scene, and so the camera started to film that scene. The indicators of this are:

411

1. The black frame separation lines are all pure black, and all equal in shape, thickness, and solid black tonality. There is no white line of a cut in the base.

2. The frame aperture remains virtually identical in the two segments, as evidence they were from the same camera.

3. The first frame of the new segment will likely have over-exposure from the camera start (caused by the shutter trying to get up to full speed, and during the first frame exposure, the shutter is not yet full speed so it's open a bit longer, resulting in a slight over-exposure as compared to frame two of the segment.) An edited segment commonly cuts off the first frames and loses the camera starts.

Fig. A3-06 shows scene changes resulting from camera starts and stops of the same roll of film, and no editing.

The two examples are taken from Roger Patterson's documentary footage reel, and both show the camera ID notch of a K-100 camera, which means the copy is a contact printed one.

These two examples meet all the above criteria for an intact film where the change of scene is just where the camera stopped, a new setup was done, and the camera restarted for the new segment, with no editing. What we see is what was on the camera original roll.

But on a film copy that has been spliced, we will find the light leak (or overlap of a glue splice) and we may find camera aperture changes or mis-alignment of frame edges as well.

In the example of a splice from a National Geographic Films documentary on Mexico, a professionally produced and edited 1 hour documentary, examination of the copy shows that the edits and splicing of the previous generation master can be found with no difficulty on the copy.

Examples of Scene Change Not Edited

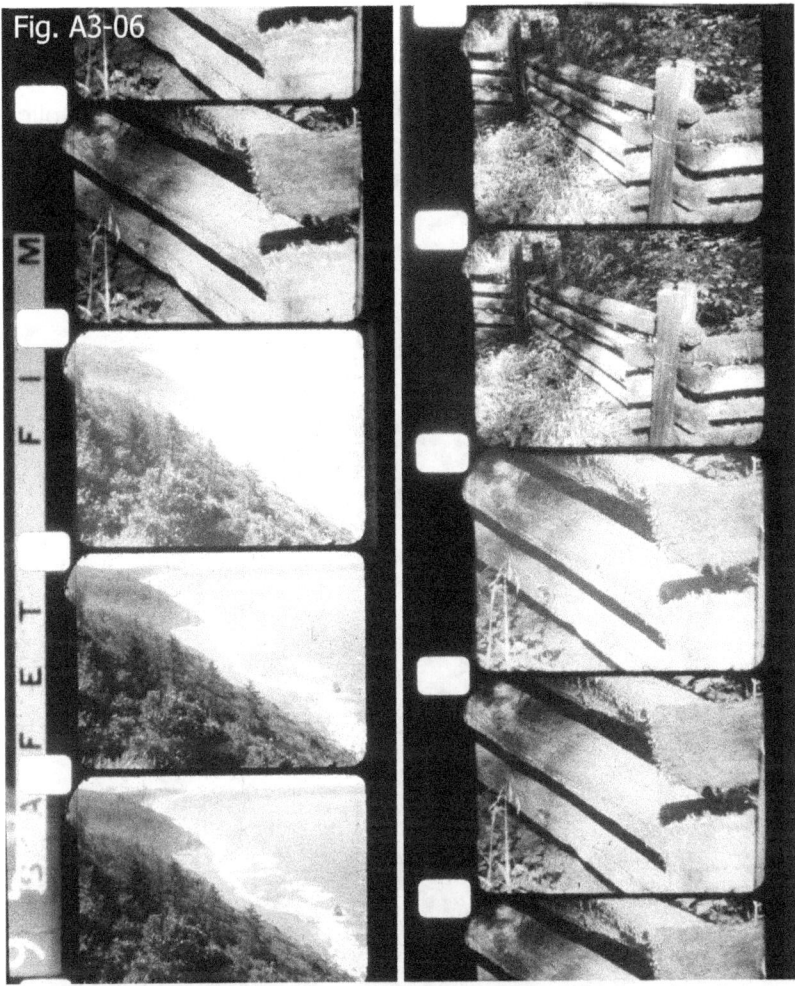

Fig. A3-06

The two above filmstrips show a scene change due to
stopping and starting the camera, with no edit or splicing.
1. Frame separation black lines are the same, with no
   light leak through the line.
2. The new first frame shows the over-exposure of
   a camera start.

On this example, the white light leak through the physical cut
in the film cellulose base is very evident.

Fig. A3-07 shows one example splice on this copied documentary.

Examples of Splice Print-through on Copy

Fig. A3-07

No Splice

Splice

No Splice

No Splice

Splice

No Splice

From the documentary "Bigfoot Man or Beast" (1971), the splice is obvious from the white line of the light leak through the cut in the cellulose film base, plus we see the change of the frame aperture shape. In the landscape image of foggy mountains, the aperture has well-rounded corners, but in the shot of the man, the frame loses those well-rounded corners completely on the right, and they are diminished on the left. On the splice line, right side, we see the top of the splice cut has the rounded aperture, and below the splice line, it is straight, no rounded corner. But the frame above has the rounded corner above and below the black frame line. So in the enlargement of the frame lines on the right side, we see the top frame line has both top and bottom rounded frame shapes, the splice line has only the rounded shape on top, and the next scene has no rounded shapes at all. Also the image of the man goes further left into the sprocket area than the landscape mountain image does, which means the two segments were taken by different cameras. All of these are indicators of editing and splicing.

Fig. A3-08 shows just the three frame lines in top section and entire frames plus frame lines below.

An enlarged image of the frame lines is also shown, and it is an excellent example of how obvious a spliced cut in the film base is when spliced together, and also shows the change of aperture shape of different footage taken with different cameras, as a supplemental verification of where the splice is, and that the change of scene cannot be explained as a camera stop and start with the same camera.

Fig. A3-09 enlarges the frame lines so you can see how a cut of the cellulose film base prints through as a white light leak.

Next is a documentary produced by the XEROX Corporation, titled "Wonders of the Sea" and we may reasonably assume it was made by competent professional film editors and producers, so what we find in it is typical of professional quality.

Examples of Splice Print-through on Copy
Fig. A3-08

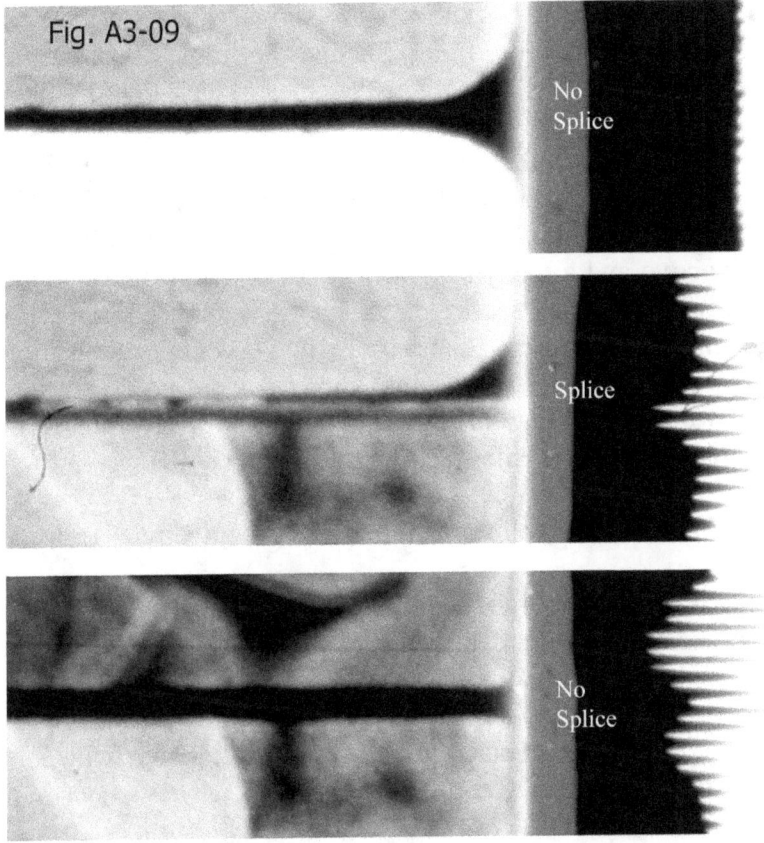

Fig. A3-09

This enlargement of the frame lines shows that above the splice, the frame line is pure balck, and has the rounded corners of the camera aperture.

The middle line shows the splice, and the light leak. It also shows the frame aperture curve is only on top, and not below the splice.

The bottom frame line is pure black again, and has no rounded corners top or bottom, from a different camera.

From "Bigfoot: Man or Beast" copy of edited show.

In the illustration, two segments are shown, and in each case we see the frame before the splice and the frame after. Where there's no splice, the frame line is a clean black line dividing the two frames. But in the splice, there is a white rough line where the light is leaking through the physical cut in the film's cellulose base. It's painfully obvious when inspected at this level of magnification. On the sample on the right (showing water above and a rocky seascape below) the left image frame is wider on the rocky seascape than on the water scene, and this difference of aperture size and shape is a secondary indication of a splice, because two different cameras were used and they have two different aperture shapes. Footage from two different cameras can only be combined by editing and splicing.

Fig. A3-10 has two splice examples on the copied release print.

In each case of these documentaries inspected, they had many more examples of very obvious and easily verified splices on the copy, and I have over 100 scans of such splices. So it is simply a fact that when a film is cut, edited, and spliced together, and if a contact print is made, then we will find splicing evidence in the copy.

With this assurance, and the fact that every frame of the PGF has been inspected for any irregularity in the frame line between frames, and no irregularities have been found, we can say with absolute certainty that the PGF was not edited and spliced when the copies were made from the original. The camera original is exactly frame-for-frame what we see in copies we study.

All one needs to do is actually study camera and film technology to verify this.

When Roger Met Patty

Professionally Produced
Documentary Release Print

Splices show on Copy,
two examples below.

In each case, the top frame
line is intact, the middle one
down has the splice, and
the lower one is also intact.

Fig. A3-10

WONDERS
OF THE
SEA

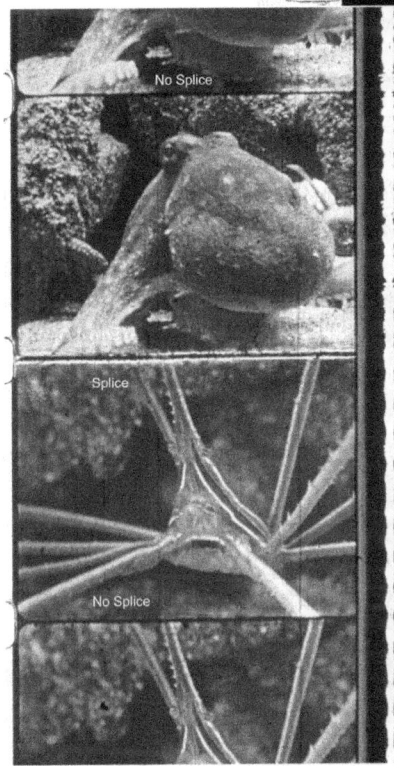

No Splice

Splice

No Splice

Above, we also see a glaring example of
a change in camera aperture, showing the
two segments of film were taken by different
cameras. Top frame has rounded corners
and lower frame has no rounded corners.
(Shows better in color)

Forensic analysis of the various copies of the PGF is conclusive that the PAC copy family (which Roger made first) are contact printed copies. The Green/Dahinden and ANE copy families are optically printer copies. We have enough samples of both to tell the difference immediately on examining one particular copy for the first time. So the fact is, some contact printed full frame copies do exist and we know exactly which copies those are. The evidence is empirical, irrefutable and has been published.

We can also appraise the generational level of copies with proper analysis.

First is image resolution, the sharpness of an image on original film. The transparencies made by Kodak for Roger from the original were greatly magnified (from 0.3x0.4" to 4x5") so the transparencies had so much finer grain than the 16mm original that no detail of the original was lost. The transparencies show an image resolution equal to the Kodak specification for an original. So the source film that made the transparency was a camera original processed perfectly according to Kodak specs for the best resolution.

Each copy on 16mm degrades to some extent from it's source film, and because we have now 20 copies of the PGF in the image database, we can appraise the image reduction of sharpness in various copies and develop a relative comparison of copy generation stages. Also, as copies build up contrast, the near white and near black tones shift to pure white and black, and we get a phenomenon called Copy Bloom, and those white and black image shapes bloom larger copy after copy. So we can also gauge copy generational level by studying copy bloom.

These processes allow us to make a fine comparative evaluation of any given copy's generational level.

Copy Quality Studies.   Fig. A3-11

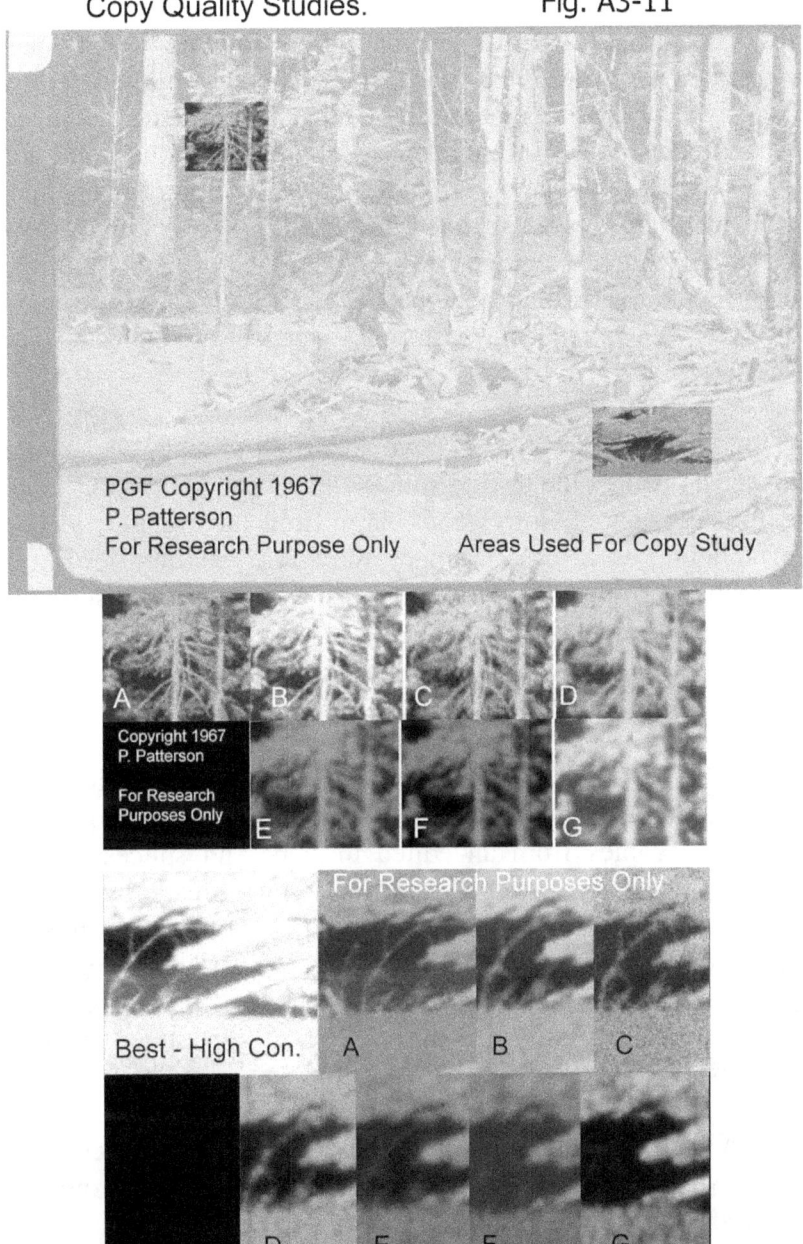

Fig. A3-11 repeats a copy generation study image (shown earlier in the book).

So, we can find splicing evidence on contact-printed copies from an edited original or master. And there is no such evidence anywhere in any PGF contact printed copy. Therefore, the PGF camera original was not edited and spliced when those copies were made. The copies are a true frame for frame record of what was on the camera original, and there were no splices! I've scanned and examined every individual frame of the 954 know PGF frame inventory, on multiple copies, and every frame line between frame images is scanned and verified as intact, with no evidence or indication of splicing.

Diehard skeptical believers have also tried to make their point about not being able to find splices on a copy, by making some wild claims about Hollywood 35mm films having no evidence of splicing, as a logical fallacy implying that they too must have been filmed in one continuous performance instead of being edited because we can't find evidence of splicing. Sadly, the people making these odd remarks obviously never looked at actual 35mm movie film. I did.

I purchased some 35mm studio trailers (the "previews of coming attractions" short film promos) and examined them. They are copies from an edited master. And splices on the master are clearly visible on the copy print, if made by contact printing. The ILLUSTRATION shows some examples I scanned. The splice technique was a glue overlap splice, but it seems narrow on the 35mm film because the film is wider and the image area is larger in relation to the splice overlap dimension, as compared to 16mm film.

Fig. A3-12 shows how obvious a splice on 35mm professional Hollywood film can be as seen on a copy made from the spliced original.

Fig. A3-12
Splice on copy of 35mm film, because of light leaking through the cut in the cell base of the film.

So we can find splice evidence on contact printed copies of Hollywood 35mm films as well. All it takes is a person who actually has sufficient technical knowledge to recognize the difference between a contact print and an optical print, and to actually examine the physical film itself.

When Roger Met Patty

Scan Configuration Two - Individual Frame Scan.

Left shows system

Projection Element frame of PGF is partially seen in blue boxed image.

The scanning computer screen showing actual scan image area, one full frame plus some sprocket area

Appendix 4 - Frame Scanning System

This is a technical discussion of my frame scanning system, and my scanning techniques as developed specifically for PGF analysis.

My System

When I first started thinking about being able to scan PGF copies myself, in 2008, I did consider purchasing a film scanner commercially made. The high-end ones like an Oxberry, I couldn't afford. The lesser priced ones, like a Viper, were too low a resolution. And I had the additional requirement that my scanning equipment had to be portable. I had to be able to take the scanner to the film copy, because the people who had copies were hesitant to loan out their copies to me.

So after evaluating the options, I decided to build my own scanning system, 4k resolution, portable, and with adaptable image carriages to scan 16mm film and other larger transparencies or photographs. The system components are:

1. The camera - I used a Canon 12.2 Megapixel digital camera, a Rebel XSi, with a 100mm macro Canon lens, and multiple lens extension tubes, to give me a magnification of 2:1 (being able to get a image area that was twice as small as the camera's image sensor measurement). The Canon system was specifically chosen for its excellence in optics (Canon lenses rate very high in optical precision) and for the EOS software utility that comes with the camera and allows the camera to plug into a computer through a USB port and the computer can run the camera and display the camera image on the computer screen (the Live View Function). This system captures an image at 4272 x 2848 pixels in the highest image quality mode.

2. The Computer - I've used two Toshiba laptops over the years, one running Windows Vista, and the newer one running Windows 7. I usually save the images captured to an external

hard drive, for ease in moving the scans over to my main workstation for analysis. The laptops are 17.3" screens, set at 1600x900 resolution.

3. The film holders - I have had to custom make my various film holders. The first device was a simple set of glass panes glued together with just enough separation for 16mm film to slip through, and guides to hold the film is a fixed path. But I didn't have any way to open the holder, and so I had to slide the film down through the top, and scan each frame continuously, with no easy option to go to any specific section. I wanted to see the images upright and film is actually set up for inverting during projection, so upright, film runs end to front. So my first scan of John Green's full frame copy was end to beginning, and the numbering of the scans is back to front (scan #1 is the last frame, and scan #941 is the first.)

I subsequently upgraded my film holder to a vintage Keystone Projector, because it had a beautiful film gate where you can put the film in at any point in the reel. I did have to modify the projector aperture, because projection apertures tend to be a bit smaller than full frame, and I needed to scan larger than full frame, so I removed the aperture plate and filed it to widen it considerably larger than the film frame. This worked well, but as film runs through a projector upside down (and is inverted right side up by the projection lens I wasn't using) my scans with this were originally captured upside down and need to be inverted in Photoshop when used for analysis.

Later still, I finally modified the projector to build a new frame and mount the projector upside down in the frame, so films scan right side up. I also added a 30 rpm motor to the projection movement so I didn't have to hand turn the movement for each frame scanned, a task which caused me a lot of hand and arm strain when I scanned the PGF fully at Mrs. Patterson's home in 2009. The motorized projection movement, which gives me a new frame every four seconds, is just about right for the time it takes for the computer to save each captured image from the camera. (If you do the math, and wonder why a 30-rpm motor allows for 4 seconds per frame, it's because the projection shutter actually revolves twice for each frame, so 30 rpm is 15 frames per minute through the film gate).

The film holder for Inventory scans I had to scratch build, with two glass sheets spaced for the film thickness, a guide the film slides through on, and spindles with reel mounts on each side of the glass holder.

The light source is a double compact fluorescent bulb set, behind a plastic sheet that was sanded so the surface is not clear but diffuses the light of the bulbs to a larger flat panel of light. It is also set back so the sanded surface is out of focus, thus diffusing the light even more, for a flat-across-the-scan image light intensity, with no hot spots.

Fig. A4-02 shows the setup configured for an inventory scan, and below, a sample thumbnail page of a group of inventory scans to index a film roll.

The EOS Software is excellent and in the Live View Mode, it allows the computer to control the camera, and display what the camera sensor is picking up. We can control settings for shutter speed, F-Stop, and sensitivity ISO, designate the folder for saving images, and click the camera shutter on the computer screen. So I see the camera's image perfectly on screen far larger than the digital screen on the back of the camera.

In General, I do three types of scans, depending on the film under study.

Fig. A4-02

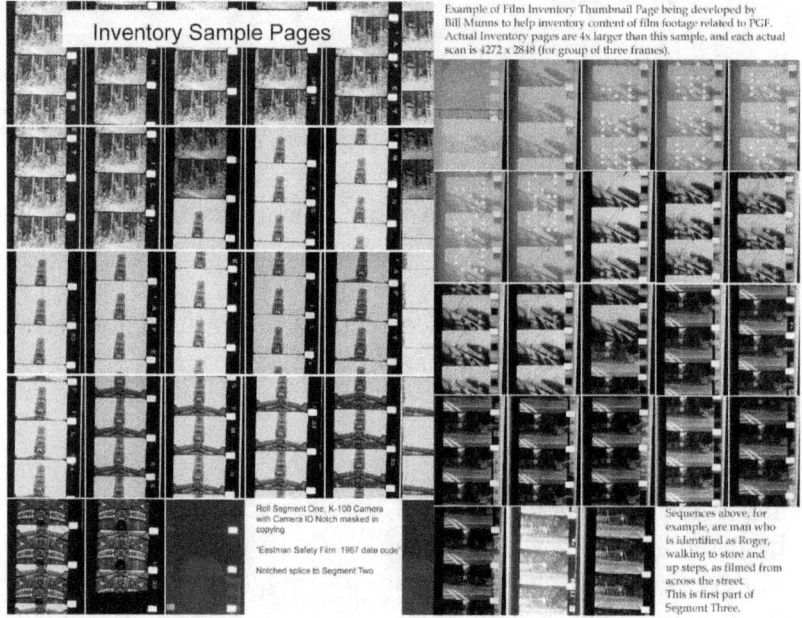

Inventory Sample Pages

Example of Film Inventory Thumbnail Page being developed by Bill Munns to help inventory content of film footage related to PGF. Actual Inventory pages are 4x larger than this sample, and each actual scan is 4272 x 2848 (for group of three frames).

Roll Segment One, K-100 Camera with Camera ID Notch masked in copying

"Eastman Safety Film 1967 date code"

Notched splice to Segment Two

Sequences above, for example, are man who is identified as Roger, walking to store and up steps, as filmed from across the street. This is first part of Segment Three.

These options are:

1. A Frame Scan - My frame scans are slightly more than a true frame, so I also get the frame separation line and even a bit of the next or previous frame image. The usual content frame scan sets the scan area to the image content alone, and that's fine for image content preservation. But for forensic studies, the frame separation line is as important as the image frame itself, because any concern about editing is found in that frame

separation line. Having some width beyond the frame edge is vital for camera ID indications as well, and trace evidence of camera runouts, such as the orangish burn through of light leaking through sprocket holes onto lower layers of film when the camera door is opened and a daylight-load roll is removed.

For those analysis reasons, I select this overscan format to capture 100% of the true frame image plus the frame separation line, some side sprocket area content, and traces of the next image frame. Then I can crop to image size only for analysis and any video assembly for rendering as an animation.

Fig. A4-03 shows a Frame Scan above, including some image frame separation line and overscan, and below, a Zoom Scan for content only.

2. A Zoom In Scan - This is for enlarged study of image content only, and is an enlarged portion of the frame. I do Zoom In Scans when I want a scan set focusing on some specific image portion of the frame, like enlarging Patty for the best analysis.

3. An Inventory Scan - I developed this type of scan specifically for study of the physical film characteristics apart from the content of an individual frame image. The film has such evidence material as latent image edgecodes (for type of film, mfg date code, possibly copied latent image data from previous copy generations), Camera ID markings, any copy device markings, any man-made edge markings, scratches or irregularities, splice studies, camera start studies, leader markings, and other filmstrip studied where I make a composite image of multiple frames to look like the filmstrip does. The inventory scan takes in the entire 16mm width of the film, including sprocket holes, and captures 3 full frames of the filmstrip.

Fig. Forensic Frame Scan, with overscan for frame
A4-03 separation line and traces of frame before or after.

Zoom Scan, only for image content and not frame lines.

For example, the indicator of a camera start is usually over exposure of frame one as the shutter is getting up to speed, and so having a scan that captures the full first frame and second frame together assures that both frames get equal exposure and any over exposure of the first frame truly is from a camera start and not from an imbalance of scan exposures.

Fig. A4-04 shows a three frame Inventory Scan to study a camera start, because the single scan of three frames guarantees all three are of equal exposure in the scanning, so any exposure difference in the frames is preserved.

Segment Indexing allows me to scan three consecutive frames and so I can get one frame of a segment ending and one or two frames of the next segment starting, and this makes it easier to later reference the order of the segments on the film. I can also see if the segments are connected by a camera start or an edit. So with a full set of these Segment Indexing Scans, I can look at a visual inventory of every segment on the roll, and their order or sequential relationship as well.

Filmstrip assemblies are easier with a three frame scan than a single frame scan, and I do such filmstrip assemblies for reading leader markings, reading film stock edge codes, studying light washouts, and copy comparisons for where each copy starts or ends.

So if my scan formats are somewhat unorthodox, they have been designed specifically for the unique forensic challenges of analysis of the PGF, and that's why they are different from the traditional scans for content preservation only.

Fig. A4-04

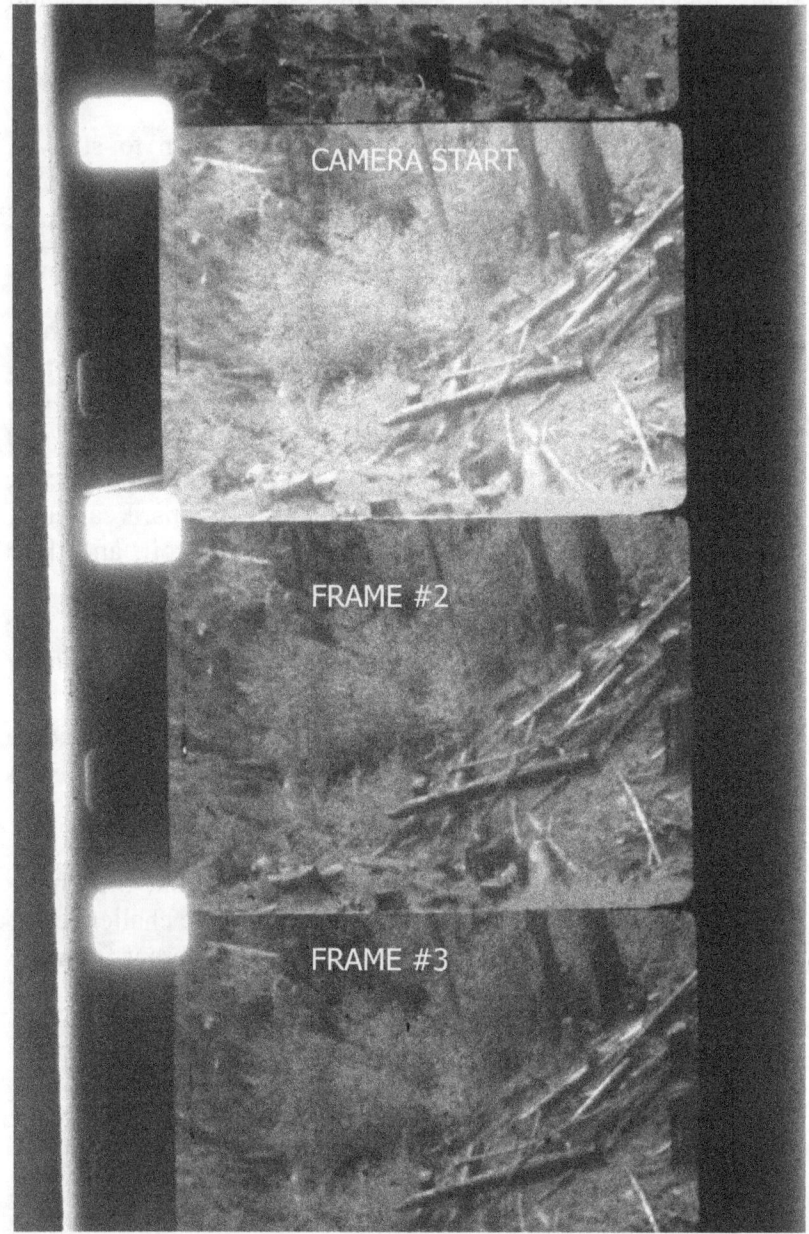

When Roger Met Patty

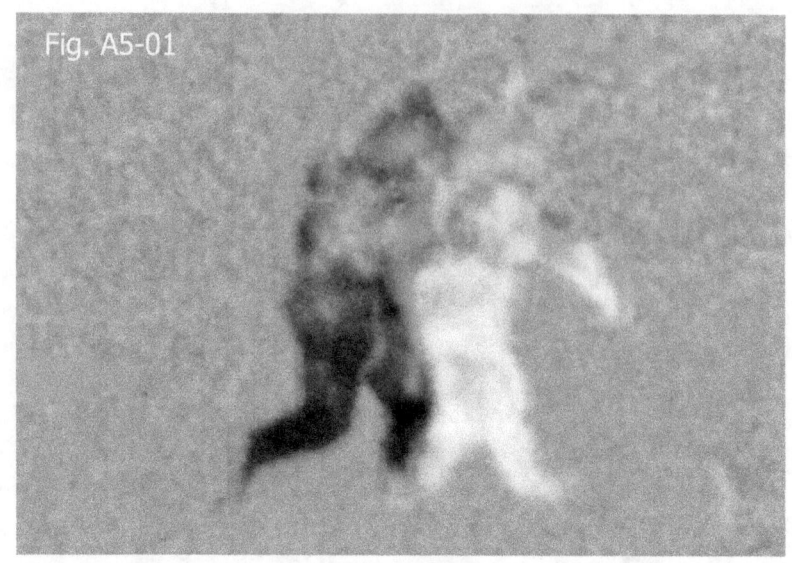

Fig. A5-01

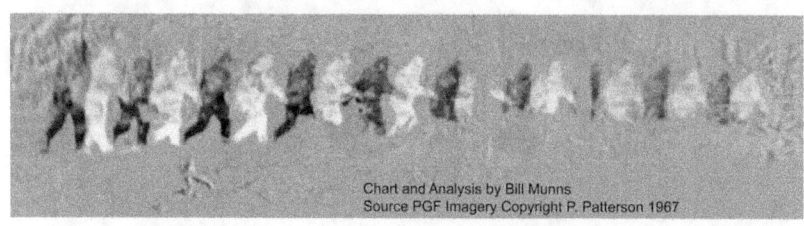

Chart and Analysis by Bill Munns
Source PGF Imagery Copyright P. Patterson 1967

Appendix Five - Image Analysis Techniques

1. Basic Image Comparison with Tonal Inversion - Computer graphics technology, such as Photoshop, has enabled us to analyze film images in ways we never could have realistically done before the computer graphics era. And these analysis techniques allow us to understand the PGF with far greater scientific accuracy and in ways Roger Patterson could never imagined in 1967. I rely heavily on these processes in my analysis, and felt that a sort of "primer" on the techniques would be valuable for the reader who wants to delve further into this study, or simply better understand why this analysis is real science and not the pseudo-science the skeptics so casually and confidently throw out as an accusation to anyone who's analysis rattles their belief of hoax.

Image comparisons allow for many forensic determinations, such as identifying motion in the filmed image (such as Patty's motion through the location), identifying motion by the person holding the camera, or motion of shadows over an elapsed time. The comparisons also show changes in lens focal length or camera position, show image distortion in bad copies, and alterations of an image as compared to another. Many of these studies are done on the computer where the two images are switched on and off and so I (or any analyst) sees the changes as the image switches from one to the next. This can't be done in a book illustration however, but a way of showing such image disparities in comparison can be represented with a tonal inversion technique I use extensively.

In its simplest form, one image is copied and pasted over the second image in a new layer, and then the pasted image layer is set at 50% opacity and the colors INVERTED. This is the digital equivalent of a film negative. When the two images are compared with the top one at 50% opacity, so 50% of what we see is the image below, the parts of the two images that were the same will appear as a solid neutral grey, and differences between the images will appear as varied light and dark tones, sometimes going to the extreme of blacks and whites. With this technique, we can easily see what is the same in the two images, and what is different.

The first example is something moving in the picture with the landscape elements more-or-less stationary. The overall image is grey, but the figure in motion (Patty walking) has distinct white and black lines indicating changes of position for the body overall, and the arms and legs in the walk cycle.

Fig. A5-02 Shows Patty walking in the top image, a motion study. Second row shows a rotational analysis of one frame to another, a part of an image stabilizing process. The third and fourth rows compare the same frame from several copies and the copy in bottom right had distortion because of the black lines of the trees when the landscapes tend to equalize.

The second example is VFC -2 Frame 613, where a light flare on one copy was mistaken for some type of gunshot "muzzle flash" used in the absurd "massacre theory". Two copies from John Green's archives were compared, (listed in the Copy Inventory as Copy #12 and #13), and both were same generation from the Ektachrome master, both full frame (minus optical printer cropping). One has a light flare in it, and the other does not. Overlaying one on top the other at 50% opacity and tonal inverted, the body of the image goes grey, and the flare stands out as a black dot, (surrounded by a white circle to help you locate it in this book illustration) because one copy has it and the other copy doesn't. Thus it could not be on the camera original, and thus was no anything that occurred when the film was originally taken at Bluff Creek. Image 2-05 in Chapter Two illustrates this.

Fig. A5-02

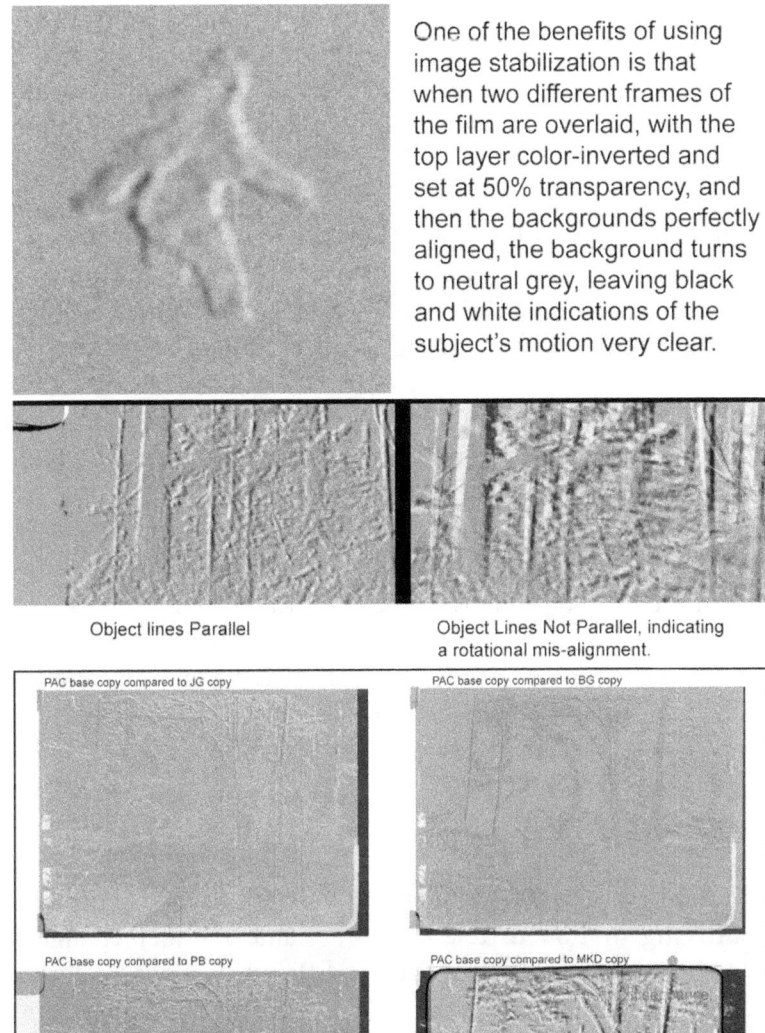

One of the benefits of using image stabilization is that when two different frames of the film are overlaid, with the top layer color-inverted and set at 50% transparency, and then the backgrounds perfectly aligned, the background turns to neutral grey, leaving black and white indications of the subject's motion very clear.

Object lines Parallel

Object Lines Not Parallel, indicating a rotational mis-alignment.

PAC base copy compared to JG copy

PAC base copy compared to BG copy

PAC base copy compared to PB copy

PAC base copy compared to MKD copy

Red dot above right references the tree showing angular distortion.

This technique was also used to verify each film frame was truly an individual frame and not a duplicate from a scanning error, and was not a TV conversion blended frame, because some frames do actually look a lot alike (such as the end of Segment Five, when Patty is mostly hidden behind a cluster of trees, and Roger is still filming, hoping she's come out from behind them and he can get a good shot again).

Same Image or Not - Because this technique is ideal for comparing two images, it is excellent when an image is suspected of being the same one, copied over, instead of just a very similar one. Some PGF frames look so much alike, this is the only way to verify if they are truly different frames.

Same Cropping - Comparing two cropped images by this technique is ideal to determine if they were cropped the same.

Distortion - I have found some examples of copy distortion, and this technique is excellent for comparing to copies of the same frame to see if one was copied with some distortion.

Fig. A5-03 shows another technique used for analysis, composite images that combine multiple frames into one composite picture, and give us a sense of time passing in a single analysis image. Patty walking and Jim McClarin walking are both shown.

2. Motion Analysis - The Tonal Inversion method is ideal for quantifying motion in relation to a frame number change, to see what moves in one frame difference, or two frames difference, etc. It allows for a very measurable evaluation of motion in a time duration context.

Fig. A5-03   Composite Image Studies

Above, Patty in PGF, below, Jim McClarin replica walk

Fig. A5-04   Locating the foot precisely

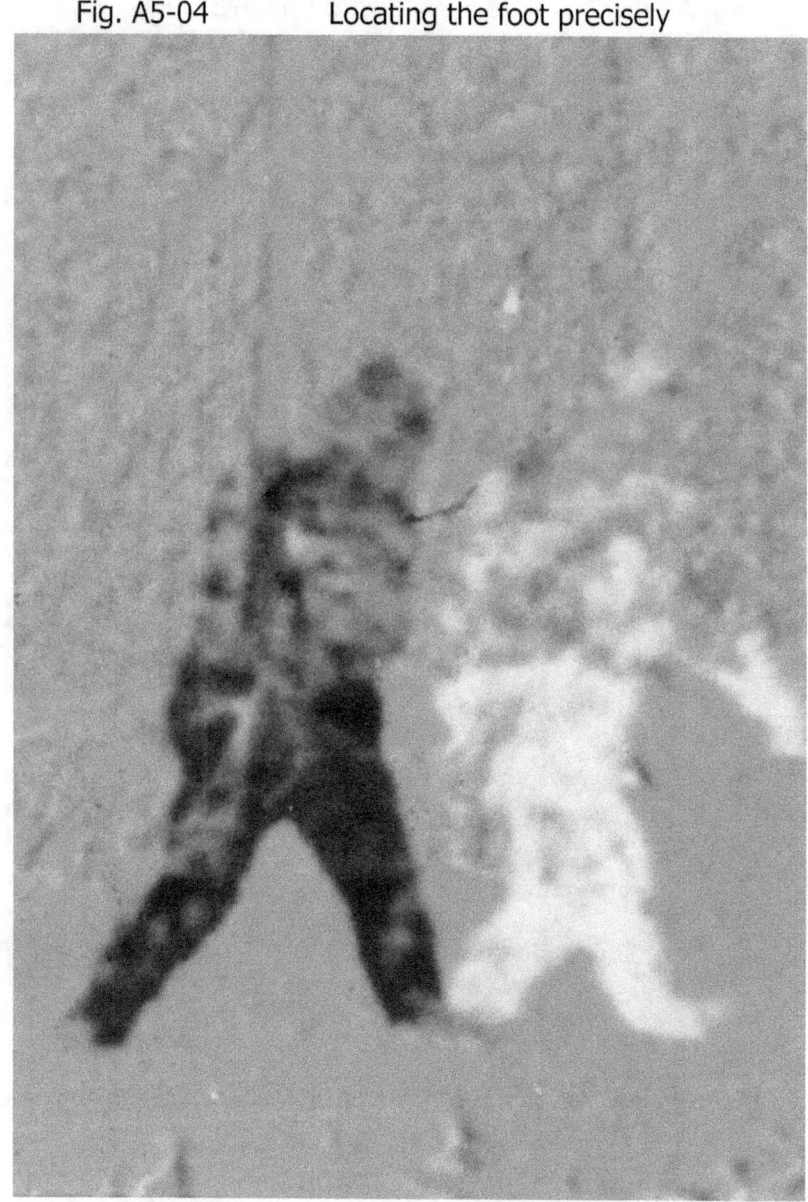

Fig. A5-04 shows one aspect of motion analysis, where we try to make an exact determination of where the foot is, and we can do so by combining two walk steps, where the same foot is the forward foot in one frame and the rear foot in the other. Where the black and white shapes of patty converge is the precise foot location.

3. Grain Nullification - The concept of Grain Nullification is rare because it requires unusual circumstances as a preliminary, and those circumstances aren't done as a pre-cursor for the grain nullification technique. It's rather an opportunistic derivative of something originally done for an entirely different purpose.

Basically, film is captured frame after frame, and projected at the same frame rate to reconstruct motion in real time as it occurred while being filmed. The optical printer allowed for the creation of a slow motion effect without needing a high speed camera to do the original filming. You can slow down the motion by simply printing each source film image two times or three times, or four times, onto the copy stock in the printer. And this was done both for the Green/Dahinden and ANE master copies. How slow the action will appear is based on how many times you repeat the frame.

Fig. A5-05 Shows four copy images reduced to two composites, and those two then reduced to one final composite.

So this multiple printing of the same frame was never intended for forensic image analysis. But each copy has a slightly different grain pattern than the source film, and that irregularity in the grain patterns causes slight image detail degradation on the copy. But if you take the copies of the same frame (each with its own distinct grain pattern, and blend these various copies of the frame, you can null out the copy grain, and get a composite that is closer to the source image in detail. It helps if the copy frames are a number in a multiple of two, like 2, 4, or 8.

Fig. A5-05 Grain Nullification Process

Copy 8 of the PGF has a great 4x zoom in study of Patty in the lookback, and the ANE people set it up with a four frame repeat of each source image, for a real slow motion effect. I take those four copies of the same frame, and blend them to null out the grain patterns of the copies, and help restore the detail level that was on the source film one generation earlier.

The technique (with four printed frames to one source frame) is to take copy frames one and two, copy two over to a layer above one, set opacity at 50% and align the two images perfectly. But unlike the tonal inversion, here, we keep both images in color correct form. Then the layer is merged down with the image beneath it, and the composite image has less grain than either one we began with. Next copy frame three and four are merged the same way, so now we have a 1-2 merged composite and a 3-4 merged composite. We copy the 3-4 composite and paste it over the 1-2 image, and set the 3-4 layer at 50% opacity, and merge the two down into a final composite. Everything which is common to these four copies of the source frame is retained, and any irregularities from the copy stock individual grain patterns is lost. The composite is now far closer to the source generation in detail. And grain noise is greatly reduced so any motion studies are not affected by random grain noise shifts and patterns (which sometimes give the false impression of motion when film is highly magnified and the grain is enlarged).

The result is superior study and analysis version, especially for aspects of Patty's body.

Fig. A5-06 shows a comparison of a single frame, with a grain-nullified result on left, and one of the source original grainy copies on right.

4. Lens Diagnostics - We can study the lens of a camera if the filming moved the camera side to side, by taking an object to one side of center, and finding it in true center of another frame, and comparing the two. This study allows us to evaluate peripheral distortion of lenses. It is sometimes quite pronounced.

Fig. A5-06

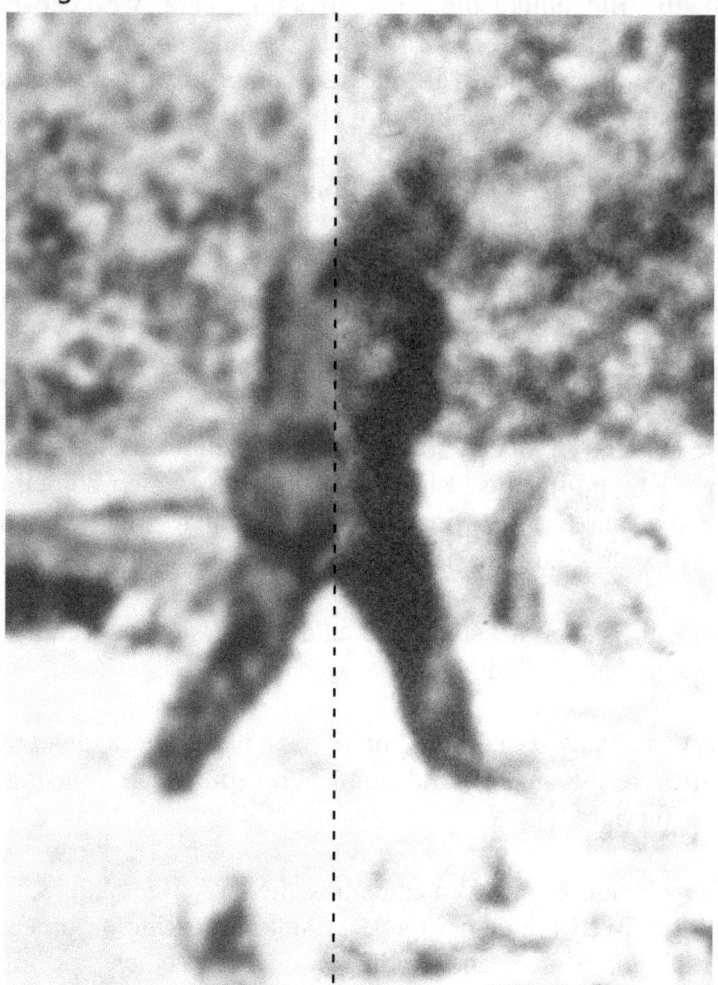

Grain reduced   Original grain sample

5. Image Stabilization - Image stabilization is the process of reversing the motions of a hand held camera, and keeping the images anchored on one designated fixed spot for analysis. For example, we can keep Patty centered in frame as she walks, or keep the landscape rock solid as she walks through it. Each frame is moved and possibly rotated, as needed to align with an arbitrary anchor point we choose, depending on the specific analysis goal of the stabilization.

Research Studies

Phase Two of the Breast Motion Studies which compare the natural breast and costume prosthetics.

Project Goal: To compare motions of natural breast mass vs motions of costume breast prosthetics to the PGF Hominid breast motions.

Appendix Six - Research Grants

I have been blessed by the support of the Cestone Foundation, which supported my research in conjunction with the research of Prof. Jeffery Meldrum at Idaho State University. And like most of what I do, the goals and intent of the research work has been misrepresented by critics (especially in the JREF forum) who had no knowledge of the actual grant proposals or goals, and because the grants were for research into the PGF, these critics just assumed the money was to "search for mythical beasts". Nothing could be further from the truth, but that's not surprising, given the JREF mentality could not be further from the truth either.

The major grant in question was one in which Prof. Jeffery Meldrum was the faculty Principle Investigator at Idaho State University, and I was the primary Research Associate and Sub Contractor. The grant funds were divided between Meldrum's work at ISU and my research work in Southern California. It was for research into the PGF, and had several goals.

First among the goals was to fund the design and fabrication of fur costumes specifically for scientific testing, to study the motion dynamics of fur suits, because the PGF is of course suspected of showing one such suit. We see the PGF figure in motion and we can identify some specific motion characteristics of the body seen in the film. So the research was intended to do the first ever systematic and methodical comparison of fur costumes in motion, as the person wearing the suit walks, and compare real human anatomy, in the form of female figure models consenting to be photographed without clothing so we can study how the human body moves in the same action and observable circumstances as the fur costumes. Them both can be compared to what we see in the PGF and we have some scientific basis to conclude that the body motions in the PGF resemble either costume motions or real anatomical motions.

For example, the PGF breasts do have a motion dynamic when the figure takes a hard step down while looking back at the camera. There is a distinctive ripple in the breast mass. Can a costume's fabricated prosthetic breast piece, made with any of the materials of the era (the 1960s) demonstrate a similar fluid dynamic motion? Or can real human anatomy demonstrate a similar fluid dynamic motion? We can theorize endlessly and debate it ad nauseum, but we can only settle the matter by a clinical test of the two options, walking the same way, photographed the same way, subjected to all the same circumstantial factors. Does the figure's breast motion in the PGF more resemble the motion of fabricated costume prosthetic breast pieces, or the motion of real anatomical breast masses?

Second, the PGF subject's body shows some shifting of the surface hair and presumably what the hair is attached to, be it furcloth base or real skin. So the research was intended to use these research costumes to study the shifting motion of the furcloth as the person wearing it walks, and then study the shifting motion of real human skin as the person walks. Does the PGF body surface shifting more resemble what costume furcloth does, or more resemble what real dermal tissue on a human does?

These are valid and responsible questions to try and make some factual determination as to what it is we see in the PGF. Is it a fur costume or real anatomy? The research goal has as much opportunity to prove the PGF is a fake as prove it real, so the research was not prejudiced toward any agenda. It simply was intended to try and provide for the first time some factual basis for comparing the PGF subject figure's body to both costumes and real anatomy. I would think anyone wanting to get to the truth of the PGF mystery would welcome such research.

The second goal was to do a site survey of the Bluff Creek location, with new survey analysis technology, simply to

450

provide a new and more precise factual specification for the dimensions and characteristics of the location. This would add to our knowledge of the site and thus to our knowledge of what actually occurred. The data could as easily prove the filming was hoaxed as prove it was authentic. It simply expanded our knowledge of the truthful circumstances of Bluff Creek, and should thus bring us closer to a truthful understanding of what actually occurred.

The third goal of the research was to acquire and scan more film footage into the research database, so we have more image data to work with. Is there anything disreputable about wanting to gather more data for analysis? Of course not. It is and always will be an admirable goal of any research program.

The fourth goal was to acquire and study 16mm camera and lens technology, so we can do actual camera tests and increase our knowledge of the 16mm film technology with is integral to the study of this film.

The grant was not for any search for a mythical beast. It was for research in comparing fur costume motion dynamics to real anatomical motion, gather survey information about a woodland location, and expand our understanding of 16mm film technology so we can better evaluate the 16mm film evidence connected to this investigation.

The grant proposal was quite reasonable, and the following describes some of the analysis of the research.

Considerations of Breast Mass Fluidity and Dermatological Elasticity In the Analysis of the Patterson Gimlin Hominid

Abstract: The Patterson Gimlin Hominid exhibits two anatomical motions during the hominid's walk which have not been previously studied in a comparative form to both real human anatomy and fabricated costume anatomy. The hominid female's conspicuous breasts (seen during a segment of the film referred to as the "lookback", frames #345-375 from the Verified Frame Count Inventory, Ver. #2) exhibit both a random fluidity during a smooth step, and a sudden vertical vibration during the landing of a hard step. Additionally, the fur/skin surface from knee to armpit exhibits a shift of position during the lookback right leg step because the subject body is seen in near true profile at that point. The motions can be clinically studied and experiments can be devised to compare both live human anatomy and fabricated costume anatomy to these identified PGF Hominid motions of anatomy. The motions are consistent with real human anatomy, and inconsistent with fabricated costume anatomy motion potential.

Introduction - The Hominid Subject seen walking through the famous 1967 Patterson Gimlin film has been analyzed for 45 years now. Many of those analysis efforts focus on two opposing lines of reasoning. One line of reasoning tries to compare and explain the anatomical appearance in relation to human or other primate physiology, to support arguments that the film's hominid subject depicts a real biological entity. A second line of reasoning tries to compare and explain the anatomical appearance in relation to a fabricated fur costume, to support the argument that the film's hominid subject depicts a falsified costume worn by a human performer, and thus claim the film depicts a fake or hoaxed entity.

Analysis of the past failed to reveal the motion dynamics herein studied because of a lack of effective computer graphic technology, as well as a failure to understand how to properly use the film image data for effective analysis. Current computer graphics software tools and applications do provide the necessary capability to identify the anatomical motions and the correct film copy image data has been identified as the appropriate data to determine the motions occurring on the PGF Hominid body. Using the PGF Hominid motions as the foundation, experiments were devised to study both live human anatomy and fabricated costume anatomy in motion to determine if one form in motion produces results more favorably compared than the other to the PGF Hominid motions.

This concept forms the Null Hypothesis - Breast motion and dermatological shifts of the PGF Hominid while walking will tend to replicated by one test alternative and will fail to replicate on the other alternative, allowing a conclusion as to whether the PGF Hominid's anatomy is live anatomy or fabricated costume anatomy.

The Alternate Hypothesis - The motions being studied will either occur on both live human anatomy and fabricated costume anatomy, or will occur on neither, and no conclusion can be drawn from the experiments.

This concept is examined herein, to see if the null hypothesis has merit or not. If we find similar motions do in fact occur on one tested alternative and not the other, the null hypothesis is supported and validated, and the tested alternative which has affinity with the PGF Hominid motions will be thus applied to determine if the PGF Hominid is biologically real or the result of fabricated costume devices.

Film Image Data (source for comparison)

The PGF consists of a 16mm movie film taken in 1967 and copied multiple times and by varied methods. Currently researchers only have the copied forms as study material, and the film medium is a technology which experiences some loss of data with each copy generation. When the copies were made, the persons making the copies produced various copy versions, but without full appreciation for what type of copy is best suited for specific types of analysis. Lacking this full understanding of film copy technology and what type of copy format is ideal for specific analysis goals, the analysis of the film resulted in considerable confusion and some aspects of the PGF Hominid anatomy were overlooked and unappreciated for their potential to make determinations about the most fundamental controversy of this film, the question of whether the hominid is biologically real or a result of a fabricated costume effect worn by an ordinary human performer.

Understanding the copy technology of film is essential as a foundation for understanding the PGF data which the experiments will compare to. So reference back to other chapters discussing film technology might be useful to the reader.

Comparative Models and Material

Comparisons are made with both real human anatomy and fur costumes fabricated for experimental purposes.

The comparison of real human anatomy included photographs of human models employed as research models. It should be noted that in some examples of human female bodies, the bodies are painted a grey coloration with black markers on the body, and this was for clinical experiments. It allowed the shifts of the dermal tissue to be studied as the body moves through a walk cycle. Similarly, on some test costumes, small patches of fur were cut out to make specific markers that could be tracked as the human wearing the costume walks in specified motions for tests.

The comparisons using fabricated fur costumes included the costumes being worn by human models and going through specific directed motions and activities, same people filmed candidly at times they were simply engaging random activity, and in some cases, experiments with the costumes mounted on mechanical body forms for specific tests of motion or fold dynamics. The costumes, when worn by human performers, were tested with no internal padding, with partial padding in some areas of the body, and complete padding over the entire body, to study how padding (or lack of same) impacts upon the costume shape and capacity to fold during motion of the performer wearing same.

Features Described and Compared

Topic A - The Breast Motions

The PGF Hominid does demonstrate some fluidity in the motion dynamic of the breast mass, as it walks and looks back at camera, thus allowing a clear view of the breast anatomy on the subject. Fig. #A6-01 (PGF Hominid Breast Motions) illustrates two film image frames which best illustrate the PGF Hominid breasts exhibiting motion dynamics and deformation of shape of the breast mass. This particular motion shift of the breast mass occurred when the PGF Hominid stepped with the right leg forward and the terrain was lower than expected, and the body drops abruptly onto the lower terrain and a shock wave of this hard step ripples up the body, causing the breast alteration of shape.

The two experimental comparisons are with human female breast anatomy, and fabricated costume breast form prosthetics. The question posed for study is, do the natural human female breast anatomy and the fabricated costume breast prosthesis have a similar potential for motion and appearance of fluidity, when both are subjected to a specific controlled and repeatable vector of downward motion and then an abrupt stop?

The soft tissue structure of the biological hominid breast mass is well described anatomically. The physical structure of costume breast prosthesis objects is not. Therefore, before the experiments can be described, a primer on costume breast prosthesis is in order.

William Munns

Immediately below is an illustration of the image analysis procedure. LEFT has a dark circle toward the left side of the grey box. CENTER has the dark circle moved to the right. RIGHT image shows the second image set at 50% transparency and color inverted so an analysis can document the shift of position, the dark being the original position, the light being the moved position.

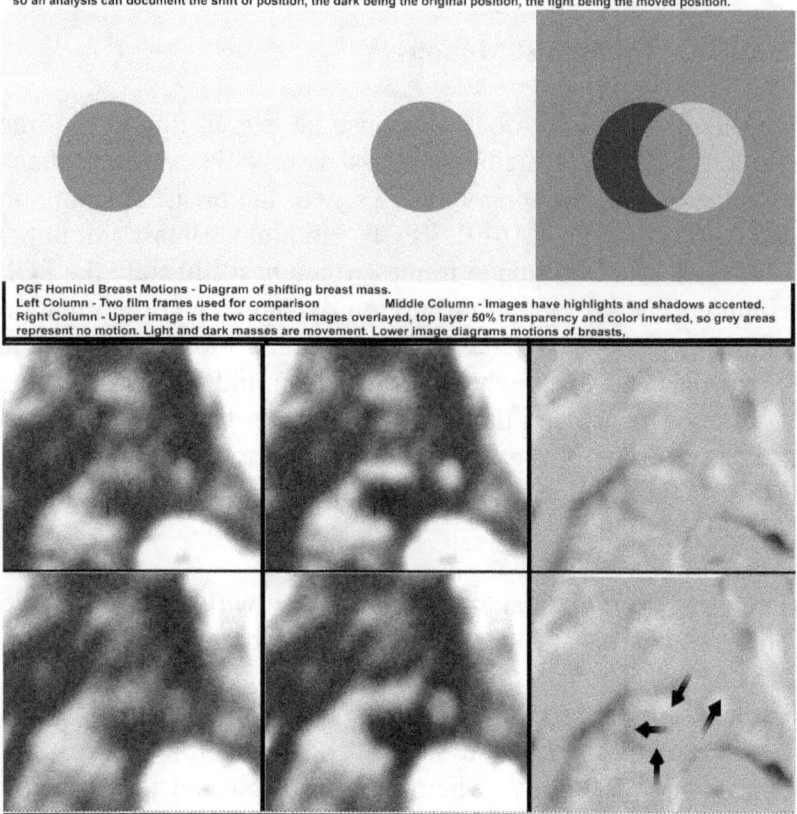

PGF Hominid Breast Motions - Diagram of shifting breast mass.
Left Column - Two film frames used for comparison   Middle Column - Images have highlights and shadows accented.
Right Column - Upper image is the two accented images overlayed, top layer 50% transparency and color inverted, so grey areas represent no motion. Light and dark masses are movement. Lower image diagrams motions of breasts.

In 1967 (the year the PGF event occurred), prosthesis devices were common and well established in the motion picture and special effects industries. Molded and cast prosthesis objects actually date back to the 1930s and were extensively used in the famous 1939 motion picture "The Wizard of Oz". So by the 1960s anatomical prosthesis for costumes was a well-established and highly perfected craft. I learned these fabrication methods from Michael Westmore, director of the makeup lab at Universal Studios (Hollywood), in January, 1969. The fabrication process was at that time in relative stasis with well-established techniques practiced consistently over the entire decade of the 1960s and well into the 1970s.

456

A prosthesis for a makeup or costume effect could be fabricated in one of two ways. A direct buildup was the more crude and old fashioned way, having been used by Jack Pierce in 1931 for the transformation of Boris Karloff into the Frankenstein monster for the 1931 film "Frankenstein". In the 1960s direct buildups were occasionally done still, usually for both expediency and lack of appropriate budget. A direct buildup technique tends to produce a prosthesis of notable rigidity, whereas molded prosthesis devices can be made of materials that are more pliable and flexible, so they are considered the material technique more likely to produce what is observed in the PGF Hominid's body. The generally regarded proper professional technique (one which consistently produces a superior result) is the molded prosthesis.

A molded prosthesis starts with an impression of the anatomical part of the human performer to wear the resulting prosthesis, and a positive plaster cast of that impression serves as the foundation for the sculpture. Either a water-based ceramic-type clay, or an oil-based plastilina clay was used to sculpt the desired anatomical feature. As we are discussing breast prosthesis, a chest impression and positive plaster cast would be made of the human performer, and the breast masses would be sculpted upon this chest plaster cast. The desired size, shape, and level of anatomical detail would be determined by the sculpture and the specifications of the job. Once the sculpture is complete, a negative mold is made of the sculpture, using a relatively hard gypsum plaster such as Ultra-Cal 30, or even dental stone. The negative mold covers all the sculpture area plus some peripheral surface of the chest impression cast, with devices called "keys" incorporated so the negative mold and the positive impression cast can be separated and reassembled in a precise interlocking position.

Once the negative mold is hardened, the two mold pieces are separated and the sculpting clay is removed, creating an empty cavity in the shape of the intended prosthesis. That mold cavity then needs to be filled with some type of prosthetic casting material to produce the finished prosthesis.

Such techniques were well established and described by Carl Dame Clarke, "Molding and Casting" (First edition, Pub. 1938, John D. Lucas Co.) and "Stage Makeup" by Richard Corson, 3rd Edition 1960

There were three options of material in general professional use in the 1960s. The simplest option was a vulcanized liquid latex commonly called "slip rubber" which could be poured into a mold and slushed around, or brushed up in the negative mold only, to build up an appropriate thickness so the prosthesis would hold its shape while still being flexible. The second option was a two part resinous polyureathane material, commonly called "flexible polyfoam". Once the two components were mixed, the compound mixture creates air bubbles which are trapped in the viscous liquid, thus chemically causing a foam expansion of the mix. It was poured into the negative mold, the positive back part was quickly closed against the negative mold, and some type of clamps or restraints were applied to hold the two mold pieces firmly together as the polyfoam continues to expand. Once the expansion ceases, the material cures and the result is a pliable resinous foam mass in the molded shape of the prosthesis. However, polyfoam has inherent strong adhesive tendencies, and will adhere ferociously to the plaster molds, so a thin coat of slip rubber was traditionally painted onto each mold surface and that thin slip rubber "skin" adhered to the polyfoam, but easily peeled out of the molds as they are separated. The primary advantages of polyfoam prosthetics were that the mixing was quite simple, and the material cured at room temperature as part of its inherent process.

The third alternative for prosthesis was natural foamed latex, and this was generally regarded as the superior prosthesis material. It was, however, the most challenging to mix and cure. It would generally be bundled together as a package of components, from 3 to 6 components (depending on a company's specific formulation), but the essential ingredients of all systems were the unvulcanized base rubber latex, a sulfur component which will facilitate the vulcanization, a foam agent which helps stabilize the foam, and a gel component which causes the foamed liquid to gel or solidity in a predictable and uniform manner, once it is put into the molds. Natural foam latex was mechanically foamed, meaning the user must use some form of whipping mixer to actually whip the air bubbles into the rubber liquid. So equipment to accomplish this was a necessary workshop tool. Once the foam was poured into the molds and gelled, it needed heat to cause the sulfur component to interlock with the rubber compound and vulcanize the rubber, so a curing oven that can hold the mold apparatus and sustain a temperature of 225 degrees for about 4 hours was also a necessary piece of workshop equipment.

These were the three options for fabricating costume and makeup prosthesis devices in 1967, the year the PGF was filmed. Additional description can be found in "The Technique of Film and Television Makeup" by Vincent J. R. Kehoe, Pub. 1969. The current experiments herein described and conducted for this analysis included three breast prosthesis pieces, one each fabricated from all three materials, so all options were tested.

Experiment #1 - Breast Motion

To compare fluidity of natural anatomical breasts with costume fabricated prosthesis breasts, the experiment was designed so the motion/inertia vectors were direct, repeatable and of consistent force applied to both tested alternatives. A platform was constructed which is supported by aluminum bar parallelogram brackets, which allow the platform to first be pulled forward and then the brackets would swing in a circular motion to a downward vector. A rigid material was placed beneath the platform as a stop, to insure that each time the platform is moved, it falls the exact same distance before coming to an abrupt stop. Equalizing the weight on the platform for both the human test subjects and the costume breast prosthesis insures that the acceleration of the platform in its fall was consistent and the impact of its abrupt stop would produce a shock wave of consistent force with every test action.

Fig. #A6-02 (Breast Motion Test Platform) illustrates the test device. On that illustration, image A shows the basic device. Image B shows the platform in its high start position. Image C shows the base platform in its lower "dropped" position. Image D shows a human model standing on the platform. Image E shows a costume mounted to a support rig on the base platform.

Three drop distances were studied, both for the human subject and the costume prosthesis (in all three material forms). Those distances were 1", 1 1/2", and 2". The 2" drop was the most severe and resulted in the strongest shock wave of force.

The result of this experiment on two female subjects demonstrated that the real anatomical breast mass will consistently shift downward relative to the general thorax at the point of platform impact and abrupt stop, because of the natural breast mass fluidity and lack of any skeletal support results in the breast mass continuing to fall even as the skeletal support of the body has abruptly stopped. This initial shift downward of the breast mass (in relation to its normal position) would eventually be opposed by the dermal tissue and breast mass tissue connectivity, and the breast mass would then be pulled upward by the elasticity of the skin and breast connective tissue, a bouncing effect. After several vertical shifts up and down, of diminishing travel, the breast mass finally settles back into an equilibrium of position on the thorax. This was to be expected, because the human breast mass is fluid-filled tissue without any skeletal or musculature structures to stiffen the mass, but has both internal connective tissue and external dermal tissue that define its shape when no motion vectors deform it. So it behaves exactly as fluid dynamics would predict.

Breast Motion Test Platform

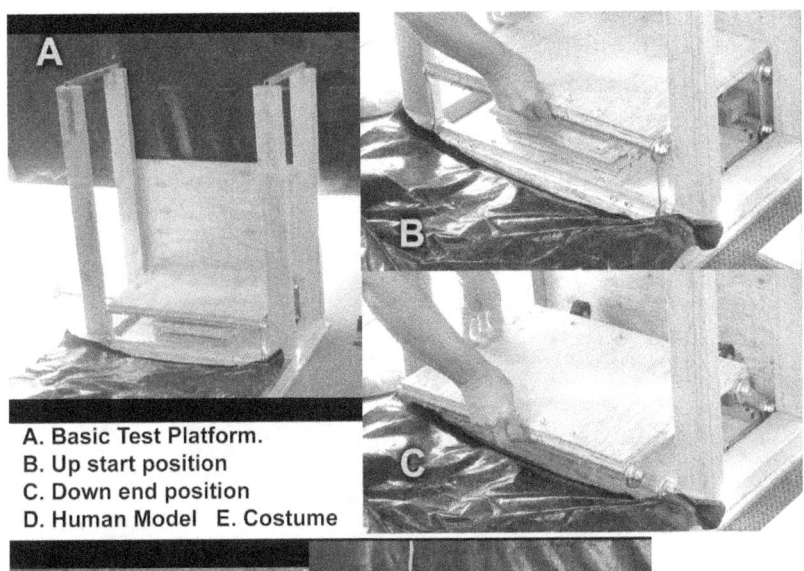

A. Basic Test Platform.
B. Up start position
C. Down end position
D. Human Model   E. Costume

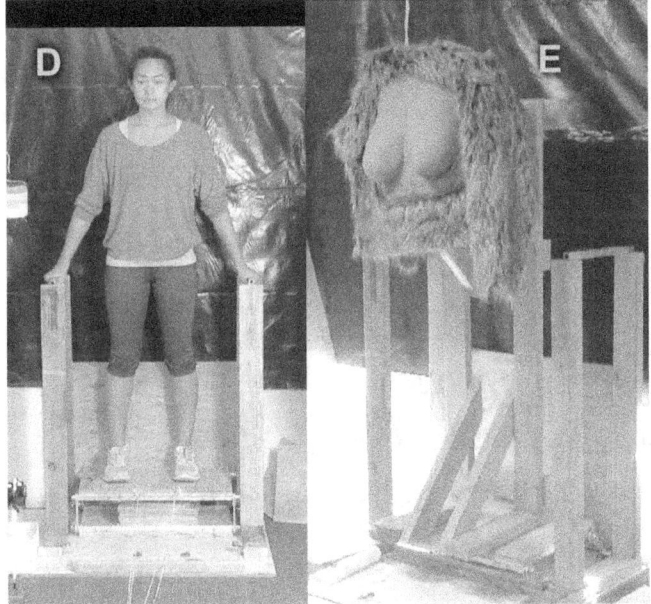

Fig. A6-02

Fig. #A6-03 (Biological Breast Mass Motion) shows two female test models and the panels (left to right) show the breast mass prior to the drop and shock wave (left column), the breast mass on it's reactive rise after the abrupt stop (center column), and finally a composite of the two images showing the displacement of the breast mass vertically and to a slight extent, horizontally shifted away from the body's medial line (right column).

Biological Breast Mass Motion - Two research figure models were placed standing on the drop platform, and experienced a vertical drop of 2" before the platform came to an abrupt stop. LEFT Column shows the breast shape prior to the drop. CENTER Column shows the upward bounce after the initial shock of abrupt stop. RIGHT Column has superimposed the second image at 50% transparency and inverted color, to show the vertical shift in breast mass and the outward horizontal shift of the areola.

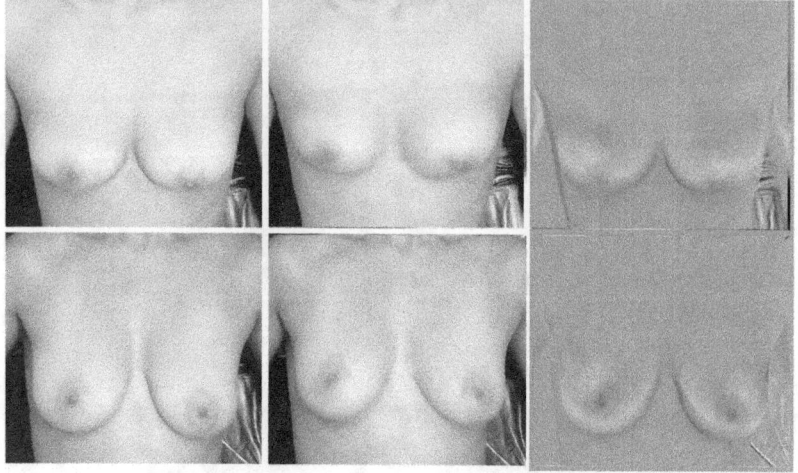

Fig. A6-03

The unpredictable element is the costume breast prosthesis objects. To what extent their molded shape would be altered or deformed by the shock wave of the drop and abrupt stop could only be determined by testing. These breast prosthesis objects were secured to a chest mold mounting plate, which was then secured to the drop platform. Additional weight was added to the platform to equal the weight of the human models gross body weight, to insure that each prosthesis test would experience the same acceleration, the same velocity while dropping, and the same shock wave force upon the abrupt stop of motion as experienced by the human models and their breast mass.

For all three types of breast prosthesis (slip latex, flexible polyfoam, and natural foamed latex), and the most severe drop of 2", these prosthesis objects demonstrated essentially no deformation of mass comparable to the human natural breast masses. In evaluating the failure to demonstrate any fluidity of motion, the slip latex prosthesis exhibited one type of physical dynamic, while the two foamed prosthesis objects demonstrated another dynamic.

The slip latex, being a hollow shell outer shape, must have enough rigidity to maintain the molded breast shape, but it has very little mass, as is expected for a hollow object. Thus the inertia of the small mass was not sufficient to overcome the inherent rigidity of the latex shell construction. Thus no vibration or bounce of the breast shape occurred in these experiments.

For both the flexible polyfoam and the natural foamed latex prosthesis, their mass did fill the breast shape but both are foamed, so their mass is largely air bubbles encapsulated in the elastomer compound. The mass is insufficient to deform the interlocking elastomer matrix around the air bubbles, so the inertia of the mass in motion (dropping) was less than the connective strength of the elastomer to maintain it's molded form.

Fig. #A6-04 (Biological Model and Fabricated Prosthesis Compared to PGF Hominid) compares the motion dynamic of the natural human female breast (upper left), the PGF Hominid breast mass shift (upper middle), and the slip latex breast prosthesis (upper right) which exhibited no measurable motion shift, all examples using a imaging technique where a "before" image is at 100% opacity, and the "after" image is overlaid at 50% transparency and color inverted, so image elements which do not move become a neutral grey tone. The slip rubber prosthesis image examples (from two separate test drops) are remarkably grey overall and thus has not altered its shape in

any substantial or measurable way. The upper and lower image
are test drop #1 and test drop #2, for each prosthesis.

Biological Model and Fabricated Prosthesis Compared to PGF Hominid -
LEFT shows the two test figure models and the observed breast mass motion, vertical compression and lateral displacement.
MIDDLE shows the PGF Hominid breast mass motions, vertical compression and lateral displacement.
RIGHT shows the Slip Latex Prosthesis, with no evident motion, deformation or shifting of shape.

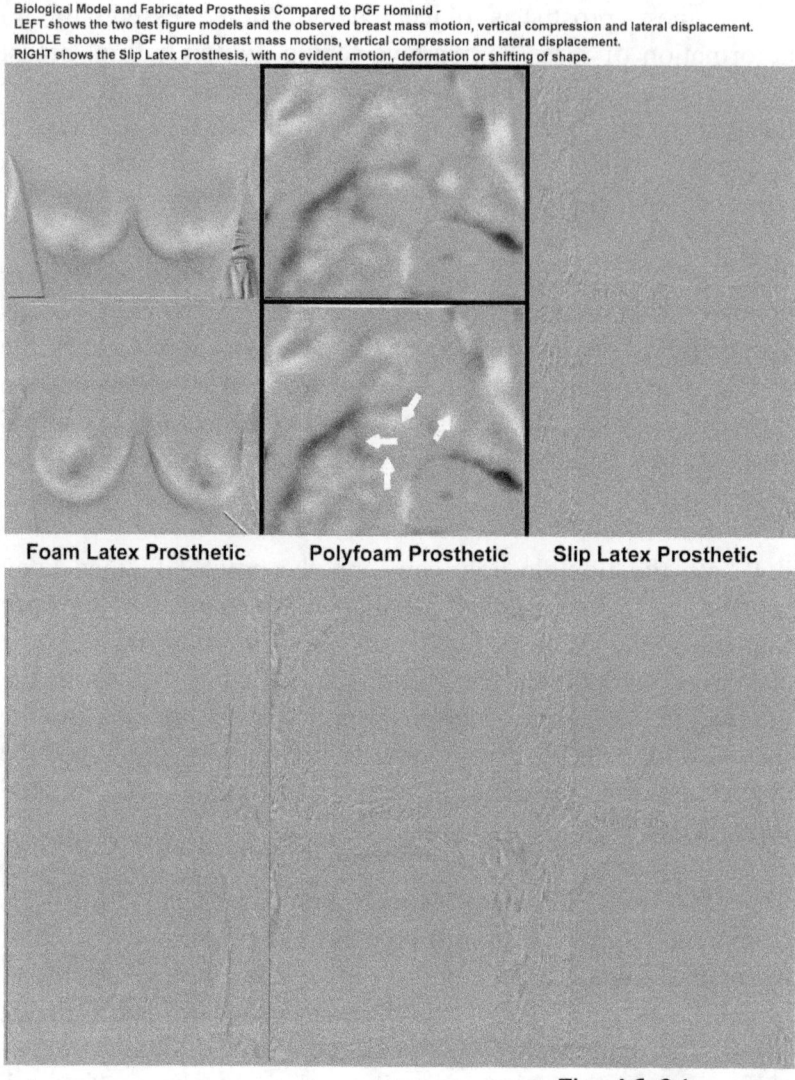

Foam Latex Prosthetic   **Polyfoam Prosthetic**   **Slip Latex Prosthetic**

Fig. A6-04

464

The lower image pairs show all three types of costume breast prosthesis compared. The natural foam latex prosthesis (lower left), the flexible polyfoam prosthesis (lower middle) and the slip latex prosthesis (lower right) all demonstrate a continuity of form indicating that they have no capacity for replicating a fluid mass motion dynamic.

The result of these experiments was a consistent demonstration that real human female breast mass does exhibit the fluid dynamic motions as seen in the PGF hominid's breasts, and that costume prosthesis breast objects consistently do not exhibit any fluidity of motion and deformation comparable to either the PGF hominid's breasts or the natural breasts of the human test subjects.

This phase of experiments thus affirmed that human organic breasts do demonstrate motions comparable to the motions of the PGF Hominid, and that costume breast prosthesis objects do not demonstrate any such capacity for fluid motion or deformation.

Experiment #2 Dermal Tissue Shifting during anatomical motion

In analysis of the PGF Hominid, particularly using Cibachrome Images which were two frames apart in the sequence and select frames from PGF Copy 8 (which has an excellent 4x zoom in sequence of the lookback), we can study some shifts of the surface features along the right leg (nearer to camera) and up the right side of the torso to the armpit. These observations show that when the right leg is extending to take a forward step, that extension of the lower leg causes a shift in dermal tissue (and the fur attached to the said tissue) from the rearward portion of the knee up through the thigh, past the pelvic region, and all the way into the thorax region below and behind the breast mass. Irregularities in the fur pattern of the PGF hominid are used as markers to chart the shift and elongation of the dermal tissue.

PGF Hominid Skin/Fur Shift Diagrammed -
TOP ROW - Skin/fur markers in contracted position
SECOND ROW - Skin/fur markers in extended position
BOTTOM ROW -   elongation shifts.

Fig. A6-05

Fig. #A6-05 (PGF Hominid Skin/Fur Shift Diagrammed) illustrates two separate phases of the PGF Lookback walk cycle and the skin shift is diagrammed. White dot markers show the identification lines in compression, the black dot markers show the same identification markers in extension, and finally the gray zone indicates the shift zone between the compression and extension positions.

The experiment is to compare both human dermal tissue and costume fabricated furcloth material (which forms the substitute "skin" of a costume intended to appear anatomically real) to what is observed in the PGF Hominid studies.

Defining the Alternatives

Human (or primate) dermal tissue (skin) is an anatomical component of a living body which has a generally consistent thickness and a shape that generally conforms to the anatomy within. The skin has an elastic connectivity which allows it to shift its position on the body as the internal skeleton and musculature move and cause some surface aspects to either expand or compress, depending on where the surface points are in relation to the skeletal centers of rotation. For example, when the leg rotates forward to step ahead, in a normal walk cycle, the skin bridging the forward aspect of the pelvic region of the torso and the skin of the front of the thigh compresses somewhat, and the skin on the rearward buttocks and the rear aspect of the thigh tends to elongate. Similarly, when the lower leg extends or flexes, the skin behind the knee elongates or compresses the skin up the rear and side of the thigh well into the pelvic region and to some extent into the skin of the thorax on the lateral regions.

This elongation and contraction of the skin during physical motions of the body is generally diffused over a large area of the total dermal structure commensurate with the size of the body part in motion. The leg rotating about the hip socket or the lower leg rotating about the knee joint will cause a fairly large mass of dermal tissue to shift, whereas a finger bending will only cause a dermal tissue shift from the finger to about the wrist, but likely no further up the arm.

The architecture of a fur costume, especially one with a breast prosthesis attached, has a totally different design, and thus should be expected to exhibit a totally different motion shift when the costume is worn by a human performer and the performer goes through such motions as walking. The principle distinction between real dermal tissue encompassing a body, and a costume encompassing a body of a human performer, is the different materials needed to produce various aspects of the costume.

A costume designed to appear as a realistic higher primate (a great ape) must address the fact that all known great apes have areas of the body covered by hair dense enough to fully obscure any view of the skin, and some areas of the body so devoid of hair as to make the skin fully apparent in it's shape, tonality and texture. And finally, there are transitional areas where the hair density transitions from very sparse to very thick. But whereas the real anatomical body has a consistent dermal tissue structure on both the bare skin, and hair covered regions, and thus behaves as one unified structure, the costume designed to appear as real does not have a single unified structure.

Traditional "creature" costume design of the 1960s divided the costume into three elements, those elements being the regular fur aspect, the bare skin aspect, and the transitional skin-to-fur aspect. The fur aspect was most commonly accomplished with some type of synthetic furcloth or real animal fur pelt, especially when a fur effect of short and dense fur was required (as seen in the PGF Hominid). Wefted human hair or synthetic hair fiber was occasionally used, the wefts sewn onto a cloth costume base structure, but this effect was only desirable when the apparent visual effect was hair longer than 6" and the shaggy quality of the hair essentially obscured any perception of anatomical specifics of body contour. As this is clearly not the case with the PGF Hominid, this option can be excluded from further consideration.

Both real fur pelts and the fabricated furcloth of the era were non-elastic. Elastic furcloth was not documented to be developed and introduced to the market for use until the mid-1980s, by a company currently known as National Fiber Technology (formerly National Hair Technology, and before that, Reed-Meredith). Non-elastic furcloth was woven on a loom just as carpets are, and has the same basic components, the base weave and the pile (which in furcloth is the hair exterior). The base weave structure has no elasticity whatsoever along either the horizontal or vertical lines of the weave, and potential stretch along the bias (a 45 degree diagonal to either the horizontal or vertical weave fibers) is determined by the tightness of the weave. Furcloths, like carpets, generally mandate a tight weave and so have minimal stretch on the bias. The pile or hair mass of the furcloth has a "lay", a directional orientation in which it lays the flattest when brushed down, and that lay is generally parallel to the vertical weave of the base fiber.

The skin areas of a costume however are generally fabricated from some rubber or synthetic elastomer compound. One might think that to merely trim the furcloth hair pile down to the base weave would result in a section appearing like hairless skin, but if done, the base weave pattern becomes quite obvious and has no realistic contour, texture, or tonality resembling real skin. So skin sections of a costume are instead fabricated by first making a sculptured piece which allows for natural skin shapes, folds and textures such as wrinkles. Molds are made and from those molds the actual skin-like prosthesis pieces are cast. The area of these prosthesis pieces is larger than the intended apparent skin area, so there is a surface where the fur can be glued to the prosthesis surface, and then the area where transitional hair can be applied to give a natural-appearing transition from thick fur to sparce hair strands to bare skin.

But it it this attachment of the furcloth to the prosthesis which creates a physical disruption of the physical density and flexibility of the costume. Furcloth alone has a consistent degree of homogenous density and transmission of motion. A prosthesis piece of consistent thickness also has a homogenous density and transmission of motion. But when the furcloth is glued to the prosthesis, there are now two different materials fused together, creating a boundary zone of far greater rigidity than either the furcloth or the prosthesis material alone possesses. Further, the glue material itself compounds the rigidity in that zone. So the costume no longer has a singular overall degree of density, flexibility or capacity to shift in motion. Instead, there is a continuous furcloth mass going up the legs into the pelvic area, and then an abrupt rigid line where the furcloth joins the prosthesis material. Any shift of the furcloth up the leg portion of the costume will abruptly buckle when it fails to shift the more rigid mass where the furcloth is glued to the prosthesis chest object.

So, in theory, when furcloth of a costume is studied to see if it can replicate a shifting of skin as observed and documented in the images of the PGF Hominid, it is expected that the shift will buckle conspicuously at the edge of the rigid line where the furcloth is glued to the chest prosthesis.

Fig. #A6-06 (Costume Fabrication Design) illustrates the concept, showing (top row) a diagram of a breast prosthesis and a furcloth surface as two discrete objects, (middle row) a diagram of how the furcloth would be glued onto the prosthesis (red line indicating the glue, area circled in dotted red circle for identification), and (bottom row) how one may expect the furcloth to buckle if shifted toward the junction of the furcloth and the prosthesis.

Costume Fabrication Design

TOP ROW - Breast Prosthesis (left) and furcloth (right)

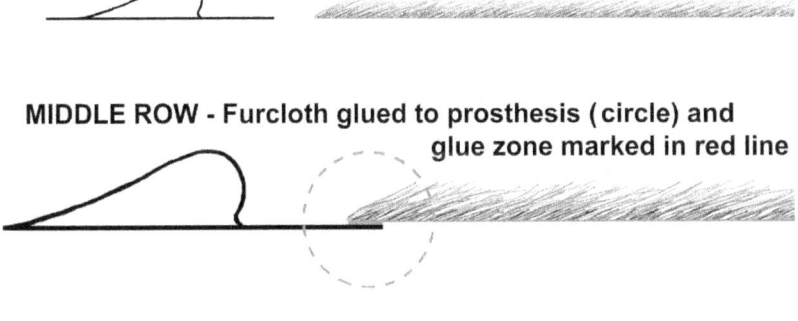

MIDDLE ROW - Furcloth glued to prosthesis (circle) and glue zone marked in red line

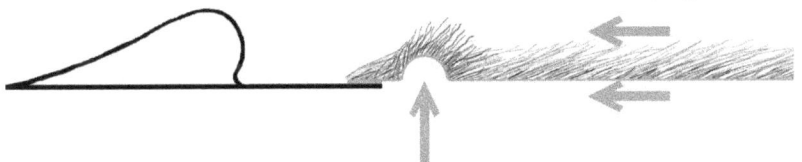

LOWER ROW - When furcloth shifts toward prosthesis, the furcloth will buckle before the rigid zone

Fig. A6-06

In actual testing of costumes, this did occur in practice and affirmed that a costume, fabricated with a furcloth body and supplemental prosthesis pieces such as a chest piece with molded breast shapes, will not replicate the continuous skin shift from knee up to the mid thorax region immediately below and beside the breast mass, which is observed in the PGF Hominid. The costume fails to replicate the motion dynamic exhibited by the PGF Hominid's skin/fur surface.

Females models were employed by this research project to study the motion dynamic of real skin. To allow for precision in evaluating and documenting the skin shift characteristics, and the motion dynamic of human anatomy, the models were unclothed so the skin area, from knee to armpit and shoulder, was unobstructed by any clothing, in terms of visual evaluation or actual motion constraint or modification. A grey body

471

makeup was applied so the body's tonality was consistent overall for each body and from one test subject to the next, and a series of black marker dots were applied from the knee up to the armpit along the side of the body, allowing shifts in position to be studied in an absolute and measurable way.

Fig. #A6-07 (Experiment Body Painting Pattern) illustrates the body paint and study marker pattern used on these figure models.

Experiment Body Painting Pattern, including marker dots on front collar line and right side of leg and torso

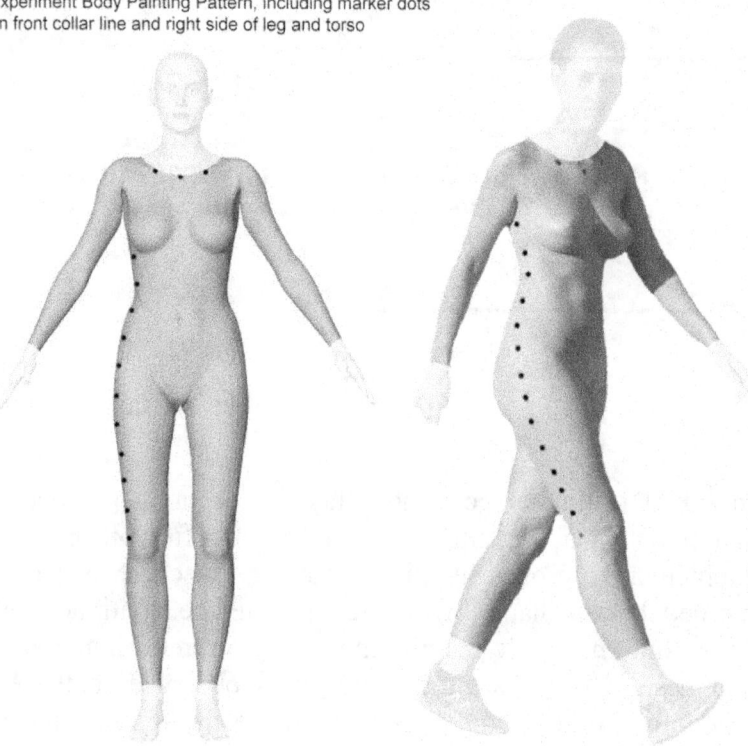

Fig. A6-07

The anatomical models employed for this were studied in normal walk cycles as well as slight step downs, as occurs in the PGF Hominid walk on uneven terrain. Individual frame images from the video documentation were compared to see how the skin would shift as the lower leg flexed and the thigh rotates about the hip forward or rearward.

Biological Skin Shift - Charts the extension of shifting skin as the lower leg flexes.
LEFT are the figure models at two different phases of the walk, and the lower section has
the overlay of one set on the other at 50% transparency and inverted color to mark shift.

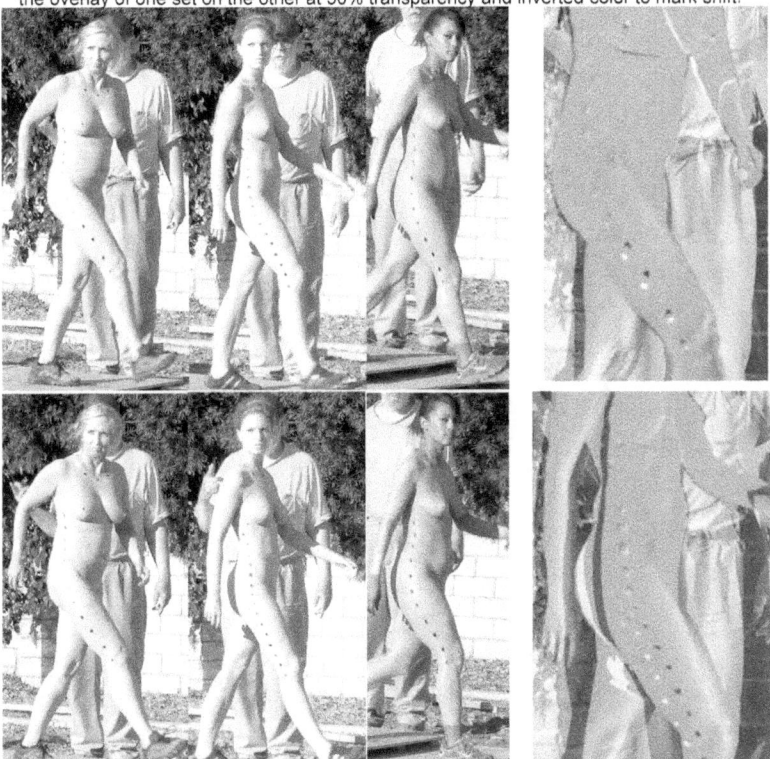

Right Column images show the shift marked in red for each figure model.

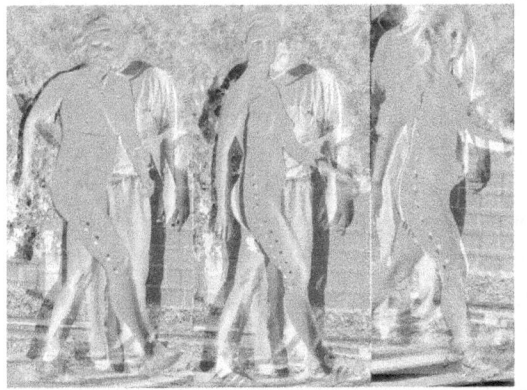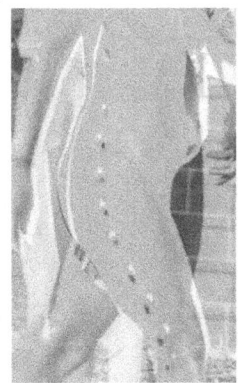

Fig. A6-08

Fig. #A6-08 (Biological Skin Shift) illustrates the actual shift
of skin markers on three figure models, and the elasticity of the

skin, from armpit to knee, was consistent without any artificial buckling.

Costume Furcloth Shifts
Top Row are two spaced frames of the walk cycle, and the shift of the furcloth on the leggings and hips buckles as predicted.

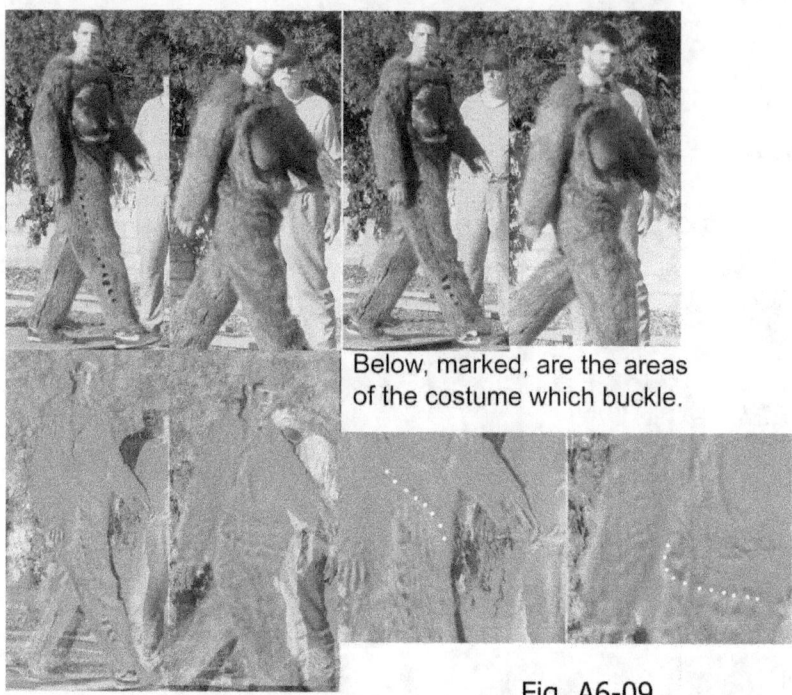

Below, marked, are the areas of the costume which buckle.

Fig. A6-09

Fig. #A6-09 (Costume Furcloth Shifts) illustrates two of the costumes tested, and the exact buckling of the furcloth as it shifts toward the rigid junction of furcloth and breast prosthesis, producing a pronounced fold as predicted, and unlike any shift of dermal mass seen in the PGF Hominid studies.

Analysis and Conclusion

Both live human anatomy and fabricated costumes resembling fur-covered primate anatomy were tested in this study of skin shift motion, and compared to what is observed on the PGF Hominid. The live human anatomy consistently and clearly demonstrated a skin shift and motion dynamic consistent with the PGF Hominid anatomy. The PGF Sampling is from several separate instances of leg motion, and the live human models were sampled from multiple test subjects with ages ranging from the mid 20s to the mid-50s, and there was no disparity in the fundamental skin shift motion dynamic observed and documented. The live biological motion dynamic of dermal tissue as the body extended and flexed the lower leg consistently replicates the PGF Hominid dermal motion dynamics.

The fabricated costume anatomy exhibited shifts of the faux dermal surface exactly as the structural theory of a flexible surface abutting a more rigid mass would be expected to do. The furcloth buckled as it approached the more rigid zone where furcloth was glued to the chest prosthesis. The buckling contour was conspicuous and occurred exactly where theory predicted it would, at the base of the flexible fur immediately before it is glued to the prosthesis. Multiple costumes and persons wearing same were tested and the test results were unambiguous and consistent. There was no continuum of motion and elasticity from knee to armpit in the costume, but rather a pronounced fold as the architecture of the costume changed its material rigidity, immediately below the baseline of the chest prosthesis. This pronounced fold does not occur on the PGF Hominid.

In the two costume examples, the variance of the angle of the fold is accounted for by the fact that one human mime in costume was 6' 5" tall and weighed 165 lbs. while the second mime was 6' 2" tall and weighed 185 lbs. so the disparity in height and weight of these two mimes resulted in a different fit of the costumes and thus a different angle of the buckling furcloth. But neither example replicates the evidence studied on the PGF Hominid.

The conclusion is that the Null Hypothesis is fully supported by both the test for breast motion and the test for skin shift. In each case, one testable option, live human anatomy, does replicate the studied motion dynamic of the PGF Hominid consistently and repeatably, while the motion dynamic of the alternative, the fabricated furcloth/prosthesis costume, fails to replicate the PGF Hominid in any way. The conclusion is that the PGF Hominid can be confidently and factually be described as exhibiting the breast motion and dermal motion dynamic of biological primates (the human test models) and equally confidently and factually not exhibiting the furcloth motion dynamic of fabricated costumes.

On this basis it can be concluded that the PGF Hominid is a biological primate fully organic in its appearance, and is not the result of a furcloth costume worn by a human mime and attempting to appear as a real biological entity.

I would like to gratefully acknowledge the assistance and cooperation of all the women who consented to be figure models, the men who wore the fur costumes, the makeup crew who assisted on the filming test day, the general film crew, the people who provided the location, and those who assisted in the costume fabrication. This research endeavor would not have been possible without the collective efforts of these fine people.

Author Contact and Resume

I have always been open to emails from people curious about my work, and have no hesitation about publishing my email addresses, just as I publish them on my websites.

If you are interested in anything about this publication and would like to either know more, or ask me a question, email to:

wmunns@gte.net
wmunns@verizon.net
2billmunns@gmail.com

Any of the above should work, and I keep several simply so that if one is phased out for any reason, the others should still be active.

Websites:

www.billmunnscreaturegallery.com (Bill Munns Creature Gallery)

www.themunnsreport.com (The Munns Report)

www.billmunnsgallery.com or .net (this digital art gallery is planned for renovation, and hopefully will be restored to its former completeness in the near future).

www.whenrogermetpatty.com

On Facebook "When Roger Met Patty"

As a gesture of full disclosure, the following is a fairly complete resume of my career thus far (I may have missed some smaller jobs, but the ones I do recall and have documentation on are listed).

William Munns
RESUME DETAIL

Born: Los Angeles, 1948
Education: Hollywood High, Los Angeles Valley College (film studies)

Advanced makeup training with Mike Westmore, and began career as makeup artist at Universal Studios, 1969, doing the Makeup Show for the studio tour.

Freelance Prosthetic Makeup work creating prosthetics for the Blackenstein monster in the movie "Blackenstein" 1970

Makeup Artist ABC TV Little Abner, Dating Game, Newlywed Game, News

1970-1972 Assorted film crew jobs as cameraman, lighting man, sound man, makeup artist, for small independent productions, documentaries, etc.

Timber Tramp, independent feature film shot in Alaska.

Started teaching makeup in 1971 and promoted to director and supervising instructor of Elegance International, the first professional makeup artist school. Director from 1973 to 1979. Taught salon, theater, film makeup, prosthetics, and high fashion.

1980 returned to private practice as makeup/prosthetics artist.

Film credits in the 80s include:

Savage Harvest
The Boogins (a minor horror cult classic)
Dead and Buried
Doc R&D
Swamp Thing (first one)

Witch
Dance of the Dwarfs
Beastmaster (first one)
New Kids
Quest for Fire
Better Off Dead
What Waits Below
Brainstorm (Chimp prosthetics for lab animal look)
Return of the Living Dead
Hugga Bunch
Where the Boys are 1984 (custom "love doll" called inflatable Dave)
Baby
No Man's Land (Charlie Sheen prosthetics)
Blind Date (Mechanically animated erotic sculptures for art gallery scene)
Micki & Maude (custom props)
The Man who Loved Women (dog stunt double)
Dian Fossey Biography (prototype work only, on making real chimpanzees look like gorillas. My work isn't in the actual film, but it's impressive experience none the less)

Cougar, spider monkey,

ABC TV show Probe Orang Suit

1985 - Alchemy 2, working with Ken Forsee, inventor of the Teddy Ruxpin doll, for the ABC TV special on Teddy that Ken's company produced. I was a foam latex suit specialist among the costume fabrication crew.

1987 - The Munsters Today - prosthetic designer and key makeup/wig designer on the pilot episode and first show of the series.

Learning Tree started courses on makeup and prosthetics 1987 to 1993

Commercials:

AST Computers - parody of 2001 Dawn of man sequence, with me designing five ape suits and servo motor animation of heads for facial expressions.

Kraft Macaroni & Cheese "Alice in Wonderland" parody (I did the prosthetics for the Mad Hatter character, faithful to the Tenniel illustrations).

Goolab Toys aliens

Chevy Metamorphous

Gold Elephant

Quest for Burger

Three Breasted Chicken

Hostess Twinkie Bear

Cat food metamorphous

Foam Latex Puppet Fabricator - working for puppeteer Tony Urbano, molding and casting foam latex puppet heads for numerous jobs for him over 15 years.

Media References - Fangoria (two-part profile), Cinefantastique (8 page Swamp Thing story)

Transition to Other Media - Having grown weary of doing zombies and swamp monsters, desiring to prove my skills were up to the challenge of excellence and realism, I started creating superbly realistic wildlife sculptures of full scale animal figures and museum quality prehistoric human ancestor models.

Displayed at Game Coin wildlife art show 1987

Displayed at LA Zoo wildlife art show, 1988

In 1988, I took the "Best in World Recreation" award at the world taxidermy championships (and won again in 1992).

Media References - Breakthrough Magazine, numerous articles, Southwest Art magazine

1988 Started working at Creative Presentations, Inc, Valencia CA, hired to help them develop a capability to enter the museum market and compete with Dinamation and Kokoro with dinosaur shows.

While there, I designed, sculpted and figure finished a Gigantopithecus figure as a museum showcase figure (and photos of me beside it now populate many Bigfoot websites), was project manager on the LA County Natural History Museum Bird exhibit, was designing sculptor and project manager on the T Rex for Knotts Kingdom of the Dinosaurs upgrade. I was designer of a realistic bald eagle figure for an Indian Heritage Visitor Center in Vancouver, worked on an ET stroller costume for Universal (that later got replaced with an animatronic figure), and oversaw the refitting of the Gigantopithecus figure into a "Bigfoot" figure for the IAAPA trade show in 1989. Elephant bird for Paris museum.

Promoted to VP (one of four) in 1989, when John March moved up to President and Gene Bullard moved up to Chairman.

Left CPI in August 1990 to resume my career as independent artist.

1991 Commissioned by the French National Museum of Natural History to create two animatronic models of Archaeopteryx, for their renovation of the Grand Gallerie of Zoology.

1992 - returned to CPI as freelance artist for the Fuji/San Rio dinosaur show, sculpting the 23' Tarchia dinosaur.

1992 - returned to World Taxidermy Competitions to take home my second "Best In World Recreation" award.

During this year I also teamed up with the San Diego Museum of Man to create a joint venture exhibit "Faces on Fossils" , that showed both the evolution of humans and how the appearance of these ancestral forms is created. It debuted in San Diego successfully, and then went on a five-year exhibit rental tour across America. The figures I created are now part of a permanent installation at that museum.

Giant panda figure commission for private collection (Bob Howard)

Multiple Dodo figures for private collections

Displayed dinosaur sculptures at the artist room, Society of Vertebrate Paleontology, annual meeting 1991

1993 - Started working at AVG as contract sculptor and figure finisher, doing numerous tigers for a theme park job of theirs, and then helping them get the bid for the 1993 Fuji/San Rio dinosaur show. I was AVG project manager as well as figure designer, lead sculptor and figure finishing lead on that dinosaur show.

Panda, water buffalo, dog.

While there, I also helped sculpt and design the 65' dragon for the Excalibur Hotel in Las Vegas.

While working at AVG, one of my innovations in molding technology so impressed Alvaro Villa that he applied for a patent for the technology, with me principle inventor. As I understand, a sudden lack of business and revenues about that time caused him to "tighten the belt" and cut non-essential costs, and the patent attorney fees were non-essential, so the application was suspended.

Lord of Sipan for Fowler Museum exhibit UCLA

In 1993 and 1994, I received a contract to outfit a new Archaeological Theme Park in Holland (Archeon, in the town of Alphen Rhine) with a full set of prehistoric wildlife figures showing the evolution of life first in the sea, then on land, and finally human evolution. Sadly, this enterprise never reached profitability, and closed a few years later.

Chimps for Educational exhibit, Japan

A commission for a Natural History Museum in Kyoto, Japan was for creating three prehistoric cats, the small Dinictus, the classic Saber-Toothed Cat, and the giant Cave Lion.

AVG hired me as animatronic skin technician and sent me to Taiwan (with two mechanical technicians) to help open a dinosaur exhibit based on both the 1992 Fuji/San Rio dinosaur exhibit (that CPI did) and the 1993 exhibit AVG did. I was the only person who had worked on both parts, and was invaluable in all the visual repairs (the exhibits had been in storage containers for several years) as well as the general assembly of the larger elements and creatures (including a 45' sauropod).

Potomac Museum Group Mammoths

A commission for the Tokyo Broadcasting Commission and the World T Rex Expo was for showing the evolution of flight with three figures, the small dinosaur Compsognathus, the proto bird Archaeopteryx, and finally a modern Peregrine Falcon.

A commission for the Ashland Oregon Museum of Natural History was for two bald eagles.

Gorilla Rentals - Buddy, Disney animation, San Bernardino Museum, Maryland Museum

Film industry work in the mid 90s included an animatronic bear for a Hostess Twinkie commercial,

Extensive makeup, wigs and prosthetics for the TBS TV series "The Chimp Channel".

Working with Animal Makers in 1996 and 1997, I created chimpanzee figures for the HBO Jane Goodall promo spot that won a Cleo, green wing macaws for a Texaco commercial, a bald eagle for an Anhiser Busch commercial, and some dog prosthetics for a Pepsi commercial.

Komodo Dragon for Wildlife Interiors

Horse prosthetics for Mcfarlane

1997 I went digital, beginning my work in computer graphics.

My work with Bryce was so unique that MetaCreations (Bryce's developer) hired me to create graphics for a product promotional campaign, "Bryce, the Eighth Wonder of the World (Because with it, you can create the other Seven)!" With that commission, I created all seven ancient wonders of the world using only the Bryce software, nothing else. 1998-1999

2000 Computer Graphics World magazine ran a four-page portfolio of my Seven Wonders artwork. Then in 2002, in their 25 year retrospective of milestones in computer graphics, my portfolio of work was recognized as one such milestone.

2000-2001 - Hired by Jester.com (a web portal site) to do VRML scenes of the Great Pyramids, the Lighthouse of Alexandria, an NFL stadium, and the Great Wall of China where website visitors could move through the scene while chatting with others online. I used 3D Studio Max, exporting my files to VRML 1 formats for Jester to use. Like many internet ventures, this one folded a while back.

Archaeological visualization for "Dos Cabesas" excavation under direction of Dr. Christopher Donnan, UCLA Department of Anthropology

2002-2014 Freelance Digital Graphics artist.

2008-2014 Researcher for the PGF

Industry Acclaim and Recognition of Excellence (Computer Graphics):

Computer Graphics World magazine, Retrospective of 25 years of computer graphics milestones, June 2002 issue on Architectural Milestones, listed my work as one such milestone.

Computer Graphics World magazine, Retrospective of 25 years of computer graphics milestones, January 2002 issue on Digital Art, listed my work as one such milestone. (one of very few people listed twice in the six-part retrospective.)

Digital Hall of Fame, nominee in 2000

SAN (Super Artist Network) featured artist 2002 Directory (listing by invitation only)

"Big Kahuna" nominee (for Architectural Visualization, 1999, 3D Design magazine sponsored awards.

Design Graphics Magazine Issue #40 (December 1998), six page portfolio and interview.

Numerous articles in 3D Artist magazine from issue 27 to 35.

CITATIONS

Academic and Scholarly Texts traditionally have numerous citations for sources of facts or information. Sometimes texts are even judged as to their merit by the volume of citations. So the fact that I am breaking with tradition will certainly be a point of contention for some readers, so I felt that some acknowledgment of this decision was appropriate to disclose.

This book is the sum of my seven years of research and analysis on the PGF. All reference to matters regarding makeup, prosthetics and creature costumes is drawn from my professional career and actual accomplishments. Information about filmmaking is also drawn from my actual experience in hands-on filmmaking activity. So if I were to do a detailed listing of citations for statements, facts, events, and such, the vast majority would simply read "Drawn from Author's research and experience".

I do acknowledge reference texts and information sources in the Bibliography, and likewise mention some in the book text as well, as sources of knowledge the reader may wish to explore further, but that is the extent of the citations I felt was appropriate for this text. I hope that readers can understand that this text is, in essence, what I think about the PGF and the solution to the mystery. I disclose that fairly by writing in the first person. And so my consideration in regard to the process of citations is governed by that decision. I would rather disclose the facts than do an abundance of cosmetic citations that actually had no scholarly merit.

INDEX

Murphy, Chris 119

Noll, Rick   31, 37

Patricia Patterson 4, 308, 388
Pierce, Jack 82
Processing Timeline 5, 275
Prohaska, Janos 89, 90, 195, 235
Prothero, Donald 273, 352

RHI 38, 40
Rugg, Mike   307, 312

Smith, Dick 205, 206, 235

Westmore, Mike 59

BIBLIOGRAPHY

On Bigfoot/Sasquatch or the PGF

Mysterious Monsters - From the "Mysteries of the Unknown" Library - Time Life Books

Mysteries of the Unexplained - Reader's Digest Editorial Staff

Unexplained - Jerome Clark

Atlas of the Mysterious - Rosemary Ellen Clark

The Bigfoot Film Controversy - Roger Patterson and Christopher L. Murphy

Bigfoot Exposed - David J. Daegling

The Making of Bigfoot - Greg Long

Bigfoot Film Journal - Christopher L. Murphy

Abominable Science - Daniel Loxton and Donald Prothero

Sasquatch - Don Hunter and Rene Dahinden

Bigfoot Sasquatch Evidence - Dr. Grover S. Krantz

Big Footprints - Grover S. Krantz

Know the Sasquatch/Bigfoot - Christopher L. Murphy

Sasquatch: Legend Meets Science - Jeff Meldrum

The Bigfoot Casebook - Janet and Colin Bord

On Cinematography/Filmmaking

American Cinematographer's Manual - Complied by Joseph Mascelli, ASC publication of the American Society of Cinematographers, Hollywood

Cine Kodak K-100 Camera manual - Eastman Kodak Company

The New Joy of Photography - Editors of Eastman Kodak Company

Basic Book of Photography - Tom Grimm

The Motion Picture Film Editor - Rene L. Ash

Film Editing Handbook: Techniques of 16mm film cutting - Churchill

Kodak Data book - Lens manual

Kodak Dtat book - Copying

SPSE/SMPTE Proceedings - Technologies in the Laboratory Handling of Motion Picture and Other Long Films

The Technique of Film Editing - Karel Reisz & Gavin Millar

Film Editing, History, Theory and Practice - Don Fairservice

Matchmoving: The invisible art of camera tracking - Tim Dobbert

Home Movies - Alan Kattelle

On Makeup and Related Crafts

Planet of the Apes revisited - Joe Russo, Larry Landsman, Edward Gross

Stage Makeup, Third Edition - Richard Curson

The Technique of Motion Picture and Television Makeup - Vincent J. Kehoe

The Westmores of Hollywood - Frank Westmore

Hollywood: Flesh and Fantasy - Penny Stalling (section on Photo Re-Touching)

Carl Dame Clarke - "Molding and Casting"

Magazines, Multiple Issues:

Famous Monsters of Filmland

Cinefex

Cinefantastique

Fangoria

Breakthrough - Taxidermy and Wildlife Art

William Munns

www.ingramcontent.com/pod-product-compliance
Lightning Source LLC
Chambersburg PA
CBHW051436170526
45166CB00001B/8